Strategic Integrated Program Delivery

This book outlines a cutting-edge form of program delivery which the authors term SIP-Form or Strategic Integrated Program delivery. Using the Melbourne Level Crossing Removal Program (LXRP), consisting of the removal of 85 dangerous level crossings throughout metropolitan Melbourne, including rail station upgrades, signalling and track work, and other associated capital works, as an exemplar, the book sets out four features that the authors argue define the SIP-form concept as follows:

- The organisation delivers a *program of projects,* many using *an IPD contract variant form* such as a Project Alliance Agreement (PAA) in Australia and numerous other countries, or the Integrated Form of Agreement (IFoA) in North America.
- The contract form adopted is used and has been *strategically designed to accommodate the project's risk and uncertainty profile*, as is the case with the LXRP.
- Projects within the program are *integrated* with some being concurrently delivered with coordination across the projects in a *coherent and highly purposeful manner*. Projects are not included that do not strategically fit the overall program delivery strategy.
- There is a strategy for learning and innovation diffusion across projects, concurrently and sequentially. *Lessons to be learned are learned through designed-in governance mechanisms.*

The LXRP is a potentially unique program of projects, and the book takes the reader on a journey through this complex program and after giving the background and relevant context covers topics such as strategy, governance, procurement, collaboration, program alliance, HRM, leadership, digital innovation, continuous improvement, community engagement, and performance measurement. This detailed analysis of such a complex program of projects makes this book essential reading for project managers, engineers, and advanced students of project delivery and management.

Derek H.T. Walker is Emeritus Professor, RMIT University, School of Property, Construction and Project Management in Melbourne Australia.

Peter E.D. Love is a John Curtin Distinguished Professor of Infrastructure and Engineering Informatics at Curtin University and is a Fellow of the Royal Institute of Chartered Surveyors and a Chartered Building Professional.

Mark Betts has held the position of Director of Project Management and Continuous Improvement with the LXRP since early 2016 and so has invaluable insights into its early days, and all projects within the program. He also had over eight years senior management experience with Metro Trains Melbourne. Metro Trains was a major partner with the Regional Rail Link Alliance (RRLA) which was a program of projects, mainly Alliance projects that significantly influenced the LXRP concept.

Strategic Integrated Program Delivery

Learning from the Level Crossing Removal Project

Edited by
Derek H.T. Walker, Peter E.D. Love and Mark Betts

Routledge
Taylor & Francis Group

LONDON AND NEW YORK

Designed Cover Image: Mark Betts

First published 2025
by Routledge
4 Park Square, Milton Park, Abingdon, Oxon OX14 4RN

and by Routledge
605 Third Avenue, New York, NY 10158

Routledge is an imprint of the Taylor & Francis Group, an informa business

British Library Cataloguing-in-Publication Data
A catalogue record for this book is available from the British Library

ISBN: 978-1-032-48454-9 (hbk)
ISBN: 978-1-032-48455-6 (pbk)
ISBN: 978-1-003-38917-0 (ebk)

DOI: 10.1201/9781003389170

Typeset in Times New Roman
by codeMantra

Contents

Editors and Contributors

Editors

The book's editors co-authored most of the chapters with colleagues from a range of universities, and with a senior member of the LXRP leadership team, to ensure that the theoretical themes of each chapter make the best sense of perceived practice and to provide a consistent, flowing, focussed narrative.

Derek H.T. Walker is Emeritus Professor of Project Management in RMIT University Melbourne, and has been researching and writing on various topics associated with Integrated Project Delivery (IPD) and Alliancing since his involvement in a longitudinal case study of the National Museum of Australia project alliance from 1999 to its completion in March 2001. Outputs from that research included co-authored, numerous conference, professional association and academic journal papers as well as a book (Walker and Hampson, 2003). He was also researcher for a joint RMIT QUT research project into alliancing and a Project Management Institute (PMI) to study IPD and Alliancing globally. This produced many papers and a book (Walker and Lloyd-Walker, 2015). Other books relating to IPD followed in 2020, co-authoring with Steve Rowlinson from The University of Hong Kong. Derek was the founding editor of the International Journal of Managing Projects in Business.

Peter E.D. Love is John Curtin Distinguished Professor of Infrastructure and Engineering Informatics at Curtin University. He is a Fellow of the Royal Institute of Chartered Surveyors. He was also a member of the inaugural Australian Research Council's Engineering and Environmental Science Panel 2010/11 Excellence in Australian Research Exercise and re-appointed for the 2018 evaluation process. He was awarded the inaugural Scopus Young Australian Researcher of the Year Award in 2010 and later in 2012 was awarded a Higher Doctor of Science (Sc.D.) by Curtin University in recognition of his substantial and sustained contribution to the field of civil and construction engineering.

Mark Betts held the position of Director Project Management and Continuous Improvement with the LXRP from early 2016 and so has invaluable insights into its early days, all projects within the program and, together with his significant Director experience in translating strategy into action. He has senior-level experience in major civil infrastructure and transport developments, procurement, and delivery of complex projects and strategies in multiple contexts.

Recent history is in the heavy rail infrastructure industry including the Victorian Level Crossing Removal Project, Regional Rail Link, and prior across a broad range of building and major infrastructure projects including the iconic MCG Southern and Northern Stand redevelopments and Commonwealth Games preparations.

He also had over eight years senior management experience with Metro Trains Melbourne (MTM). MTM was a major partner with the Regional Rail Link Alliance (RRLA) which was a program of projects, mainly Alliance projects that significantly influenced the LXRP concept. Thus, he has highly valuable Alliancing practitioner experience to draw upon for this book.

Contributors

Kirsi Aaltonen is an Associate Professor of Project Management and Complex Systems at the University of Oulu, Oulu, Finland, where she heads the Project Business Research Team. Her current research interests include stakeholder management in complex projects, governance of inter-organisational projects, game-based learning methods, and institutional change in project-based industries.

Andrew Davies is RM Phillips Freeman Chair and Professor of Innovation Management at the Science Policy Research Unit, University of Sussex Business School. Previously he was Chair of the Management of Projects (originally established for Peter Morris) and continues as Honorary Professor in the School of Sustainable Construction in the Bartlett Faculty of the Built Environment, University College London. His research focuses on innovation in project-based organisations and large, complex projects in the UK – Heathrow Terminal 5, London 2012 Olympics, Crossrail and the Thames Tideway Tunnel projects.

He has undertaken pioneering research with major programmes created a 'movement for innovation' in infrastructure project delivery in the UK. His engagement with major projects has helped to create a collaborative and flexible project delivery (Heathrow T5), a model of delivery systems integration (London 2012 Olympics) and world's first innovation strategy for a megaproject (Crossrail).

He has published in a range of management journals such as Journal of Operations Management, California Management Review, MIT Sloan Management Review, Research Policy, Organization Studies, Industrial and Corporate Change, International Journal of Project Management, and Project Management Journal. His book Projects: A Very Short Introduction was awarded the Project Management Institute (PMI) 2018 David I. Cleland Literature Award for contribution to knowledge.

He is a Senior Editor of the Project Management Journal, a member of the Editorial Advisory Board for International Journal of Project Management, and Associate Editor of Industrial and Corporate Change. He is an Honorary Fellow of the Association for Project Management (2021).

Juliano Denicol is the Director of the Megaproject Delivery Centre at University College London (UCL) and Director of the UCL MBA Major Infrastructure Delivery. His research explores the management of megaprojects, including several iconic UK megaprojects: High Speed 1, Heathrow Airport Terminal 5, London 2012 Olympics, Crossrail, Thames Tideway Tunnel, and High Speed 2. Research areas include Procurement and Supply Chain Management, Organisation Design, Governance Structures and Project-based Firms. Dr Juliano Denicol is the Director of the Megaproject Delivery Centre at University College London (UCL) and the founding Programme Director of UCL's MBA Major Infrastructure Delivery. Before joining UCL, Juliano has worked as a supply chain management consultant at High Speed 2, the largest infrastructure project in Europe, and advisor to the European Commission on public procurement policies. As Global Head of the IPMA Megaprojects SIG, Juliano coordinates a global platform with more than 70 countries to advance our understanding of megaproject delivery. He is the founder and director of the IPMA-UCL The Megaproject

CEO and IPMA Megaprojects Book Club, two global platforms to discuss concepts and practices with leading megaproject authors and CEOs. He was Co-Investigator of Project X, a major research network that aims to improve major project delivery in the UK, established by nine universities in collaboration with the Infrastructure and Projects Authority (IPA), and the Cabinet Office. Previous research included several iconic UK megaprojects: High Speed 1, Heathrow Airport Terminal 5, London 2012 Olympics, Crossrail, Thames Tideway Tunnel, and High Speed 2. Juliano's work on megaprojects has been regarded of high global impact receiving multiple research awards, including the prestigious: PMI Young Researcher Award – Project Management Institute (PMI) (2023); Project Management Research Paper of the Year Award - Association for Project Management (APM) (2022); Most Cited Paper of the Year Award – Project Management Journal (2021); Most Downloaded Paper of the Year Award – Project Management Journal (2021); Global Young Researcher Award – International Project Management Association (IPMA) (2019).

Nathalie Drouin is Chairholder, Research Chair INFRA-S on Social Value of Infrastructures, and Professor at Université du Québec à Montréal. She is co-author and co-editor on numerous recent books on infrastructure projects and project organising, and is the editor-in-chief of the International Journal of Managing Projects in Business. She has won several research excellence awards including the PMI 2022 Research Achievement Award, the 2021 PMI David I. Cleland Project Management Literature Award for the book Organizational Project Management: Theory and Implementation, the Walt Lipke Project Governance and Control Excellence Award and the International Project Management Association IPMA Research Award in collaboration with her colleagues Ralf Müller and Shankar Sankaran.

Mattias Jacobsson, PhD, is a Distinguished teacher and Associate Professor of Management and Organization at Umeå School of Business, Economics, and Statistics, Umeå University, Sweden. Since 2016, he has, in parallel with his position in Umeå, been a part-time researcher at the School of Engineering, Jönköping University, Sweden. His main research interests are in projects, practice, and temporary organizations, and on four occasions, he was a prize-winner at the Emerald Literati Network Awards for Excellence. Over the years, he has, on several occasions, been a guest researcher at RMIT University in Melbourne, Australia, and at Virginia Tech in the US. His current research focuses on digital ambidexterity and strategic innovation in construction, the projectification of higher education, and meaningful routines.

His work has been published in many journals, including Business Horizons, Management Decision, Project Management Journal®, Services Marketing Quarterly, the International Journal of Project Management, the International Journal of Managing Projects in Business, and Construction Management and Economics. For more information, see http://mattias-jacobsson.me

Beverley Lloyd-Walker is a research fellow at RMIT University, Melbourne. Her recent research has concentrated on 'people in temporary organisations.' This research has been on the project-based Australian economy, new project procurement approaches, and careers in project management. More recently issues around human resource management policies and practices in projectified workplaces, and especially within more collaborative forms of project delivery alliances and integrated project delivery, have become a major focus of her research. Her research has explored the professional skills required for successful project leadership within alliances and other relationship-based procurement forms. Her interest in collaborative forms of project procurement led to her researching the range of forms used globally, and thus to her role in researching and writing the book *Collaborative Project Procurement Arrangements* published in 2015. She has had over 50 refereed items published, has co-authored three editions of a human resource management book, and has co-authored an industry report. She has supervised 11 doctoral theses; one masters by thesis, and eight masters theses to successful completion covering topics including women CEOs in local government, emotional intelligence in project teams, performance based rewards systems in the public sector, and expatriate project roles.

Hadi Mahamivanan is a PhD scholarship holder at Deakin University, Geelong and has been engaged in research on the LXRP, while working for McConnell Dowell, in digital engineering and the use of drone technology.

Jane Magree is an advisor on strategy and culture transformation to the Victorian Infrastructure Delivery Agency (VIDA) and has been associated with Level Crossings Removal Project (LXRP) since 2016. Jane brings deep experience and actionable insights on how the leadership of LXRP built a sustainable ecosystem of high performance that can be adapted to other delivery agencies and projects. As a former partner of Deloitte and currently CEO of her advisory company Jane Magree & Associates, Jane has worked with executive leaders in major civil infrastructure and transport projects in both public and private sector for over 20 years. With a background in organisation psychology and human resource management, Jane's focus has been on enhancing productivity and resilience in organisations managing high risk. Past projects include the Regional Rail Link program, WestGate Bridge Strengthening Alliance and Water Resources Alliance. Jane has designed innovative strategies and culture programs that delivered repeatable gold standard outcomes. Jane's strategic approach and award-winning programs generated mindsets shifts and embedded systems and practices for high performing cultures that enable better work, better working and better workplaces leaving a legacy for industry change.

Jane Matthews is Professor of Digital Construction at Deakin University, Geelong. She is an experienced academic with a demonstrated history of working in the higher education industry. She is from a multi-disciplinary construction background, holding a PhD and a first-class honours degree in Architecture and post-graduate qualifications in mathematics, computing, and environmental studies. She has specialist expertise in Digital Engineering and is currently a Chief Investigator on an ARC Discovery grant (DP210101281), exploring mechanisms for containing and reducing rework in mega transport projects. Her role on this project is to analyse and propose data structures and digital information sharing processes across design and construction to enable better management of rework.

Christina Scott-Young is an Associate Professor at RMIT University, Melbourne. She is a clinical psychologist and an experienced academic, having worked at Penn State University, US and at three different universities within Australia. With a Doctor of Philosophy (PhD) in Management from the University of Melbourne, Christina has a demonstrated history of teaching, leading and researching on project management topics. She has published over 80 refereed research articles in leading international journals, as book chapters, and in conference proceedings. Her research focuses mainly on project teamwork; diversity and inclusion; work readiness of early career project professionals; shared leadership in engineering project teams; career resilience; as well as personal resilience and mental health in individuals and workplaces. She is currently completing two research projects funded by leading international project management professional bodies. The first project conducted for the Association for Project Management UK is a cross-country comparative study on the experiences of gender and ethnically/racially diverse project management professionals working in different industry sectors. The second grant is from the Project Management Institute US to explore the project-based organisation's role in skill-building (decision-making, critical thinking, and confidence) in the next generation of project managers.

Foreword

LXRP seeks to leave a legacy for industry and the people of Victoria on the why, what and how of the LXRP model of major project infrastructure delivery.

This resource is a useful tool for those wishing to learn from and adopt or adapt the principles of the LXRP model to their delivery agency.

Matthew Gault, CEO of the LXRP

Preface

This book has been written about the role of the LXRP and explains how it was formed, evolved, and performed in undertaking the A$19.8 billion 15-year program of projects. The book editors and chapter authors carefully presented an independent account of their research. They were provided with access to internal LXRP documents and to interview 24 key LXRP staff with broad knowledge of the program's operations across the range of activities at the project and program level. Findings from these interviews were discussed with key LXRP staff and analysed to ensure that content accurately reflected current practice and strategic direction. The book also provides an unbiased, realistic, and fair picture of the LXRP's purpose, inception, and operations.

The chapters' authors have drawn upon their theoretical knowledge and insights to support their research findings and provide a picture of what the LXRP is and how and why it performs its role. They explain the what-type content, the LXRP's purpose, its infrastructure delivery project operational functions, and how this compares to current world project delivery practice. More importantly, they explore why and how chapter content provides a comprehensive contextualised theory and practice explanation of the LXRP delivery model. A key part of LXRP's vision is to facilitate a transformational industry learning legacy to its program participants and their home organisations that may be adapted, industry-wide, to improve productivity.

Key components of the what-why-how questions are explored, analysed, and explained. The 12 chapters answer the book's guiding purpose of studying the LXRP as an example of a key program-of-projects delivery mechanism initiated within a specific context. It answers questions about why its program delivery choice was made and how it was undertaken to deliver this transformational urban transport infrastructure project to benefit society.

The 12 chapters explore: the program context; the program delivery strategy and governance mechanisms; available delivery choices and why the LXRP model was developed; why and how it created a highly integrated pipeline of AWPs, how cross-project collaboration was designed-in; how it cultivated a high-performance team workplace culture; how this shaped and defined the leadership approach to facilitate its workplace ecology; how innovation diffusion was strategised and effectively built into the governance mechanisms to support continuous improvement; why and how digital engineering and technology innovation became a core LXRP designed-in aim; why and how the LXRP engaged with its internal and external stakeholder communities; and why and how the LXRP conceptualised program performance and long-term legacy through a value and transformation lens. Chapters also reflect on global practice with valid comparisons from Canada, the UK, Finland, Sweden, the USA, and other countries undertaking similarly collaborative-based delivery approaches.

The book extensively delivers fresh and original knowledge and insights about infrastructure megaproject delivery, and its main contribution is that it focuses on a very special and innovative program/project delivery approach by explaining the what aspect, but also the why and how issues. It should be a valuable learning resource for the academic world and a key go-to source for practitioners wishing to understand and adapt the LXRP model to suit their context.

Derek H.T. Walker (Melbourne),
Peter E.D. Love (Perth) and Mark Betts (Melbourne)

Abbreviations

Throughout this book the following abbreviations are used:

A$	Australian dollar
AI	Artificial intelligence
AIA	American Institute of Architects
ALT	Alliance leadership team
AM	Alliance manager
APT	Alliance project team
AR	Augmented reality
AWP	Additional work package
BAA	British Airports Authority
BAU	Business-as-usual
BIM	Building information modelling
BOOT	Build Own Operator Transfer
BOT	Build Own Transfer
BV	Best value
CAD	Computer aided design
CARES	5-Cares: Creativity, Accountability, Relationships, Empowerment and Safety
CBA	Community benefits agreement
CD	Competitive dialogue
CDBB	Centre for Digital Built Britain
CEO	Chief executive officer
CPDM	Collaborative project delivery models
CoBie	Construction Operations Building Information Exchange
COP	Community of practice
CorEX	Correlation Explanation
CRL	Crossrail Limited
D (2/3/4 etc.)	Dimension 2D = two dimensional etc.
D&C	Design and construct
DBB	Design-bid-build
DE	Digital engineering
DoT/DTP	Department of Transport and Planning
Drone	Uncrewed aerial vehicle (UAV)
EI	Emotional intelligence
EPC	Engineering Procurement and Construction
FO	Facility operator

GIS	Geographical information systems
HCS	High-capacity signalling system
HRM	Human Resource Management
HS2	High Speed 2 (UK fast train project)
IDO	Integrated development opportunity
IFC	Issued for Construction
IFoA	Integrated Form of Agreement
IoT	Internet of Things
IPD	Integrated project delivery
IVxx	Interviewee xx (01NN) e.g., IV14 = interviewee 14
IWP	Initial work package
JCC	Joint Coordination Committee (note it also has many special interest group/ discipline sub-committees)
KBES	Knowledge-based engineering system
KPI	Key performance indicator
KRA	Key result area
KSA	Knowledge, skills, and attributes
KSAE	Knowledge, skills, attributes and experience
LDA	Latent Dirichlet Allocation (algorithm)
LOD	Level of Development
LXRA	Level crossing removal authority
LXRP	Level Crossing Removal Project
ML	Machine learning
MM	Metropolitan Melbourne
MMRA	Melbourne Metro Rail Authority
MMRP	Melbourne Metro Rail Project
MTIA	Melbourne Transport Infrastructure Authority
MTM	Metro Trains Melbourne
NCR	Non-conformances reports
NEC	New engineering contract
NLP	Natural Language Processing
NOPs	Non-owner participant(s) in an alliance
OVGA	Office of Victorian Government Architect
PAA	Project Alliance Agreement
P-E	Person environment (fit)
P-J	Person-job (fit)
PM	Project management
PMI	Project Management Institute
P-O	Person-organisation (fit)
PO	Project owner
POR	Project owner representative
PPI	Public Private Initiative
PPP	Public Private Partnership
P-T	Person-team (fit)
QC	Quality control
R&D	Research and Development
RFI	Request for information

RFID	Radio Frequency Identification
ROI	Return on investment
RPV	Rail projects Victoria
RRLA	Regional Rail Link Authority
SEM	Structural equation modelling
SIM	Systems information model
T2	Terminal Two London Heathrow Airport
T5	Terminal Five London Heathrow Airport
TAE	Target Adjustment events
TCE	Transaction cost economics
TPB	Theory of planned behaviour
TOC	Target outturn cost
TTM	Alliance management team
UAV	Uncrewed aerial vehicle (drone)
UDAP	Urban Design Advisory Authority
UDF	Urban Design Framework
UK	United Kingdom
USA	Unites States of America
VAGO	Victorian Auditor General's Office
VDAS	Victorian Digital Asset Policy
VfM	Value for money
VIDA	Victorian Infrastructure Delivery Authority
VR	Virtual reality
WoW	War on Waste

Acknowledgments

The journey to craft this book has depended on us building collaboration and establishing trust with our stakeholders, much like an alliance project. It has been a learning experience for us as we uncovered new insights into how a project works in a real setting. The insights we have garnered from working closely with our stakeholders have been an invaluable experience; without them, this book could not have been delivered. Thus, we would like to thank and acknowledge the support and participation of the interviewees and thank them for their constructive feedback. Most interviews took one hour, and that is one hour out of their already busy life that they can't 'claim' back. This is not part of their work plan, so these interviews were a substantial gift to us. Also, interviewees generously gave us the benefit of their expertise, knowledge, and insights. All the interviewees are senior members of their organisations. Their open and honest responses to our questions and their voluntary and spontaneous discussions provided a deep measure of data reliability and validity. They could speak about their experiences and reflections on their workplace colleagues' conversations and insights. The recorded and analysed interview transcripts, and referral to the sound files, provided a wealth of data for this book and several supplementary papers.

The role of chapter authors and the co-editor role is primarily one of historian. We all bring our experiences as practitioners and researchers to make sense of the data we gathered, but we could not write this book without the generous donation of our interviewees' time and insights.

As historians, we draw upon the salient literature (standing on the shoulders of giants) to make sense of data gathered from interview transcripts, in-house documents and public access documents. By comparing and contrasting data from the Level Crossing Removal Project (LXRP) with other similar megaprojects, we can make sense of data to draw credible conclusions that, until refuted, can be considered reliable and factual. Therefore, the work of many other researchers and practitioners is vital and must be acknowledged.

Our co-authors are also a vital part of the team producing this book. As co-editors, we drew upon our network of expert researchers and practitioners to select those we believed could best contribute to the chapters. We acknowledge and thank them for their insights and expertise.

Writing and editing a book such as this is considerably time- and energy-consuming, and this considerable and often un-recognised cost often falls in no small part on our family and close friends. We thank them (including pets) for their patience, understanding and indispensable support.

We also thank two institutional supports for this book 'project.' The LXRP has been highly supportive of this venture and stood by its commitment to providing us with in-house materials and insights and legacy key result area (KRA). We hope this book offers in-depth 'how and why' answers to help guide and provide a straw-man model of delivering complex projects such as LXRP and how that model may be adapted. Our universities, RMIT University, Curtin

University, and our co-authors' universities provide key support through their library facilities and access to their data and information systems.

Our social and professional networks are critically important for the many conversations about this book during its writing and development and the valuable feedback and insights they have provided us.

Finally, we include this LXRP's acknowledgement. In the spirit of reconciliation, the Level Crossing Removal Project (LXRP) and its partners acknowledge Aboriginal people as Australia's first people, and as the Traditional Owners and custodians of the land on which we work and live. The level crossing removal program is being delivered on Bunurong, Wurundjeri Woi-wurrung and Wadawurrung Country. We pay our respects to their spirit and passion in their past and present custodianship of this Country.

We also pay our respects to the Elders of the Kulin Nation, past and present, and extend that respect to other Aboriginal and Torres Strait Islander peoples. The LXRP remains committed to be led by the Victorian Aboriginal people to deliver great change on Country and progress their social advancement.

Derek H.T. Walker, Peter E.D Love, and Mark Betts

1 Introduction

Derek H.T. Walker, Peter E.D. Love and Mark Betts

1.1 Introduction

What is the aim and rationale for this book? What does the book contain and how is it organised? What expertise contributed to this book?

This book is valuable for researchers and tertiary level students of project, construction, or engineering management. It should be of particular value to infrastructure management practitioners in designing a high-performance project delivery strategy, governance mechanism and delivery team recruitment. Each chapter explains this case study's findings, referencing salient academic literature. The book explains how complex infrastructure engineering megaprojects can be more effectively delivered using an innovative project alliancing integrated project delivery (IPD) approach.

Reflective practitioners, how might *you* envision a pathway to involvement with this novel IPD relationship-based way of delivering complex large projects? This book provides valuable insights to you and senior management teams presently involved in very large-scale project delivery, or those aspiring to do so. While this readership group may skim over 'theoretical' explanation chapter sections to focus on the '*how*' and '*what*' book content, they may find it useful to refer to the '*why*' supporting literature content that helps make sense of chapters. Each chapter links purposefully to explain how this innovative IPD approach may be achieved and what is expected of delivery team participants if they wish to be involved in project alliancing and similar programs of IPD projects.

The case study organisation and its activities provide a rich source of insights into how megaprojects are being delivered through a more collaborative, integrated, and sophisticated relationship basis. Key features of the Level Crossing Removal Project (LXRP) exemplify a stand-out model. The model contrasts with delivery models used for many megaprojects through its:

1. Decision to adopt an integrated collaborative program of projects delivery strategy and the evolution of its alliance work packaging alliance project team recruitment process (see Chapter 5 for more details).
2. 5-Greats program outcome strategy (great network, great places, great partnership, great engagement, and great people) supported by 5-values (creativity, accountability, relationships, empowerment, and safety) that framed a sophisticated program benefits vision (see Chapter 12 for more details).
3. Cross-project organisational program integration that was designed to achieve synergies between projects and participants with continuous improvement and innovation diffusion as focused sophisticated strategic mechanisms (see Chapters 5 and 9 for more details).

DOI: 10.1201/9781003389170-1

4. Recognition of the LXRP *purpose* as an opportunity to not only fix a problem, neglected for many decades, but to use this occasion to provide a human, social capital, social amenity, and urban transformation legacy (see Chapter 3 for more details).
5. Actualisation of the program through concerted, diligent and considered adherence to disciplined accountability and responsibility governance arrangements that was supported and underpinned through its humane orientation towards internal and external stakeholders (see Chapter 11 for more details).

1.1.1   The LXRP context

This book investigates an intriguing case study worth exploring. Eisenhardt (1989) legitimised single case study investigation for unique, or very rare phenomena. The LXRP is rare megaproject undertaken as a program of five alliance projects with multiple additional alliance work packages, therefore, a sound candidate for study to better understand how this remarkable strategic infrastructure megaproject was conceptualised and delivered. It explains in 12 chapters how major project facets were undertaken, thus, making a novel and valuable contribution to the project management and construction engineering discipline.

While programs of large-scale Alliancing-type infrastructure projects have been delivered for several decades, little has been written about these as *programs of alliance projects*. This book seeks to bridge that gap. We provide a realistic and honest history of the LXRP delivery. The book provides novel insights from several perspectives, answering questions that facilitates a better understand of how the LXRP evolved and is delivering 110 level crossing removals, together with associated rail station rebuilds, and transforming and revitalising rail station precinct and rail signal systems over a 15-year period.

1.1.2   Justifying writing this book

The LXRP is a rare form of project delivery organisation established in May 2015 by the Labor Victorian Government in Australia to oversee the removal of 50 dangerous level crossings through metropolitan Melbourne, including rail station upgrades, signalling and track work, and other associated capital works. Its oversight included taking over two crossing removal alliances managed by VicRoads at the program's start. The newly elected state government established the Level Crossing Removal Authority (LXRA) in May 2015 to deliver the program in 2022. Its remit was expanded in 2018 to 85 level crossings, budgeted at A$16.3 billion, to be completed in 2025. During the 2022 election this committed was extended to 110 crossing removal at a by 2030 totalling A$19.8 billion. The first phase of 50 crossings was completed ahead of schedule and below its program target outturn cost (TOC). The scope of this program of projects is currently referred to as the Level Crossing Removal Project (LXRP). The LXRA was disbanded as an independent organisation in 2019 and subsumed into a larger integrated infrastructure program of projects and programs to become the Melbourne Transport Infrastructure Authority (MTIA) on 1 January 2019[1] within the Victorian Department of Transport and Planning (DTP).

There is a dearth of information about this remarkable project-based organisation, its origins, evolution and how it has functioned. Indeed, there are credible and accessible sources, such as government reports and several academic publications but they are often difficult to access and do not provide a holistic overview of the project's organisation, nuances and best-practice initiatives that make it such an exemplar strategy for procuring public infrastructure.

The LXRP's organisational story, from inception to realisation, provides an original contribution to the project management (PM) and public infrastructure delivery literature. It provides more than just another PM or organisational theory text. The core of this organisational inno-vation is its strategically designed infrastructure relationship-based integrated-team delivery mechanism that coherently and systematically links a program of projects to optimise synergies between them.

The management of projects body of literature has been steadily growing since the early 1970s from the Morris and Hough (1987) seminal work on major infrastructure projects from the defence industry and construction industry. Morris later wrote two more influential books on major project delivery extending the scope of available literature. His first (1994), maintained a strong focus on major construction infrastructure project and later he broadened project delivery literature's scope (2013). Also, during that time, there had been a strong focus on major project delivery, including public sector infrastructure and commercial construction projects from a construction management perspective (Harris and McCaffer, 1977) with overlaps that explained the procurement process and roles of the actors delivering construction projects (Walker, 1985; McGeorge and Palmer, 1997; Winch, 2002).

However, government and private industry clients began to adopt more relationship-based major construction infrastructure projects procurement and delivery approaches with books ap-pearing that focussed on integrated project delivery (IPD) highly collaborative team delivery forms that used an Alliancing approach, for example on the National Museum of Australia (Walker and Hampson, 2003) and in the UK, London's Heathrow T5 Airport Terminal that use a similar model based on the New Engineering Contract version 3 (NEC3) that later developed into NEC4, even closer to the Australian and European Alliancing contract agreement forms. The T5 example was documented as an informative case study book (Doherty, 2008). While these examples are major projects, they tended to be treated as one-off projects and not con-sidered in the context of continuous delivery of a set of projects, coherently delivered within a program of projects. Interestingly, a recent trend has developed in the UK with the Institute of Civil Engineers' Project 13 initiative (see www.project13.info/). This initiative encourages and enables organisations to collaborate on not only projects but also on programs of collaborative projects. It offers information about their Project 13 framework document to guide project and program collaborative delivery and cite examples of rail[2] and water delivery[3] government or-ganisations adopting both project and program alliances using NEC3/4 contracts.

According to the American Institute of Architects – AIA California Council (2007), IPD in-tegrates teams, systems and business routines and practices to facilitate high level collaboration that increases value and optimises project output results. Project alliancing evolved over the past two to three decades as a seminal IPD form across the globe (Lahdenperä, 2012) and according to the Australian Commonwealth Government Department of Infrastructure and Transport it is a highly collaborative form of integrated form of project delivery in which the project owner's representative, design and construction contractual parties engage in a Project Alliance Agree-ment (PAA). They further state:

> All Participants are required to work together in good faith, acting with integrity and mak-ing best-for-project decisions. Working as an integrated, collaborative team, they make unanimous decisions on all key project delivery issues. The alliance structure capitalises on the relationships between the Participants, removes organisational barriers and encour-ages effective integration with the Owner.
>
> (2011, p. 9)

NEC3/4 and PAA projects are examples of megaprojects delivered through IPD-Alliancing forms of contract.

The effective management of infrastructure project (and program) delivery has been argued to be a process of realising its value-generating strategy and understanding the purpose of its undertaking with an institutional context – one that fulfils a pressing need for a social good (Morris and Geraldi, 2011). Few comprehensive literary sources exist that explain, or demonstrate, *how* a project delivery organisation can shape itself and focus on its technical delivery at the strategic and institutional levels. Morris and Geraldi explain this third institutional level as being "primarily concerned with improving success not of a specific project, but of projects within the enterprise's own organisational environment" (p. 23). In this case, LXRP integrates a series of level crossing removals and new stations to improve the effectiveness of a transport network (train, bus, tram, and road interfaces) by managing a program of projects within a portfolio (such as a government transport infrastructure department or entity). We argue that the LXRP also facilitates transformational change through its strategic approach to program management within the context of government, providing enhanced and improved infrastructure for social good. This reflects Morris's concept of reconstructing project management (Morris, 2013, particularly Chapter 21).

1.2   Aims, Objectives and Logic of this Book

Frequently, lessons that should be learned from megaprojects are not documented. Several academic papers, research reports and Victorian Auditor General's Office (VAGO) reports were published, but each focuses on a limited scope of the LXRP story. A book, rigorously researched and written, has great potential to provide a credible and valuable knowledge legacy.

This book is written with a set of chapters, co-authored by academic experts in chapter-related theoretical fields with specific insights from LXRP expert personnel, and other practitioners engaged with the LXRP. The purpose of this contribution is to ensure that approaches, processes and modes of working adopted by the LXRP can answer crafted *how* and *why* research questions using rigorous sense-making so that action-consequence issues can be better understood in context.

The purpose of LXRP personnel and other practitioner contributions to these chapters is to provide an honest and realistic basis to the story. Figure 1-1 illustrates this book's design logic.

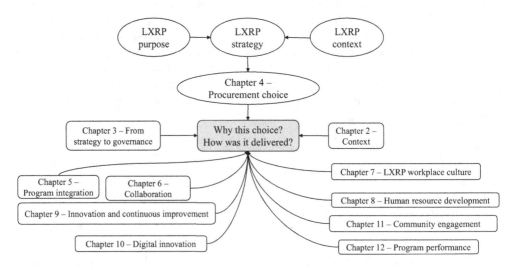

Figure 1-1 The logic of this book

The process starts by Chapter 1 introducing the book to enable readers to understand the purpose for this program of projects case study. Every project and program is subject to contextual issues that vary in intensity over time. Chapter 2 explains the strategic, complex and historical-political context influencing the alliancing approach decision and how constraints and opportunities shaped this context. Chapter 3 explains the LXRP strategy with Figure 3-1 illustrating four questions beginning with 'What outcome do we need/want?' and ending with 'What does the organisational and government design require?' The Victorian Government's program of alliance projects choice rationale is explained and discussed in depth in Chapter 4. Moving from a strategic project delivery choice to a deliver strategy required designing and implementing a fitting governance system. Chapter 4 explains how that was rationalised and understood by LXRP participants.

The remaining book chapters explain why the LXRP chose alliancing and how it was delivered. Central to the choice is alliance project team (APT) integration, and how well they collaborate to engage in genuine dialogue (Walker, Love and Matthews, 2022). Chapter 5 explores how the APT achieved integration, explains how the APT was established, their recruitment strategies for participants considering joining an APT and how APT cohesiveness was maintained, while APT collaboration is discussed in Chapter 6.

Differences between good and mediocre alliance performance is influenced by several factors. Delivering project alliance benefits relies upon high-performance people-oriented teams being committed and motivated to collaborate as an integrated team. Chapter 7 explores the LXRP workplace culture transformation strategy to effectively deliver the program. This chapter explains the specific mechanisms used to define the desired workplace culture and how a strategy was developed and enacted to support the program's purpose and vision.

LXRP participants need effective leadership to facilitate the necessary systems and mechanisms for their project integration and collaboration within the program. Requisite leadership and program participant qualities are in short supply, and so it was logical for the LXRP to develop a uniting strategy to develop necessary participant capability qualities. This requires a purposefully developed human resource capital process. Chapter 8 focuses on how the LXRP managed this challenge and how it identified, recruited, developed, and developed its key staff taking both the LXRP and individuals' perspective to make a positive impact on the program's delivery.

A key program objective was to benefit from efficiency and effectiveness dividends through continuous improvement and innovation diffusion – both within projects and across the program. Chapter 9 explains how this key objective was addressed. Associated with this, the adoption of digitisation innovations that resulted in continuous improvement is explored and explained in Chapter 10, comparing the LXRP experience with emerging international trends.

Value can be co-created with a local community impacted by a project and stakeholder engagement can facilitate much of the urban improvement and legacy goals articulated in the LXRP strategy and business case (Victorian State Government, 2017). Chapter 11 explores and explains their community engagement approach. LXRP engagement with internal stakeholders is covered in Chapters 3, 5 and 6.

Chapter 12 deals with project value delivery performance, how it is conceptualised and the extent to which it was delivered through the program of projects. It is important to understand that this book is focused on the LXRP *program* of projects. The success or otherwise may be indicated by individual *project* success but the final assessment, yet to be made, will depend on the final *program's* achievement record. At the time of writing, 75 rail crossing removal and associated works were nearing completion. Therefore, we are confident that we can make realistic and fair claims about how effectively this program is being delivered.

1.3   Research Design

This book is intended for infrastructure project delivery students and practitioners. Any research project begins with an interesting question. While presenting research papers on project alliancing at numerous research events where we mentioned the LXRP, we consistently received audience questions about how this program of projects evolved, how it was conceived and what prompted its existence. Trying to answer the consistent question of where people could search for authoritative and reliable data and findings about the problem, led us to consider writing this book. Publicly available information about the LXRP is limited. There have been two Victoria Auditor General's Office (VAGO) reports published at the initial stages of the program and other short progress reports are buried within government annual reports such as those available from the Victorian Department of Transport and Planning (DTP). A limited number of research reports and academic papers contain references to, and report to various specific aspects of the LXRP some that have mentioned it by name, others anonymise the project (some co-authored by Love or Walker).

In deciding on writing a book about the LXRP, we took some questions asked of us as a starting point and developed a framework based on 12 main questions including the first, why write this book?

This book is based on original research and reflection on previous research into IPD/Alliancing undertaken on the LXRP by the co-editors. Co-authors joining us on several chapters also reflected upon their past research and current research in their specialist areas associated with IPD/Alliancing.

Figure 1-2 illustrates the main research question answered by this book together with a series of sub-questions that relate to pertinent research perspectives answered through the book chapters.

We obtained ethics approval for the research from our universities and abided by ethics submission requirements. The purpose of the research, and book, is to predominately answer *how* and *why* questions and so we chose a qualitative research design approach (Saunders,

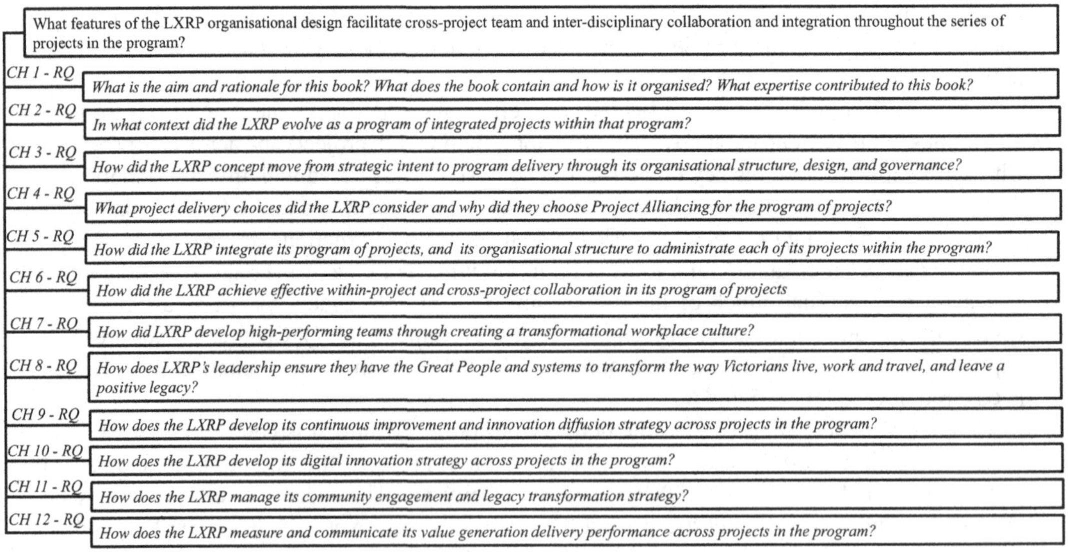

What features of the LXRP organisational design facilitate cross-project team and inter-disciplinary collaboration and integration throughout the series of projects in the program?

CH 1 - RQ   *What is the aim and rationale for this book? What does the book contain and how is it organised? What expertise contributed to this book?*

CH 2 - RQ   *In what context did the LXRP evolve as a program of integrated projects within that program?*

CH 3 - RQ   *How did the LXRP concept move from strategic intent to program delivery through its organisational structure, design, and governance?*

CH 4 - RQ   *What project delivery choices did the LXRP consider and why did they choose Project Alliancing for the program of projects?*

CH 5 - RQ   *How did the LXRP integrate its program of projects, and its organisational structure to administrate each of its projects within the program?*

CH 6 - RQ   *How did the LXRP achieve effective within-project and cross-project collaboration in its program of projects*

CH 7 - RQ   *How did LXRP develop high-performing teams through creating a transformational workplace culture?*

CH 8 - RQ   *How does LXRP's leadership ensure they have the Great People and systems to transform the way Victorians live, work and travel, and leave a positive legacy?*

CH 9 - RQ   *How does the LXRP develop its continuous improvement and innovation diffusion strategy across projects in the program?*

CH 10 - RQ   *How does the LXRP develop its digital innovation strategy across projects in the program?*

CH 11 - RQ   *How does the LXRP manage its community engagement and legacy transformation strategy?*

CH 12 - RQ   *How does the LXRP measure and communicate its value generation delivery performance across projects in the program?*

Figure 1-2 Research questions answered in this book

Lewis and Thornhill, 2016). Our, research paradigm, following Somerville (2012), is critical realism. Somerville explains this as

> characterised as a combination of a realist ontology (or theory of being, of how the world is) with a fallibilist epistemology (or theory of knowledge, of how the world is known). Realism here means the belief that reality exists (in this case the way the project is organised and delivered) independently of the human mind, with human consciousness being typically understood as perception (the observer), cognition (the knower), or thought (the thinker). Fallibilism is the belief that our knowledge of reality always has a chance of being proved wrong, that is, our knowledge has no sure foundations.
>
> (2012, p. 291)

We therefore report on what we see occurring through tangible evidential forms (e.g., authoritative documents) and evidence transmitted through interviews (where we analyse transcripts and make sense of what we were told). We adopted a grounded theory approach using extant theory as a basis for abductively analysing and making sense of the data using NVivo as an analytical tool.

The LXRP evolved over its program period and so we are aware that processes, and systems that we describe as if they are 'real things' do change, therefore, while accuracy and truth is sought, we controlled for this through validation and triangulation of evidence. We use the salient literature to establish a general or specific framework for analysing evidence. Interview evidence was gathered through sound-recorded interviews that were transcribed and sent to interviewees to validate. Internal documentation, where available and external reports, audits, and annual government reports etc. were also gathered for analysis. We also discussed our analysis extensively with LXRP senior staff to ensure that our research had not misrepresented any content or that processes, systems, and routines that we describe in the book did not represent outdated approaches to ensure that at the time of writing (2023). Two of our chapter co-authors were also members of the LXRP and so we also adopted an Autoethnography research design. Anderson (2006) describes Autoethnography as:

> the researcher is (1) a full member in the research group or setting, (2) visible as such a member in the researcher's published texts, and (3) committed to an analytic research agenda focused on improving theoretical understandings of broader social phenomena
>
> (p. 375)

and

> key features of analytic autoethnography that I propose include (1) complete member researcher (CMR) status, (2) analytic reflexivity, (3) narrative visibility of the researcher's self, (4) dialogue with informants beyond the self, and (5) commitment to theoretical analysis.
>
> (p. 378)

An outline of those interviewed are presented in Table 1-1. Most interviews were undertaken for around 60 minutes, sound recorded, transcribed and interviewees sent the transcript to correct, to retract anything or add further details.

Sources of evidence and reference material comprises independent audit reports produced by the Victoria Auditor General's Office (VAGO), together with data, either published through

Table 1-1 Interview information

ID	Relevant expertise	Comments
IV01	AM – extensive alliancing experience over many projects	Interviewed in 2018 for previous LXRP research relevant to this book. Current senior project alliance role
IV02	AM – extensive alliancing experience over many projects	Interviewed in 2018 for previous LXRP research relevant to this book. Current senior project alliance role
IV03	ALT – extensive alliancing ALT and direct team experience over many projects in rail and road sector	Interviewed in 2018 for previous LXRP research relevant to this book
IV04	ALT – extensive alliancing ALT and direct rail operator provider team experience over many alliance projects	Interviewed in 2018 for previous LXRP research relevant to this book. Current senior project alliance role
IV05	ALT and AMT project alliancing experience from an alliance road operator perspective as well as involvement as rail operator adviser and ALT Chair	Interviewed in 2018 for previous LXRP research relevant to this book. Current senior project alliance role
IV06	ALT – extensive alliancing ALT and project director experience over many alliance projects in rail and road sector	Interviewed in 2019 for previous research with LXRP experience relevant to this book
IV07	Rail project director – extensive alliancing experience over several LXRP alliance projects	Interviewed in 2019 for previous research with LXRP experience relevant to this book
IV08	Strategy and culture transformation	Interviewed in 2023. Current senior specialist program integration role
IV09	Strategic communication and stakeholder engagement	Interviewed in 2023. Current senior specialist program integration role
IV10	ALT and strategic continuous improvement	Interviewed several times throughout 2023. Senior specialist program integration role
IV11	Stakeholder engagement extensive project alliance and program culture transformation experience	Interviewed in 2023. Current senior specialist program integration role
IV12	Digitisation, digital engineering expertise and information communication technologies	Interviewed in 2023. Current senior specialist program integration role
IV13	Digitisation digital engineering expertise and information communication technologies	Interviewed in 2023. Current senior specialist program integration role
IV14	Digitisation and digital engineering expertise and information communication technologies	Interviewed in 2023. Current senior specialist program integration role
IV15	Strategy, Program Director Engineering experience	Interviewed in 2023. Current senior program integration role
IV16	Strategy, ALT Engineering Experience	Interviewed in 2023. Current senior program specialist leadership role
IV17	Strategy Program Director	Interviewed in 2023. Current senior program leadership role
IV18	Project Director Engineering experience	Interviewed in 2023. Current senior program leadership role
IV19	Culture Transformation Program Director	Interviewed in 2023. Current senior program specialist role
IV20	Program Director. Culture transformation experience	Interviewed in 2023. Current senior program specialist role
IV21	AM	Interviewed in 2023. Current senior project leadership role

(Continued)

Table 1-1 Continued

ID	Relevant expertise	Comments
IV22	AM	Interviewed in 2023. Current senior project leadership role with being project director on several alliance work packages
IV23	Director Strategy and culture transformation	Interviewed in 2023. Current senior program specialist leadership role
IV24	Strategy, ALT Engineering Experience	Interviewed in 2024 over several separate interviews with notes taken

AM = Alliance Manager, ALT = Alliance Leadership Team member

research reports, or through peer-reviewed published papers and other articles. Considerable information was also found on project teams' organisation websites and company reports. Numerous people involved through the LXRP and associated organisations were interviewed to gain insights and varied perspectives. Chapters also draw upon research data that chapter authors have already gathered through several research studies undertaken on LXRP alliance work packages. Insights gained from rigorous academic works internationally published on projects delivered in a similar manner were also analysed for chapter content. The book provides a valuable seminal reference source for governments, practitioners and students of the delivery of megaprojects.

1.4   Conclusions

This chapter details how the book provides a model for how alliance projects may be effectively delivered to meet their objective and purpose. It also discusses and analyses how this may happen within the LXRP context and summarises this book's 12 chapters. It also explains how the book moves beyond the project focus to include a program perspective and explains how the LXRP strategy differentiates itself from many infrastructure program delivery agencies within a global context drawing upon literature and co-authors with global alliancing/IPD research insights and experience.

The LXRP has special and rare characteristics that should be understood to fully appreciate its vale as a project/program delivery model. It is a program of alliance work packages (AWPs) released and integrated into a program of A\$19.8 billion over 15 years and comprises many elements that can be adapted from one AWP to the next. It is a highly publicly visible program that directly affects most Melbourne residents and visitors. It also has a core of senior project and program staff that have extensive experience in and alliancing/IPD approach. Thus, it moved beyond being an interesting case study to a potential model that is adaptable to suit a range of major infrastructure delivery projects or programs. Its context as a rare infrastructure project, demands that consideration of how this model may be used in future requires it to be adapted rather than adopted to suit a project/program's context.

Notes

1 Victorian Government Gazette No. S 580 Friday 21 December 2018 with the State Premier as the responsible Minister.
2 See   www.project13.info/news/network-rail-southern-region-announces-partners-for-revolutionary-southern-integrated-delivery-alliance-r48/
3 See www.project13.info/library/case-studies/collaborative-working-at-anglian-water-r1/

References

American Institute of Architects – AIA California Council (2007) *Integrated Project Delivery: A Guide*, Sacramento, CA, American Institute of Architects.

Anderson, L. (2006). "Analytic autoethnography." *Journal of Contemporary Ethnography*. 35 (4): 373–395.

Department of Infrastructure and Transport (2011). *National Alliance Contracting Guidelines Guide to Alliance Contracting. Department of Infrastructure and Transport*, A. C. G. Canberra, Commonwealth of Australia. www.infrastructure.gov.au/infrastructure/nacg/files/National_Guide_to_Alliance_Contracting04July.pdf

Doherty, S. (2008) *Heathrow's T5 History in the Making*, Chichester, John Wiley & Sons Ltd.

Eisenhardt, K.M. (1989). "Building theory from case study research." *Academy of Management Review*. 14 (4): 488–511.

Harris, F. and McCaffer, R. (1977) *Modern Construction Management*, Oxford, Blackwell Science.

Lahdenperä, P. (2012). "Making sense of the multi-party contractual arrangements of project partnering, project alliancing and integrated project delivery." *Construction Management and Economics*. 30 (1): 57–79.

McGeorge, W.D. and Palmer, A. (1997) *Construction Management New Directions*, London, Blackwell Science.

Morris, P.W.G. (1994) *The Management of Projects A New Model*, London, Thomas Telford.

Morris, P.W.G. (2013) *Reconstructing Project Management*, Oxford, Wiley-Blackwell.

Morris, P.W.G. and Geraldi, J. (2011). "Managing the institutional context for projects." *Project Management Journal*. 42 (6): 20–32.

Morris, P.W.G. and Hough, G.H. (1987) *The Anatomy of Major Projects – A Study of the Reality of Project Management*, London, Wiley.

Saunders, M.N.K., Lewis, P. and Thornhill, A. (2016) *Research Methods for Business Students*, Harlow, Pearson Education Limited.

Somerville, P. (2012). "Critical realism." *International Encyclopedia of Housing and Home*. Smith, S.J. San Diego, Elsevier: 291–295.

Victorian State Government. (2017). *Level Crossing Removal Project – Program Business Case*, Government V. Melbourne.

Walker, A. (1985) *Project Management in Construction*, London, Granada.

Walker, D.H.T. and Hampson, K.D. (2003) *Procurement Strategies: A Relationship Based Approach*, Oxford, Blackwell Publishing.

Walker, D. H. T., Love, P. E. D. and Matthews, J. (2022). "The value of dialogue in alliancing projects." In *The 9th International Conference on Innovative Production and Construction*, Melbourne, Australia, CIB. https://virtual.oxfordabstracts.com/#/event/public/1970/program?session=35425&s=600

Winch, G.M. (2002) *Managing Construction Projects*, Oxford, Blackwell Publishing.

2 Context of the LXRP

Derek H.T. Walker, Peter E.D. Love and Mark Betts

2.1 Introduction

In what context did the (LXRP)[1] evolve as a program of integrated projects?

Figure 2-1 illustrates key events influencing LXRP's development and eventual incorporation as a unit within the Major Transport Infrastructure Authority (MTIA).

2.2 LXRP/LXRA Strategic Evolutionary Context

In 2014 the Victorian State Labor Party committed, when elected, to remove 50 of Melbourne's most dangerous and disruptive level crossings and proceeded to do so upon winning government. Why do this? What triggered this perceived need?

Metropolitan Melbourne (MM) is Australia's fastest growing and second most populous state capital city. A recent Australian Government Centre for Population states that, in 2030–31, Melbourne's population will be 5,924,700 – surpassing Sydney's.[2] According to the Victorian government (as of 2021), the Greater Melbourne area covers 9992.5km² with a population of around 4.96 million.[3] Projections vary, but MM is expected to reach 7.8 million people by

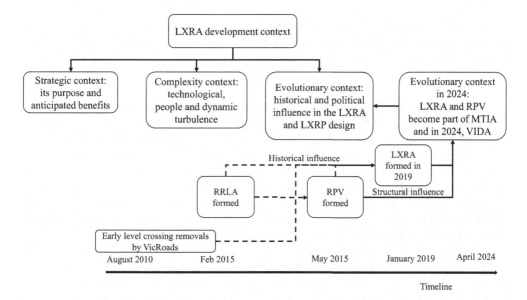

Figure 2-1 Patterns over time: Government authorities influencing the LXRP's development

DOI: 10.1201/9781003389170-2

2051, exacerbating its already congested roads due to boom-gate disruptions on its 178 level-crossings across its electrified rail network (Victorian State Government, 2017, p. 1).

This rapid population growth, and history of procrastination by all state governments over many decades, has placed severe pressure on Melbourne's transport system. Historically, changes to the networks and any improvements in level crossing removal were mainly band-aid solutions not strategic asset management decisions. According to Woodcock and Stone, "most of the important engineering work in building such a large rail system was done in the first fifty-odd years after the railways began in 1863." They add that due to Melbourne's generally flat terrain that most road/rail separations were at-grade and that "in the last hundred years, maintenance, upgrade and grade separation work slowed down, with only around 40 level crossings being removed. Today, Melbourne has over 170 level crossings" (2016, p. 13).

Thus, with procrastination in dealing with the road-rail grade separation for over a century, and rapid growth in Melbourne's population, an untenable situation emerged of boom-gate delays at level crossings caused train flow inefficiencies, danger (frustration about waiting leading to people, bikes and cars 'jumping' the boom-gates) and consequential inefficiencies in pedestrian, car, tram and other transport modes all being delayed while the boom-gates were down. Congestion, both car and rail traffic, was also a key election issue going back many years. The 2014 Victorian election proved a litmus test highlighted by a particularly contentious debate about the need for further road infrastructure projects to link existing freeways and tollways.[4] The 2014 Victorian electorate was keen to improve road traffic, rail link and public transport.

A radical solution for rail network improvements and road-rail grade separation was needed. When Labor won office in 2014, the Premier prepared for fulfilling the election promise by forming the LXRA in May 2015 as the LXRP authority charged with its delivery. Its business case clearly identifies the problem and proposed solution of initially removing 50 of the most dangerous road (and rail) disruptive level crossings (Victorian State Government, 2017). An opportunity to upgrade ageing and outdated stations, signalling and communication technology, and improve the community and local business amenities were also identified.

According to the LXRP business plan, there are three core delivery objectives:

1. Improved productivity from more reliable and efficient transport networks
2. Better connected, liveable and thriving communities
3. Safer communities

(Victorian State Government, 2017, p. 5)

These shaped the delivery strategy, project procurement approach and governance arrangements.

Previous, somewhat ad hoc, delivery of road-rail grade separations had been undertaken before the establishment of the LXRP. From 2000 to 2014 there had been 16 level crossing removals, four of them started under VicRoads in 2014 (VicRoads, 2014, Appendix B). However, while these had successfully achieved grade separation, and some rail tram-line separation, they had not accomplished satisfactory connectivity for intermodal transport transfers such as the train to bus, tram or taxis. Also, station refurbishment/refit/renewal as part of these projects still fell short of commuter expectations about their experience of leaving the stations to safely access nearby transport modes.

A realisation evolved that rail crossing removal was not so much a least-cost/time problem to solve but a problem enmeshed in strategic issues such as intermodal connectivity, rejuvenation of sadly neglected urban assets, and an opportunity to leave a rich community legacy while providing a positive travel experience for rail passengers and road users. True, a simple rail-road grade separation may make it possible for train/car/bus/tram/truck transport flow to improve. Still, it missed meeting many opportunities for positive urban transformation. In some countries, such as Hong Kong and Singapore[5] (Yang and Lew, 2009), train stations are integrated into the commercial fabric of a neighbourhood. This integrated development opportunities (IDOs) for value capture is favoured by the Department of Treasury and Finance Victoria.[6] IDO present challenges in some areas that make them difficult to realise such as difficulties involving local development and community politics where backlash become a reality, as well as a lack of whole of Government/Beuracratic support for the concept. In particular there is, as Johansen et al. observe, "Uncertainty as to whether the new opportunity will succeed and who will share the additional value and benefit" (2016, p14). The LXRP, however, evolved as a strategic transformational program of projects (Chapter 11). Understanding the strategic dimension to the program is an important consideration when judging the issue of delivering value for money (VfM) or best value (BV) to ensure that the best use of committed public resources was achieved.

The wisdom of several major infrastructure projects being announced and procured with no business case support for fear of being considered uneconomical, has been criticised. The East-West Link, for example, was highly criticised for being procured following "flawed advice" (Victorian Auditor-General's Office, 2015). Murphy (2022) studied the East-West Link major Melbourne infrastructure project from the perspective of political and other influences. This project was stopped, post contract signing, after Labor's 2014 election victory. Murphy details one critical factor leading to the project's cancellation, its lack of business case support and tactics used to initially hide critical evidence surrounding the project's procurement that was particularly important in that project's demise. This highlights the need for a business case to underpin the argument for a project's existence and prioritise it over other potentially equal or superior options. However, political opposition parties lack a governing party's access to expert advice and essential data. Therefore, it is difficult for election promises to be fully costed and their business cases fully presented. The LXRP fell into this category. The electorate considered this infrastructure project in their voting patterns, and the state premier cited public transport rail project promises as a justification and rationale for their project strategy.

The Victorian Auditor General's Office (VAGO) published an initial report on the LXRP concluding that the program was ahead of schedule but questioned the business case citing inconsistencies in the business case "[w]eaknesses in the business case undermine its purpose and its value as a basis for the government's decision to commit to the investment" (2017, p. 8). The priority logic of early level crossing removals was questioned, but later the logic of prioritising some of the most dangerous first was clarified with the logic of linking these to less dangerous adjacent crossings to allow the train network on a line to operate more safely and smoothly. In 2020, VAGO reported that their 2017 recommendations report had been addressed that related to the LXRP within the MTIA that reported to the Department of Transport and Planning (DTP).[7] Addressed recommendations (2020, p. 3) clarified the strategic intent, performance expectations, and governance structure (Chapter 3). They highlighted the strategic goals of integrating and improving the transport network, improving safety at the rail crossing locations, and delivering other social benefits such as local steel procurement, job skill and inclusive people hiring policy and other social procurement targets. The successful start to the program in terms of project delivery time and transport network improvements prompted an expansion of the program from 50 to 75 level crossing removals with an increased scope budget from A\$8 billion to A\$14.8 billion. A later announcement brought the scope to 85 crossings at A\$16.3 billion by 2025 and a further announcement in late October 2022 brought the scope up to 110 removals and A\$19.8 billion by 2030. This represented significant scope as well as value increases.

Central to project business cases is the delivery of demonstrated value that justifies the investment (Volden, 2019). Value is often articulated in VfM term but this term has been criticised as potentially problematic and confusing. When undertaking a cost-benefit analysis, costs are compared to the value generated. However, value includes items that can be monetised as well as those that are implied or best guessed. They are usually rigorously assessed but are based on contestable assumptions. For example, what cost can be attributed to injury or death in safety assessment? MacDonald (2011) argues that BV is more appropriate than VfM when monitoring, auditing and benchmarking projects. He concludes that BV may be less deterministic than VfM as it captures value more clearly and comprehensively. We continue to use the term VfM/BV to emphasise this.

It can be concluded that the LXRP's prioritised and planned sequence of level crossing removals were aligned with a VfM/BV delivery strategy – balancing safety with efficiency/effectiveness – obviating the critical problem of dangerous rail crossings while achieving safer and improved train and road traffic flow logic, combined with other considerations about the logistics and disruption impact. The LXRP instigated a significant degree of community consultation (Chapter 12) required to develop the rail crossing removal methodology. Urban planning issues were resolved to meet the strategic intent of enabling connectivity between transport modes (e.g., train, bus, tram, and taxis) and considering legacy impact key result areas. The strategic intent and removal logic process was far more complex than merely aiming to achieve the cheapest direct cost and quickest outcome. Chapter 3 discusses governance and key result areas (KRAs) more fully.

The LXRP context is far removed from an objective of merely achieving rail-road grade separation of the 'iron triangle' (i.e., the lowest cost and shortest project time). Therefore, comparing rail-road grade separation projects with a limited performance focus with the LXRP strategy that also embraced an urban transformational outcome is misleading.

We can conclude that the LXRP scope encompasses:

- A focussed program of alliance projects that has been fashioned into a series of integrated Project Alliances.
- An overall program's scope of removing 110-level crossing through rail-road grade separations in a series of coherent Project Alliance work packages by 2030.
- The renewal of existing rail stations and their precincts to vastly improve station amenities (compared to the minimum cost of existing stations). It links rail passengers entering and leaving stations to other transport modes more effectively. This is designed to enhance the efficiency and effectiveness of connecting transport systems – tram, bus, taxi, pedestrian, or bicycle.
- Addressing social and environmental public expectations through improving the urban quality of the rail line and station precincts, as well as social and commercial legacy improvement.

Central to the delivery philosophy and practice is team integration that melds the project owner (PO) representative team, the design team, and the delivery team (often a joint venture of several contractors and key sub-contractors where there the mechanical and electrical engineering components comprise a significant part of the works), and the infrastructure operators (both road and rail). These teams effectively collaborate as a united team to share their expertise through authentic dialogue to widen the perspectives to develop and deliver the project packages. This results in a deeper and more sophisticated understanding of the project package and program context so that risk and uncertainty assessment, cost/time targeting and project delivery are more effective than conventional project procurement forms (Walker, Vaz Serra and Love, 2022).

2.3   LXRP Complexity Context

Another contextual element shaping the LXRP was being inherently highly complex. Snowden (Kurtz and Snowden, 2003; Snowden and Boone, 2007) identified situations across a complexity spectrum as being considered simple (standard and following a well-understood approach characterised by known knowns), complicated (where standard professional expertise is required to solve difficulties and challenges characterised by known unknowns), complex (where there are considerable difficulties and challenges to face from dynamic and turbulent situations characterised by unknown unknowns) and chaotic (where there is no pattern evident and a situation that is highly volatile and unpredictable characterised by unknowable unknowns).

What made the LXRP complex? Complexity is a contested concept, influenced by what is known, unknown and unknowable. With clear and perfect foresight, knowledge and information, little would be unknown or unknowable, but that is a naïve and unrealistic expectation. Not only is it almost impossible to gain access to perfect knowledge, but human behaviour and exogenous events also intervene to create turbulence and uncertainty. Climate change fluctuations in weather, plague and other crises, for example, intervene.

What may be considered as technically complex may also be viewed as complicated. Technical complexity can often be obviated by gaining sufficient expertise and advice about technical issues. Therefore, much technical 'complexity' confronting well-integrated, multi-disciplinary open-minded teams can be solved. So, these issues can be categorised as highly complicated. However, complexity can be seen from a system of systems interface perspective. The number and nature of these interfaces have long been considered a complexity characteristic (Baccarini, 1996). Perceiving technical or organisational complexity through a system thinking lens may be useful. The LXRP confronted this form of complexity to ensure cross-team integration and collaboration so that these teams' multiple perspectives could help identify system interfaces that may otherwise have been missed. Their highly valued cross-disciplinary knowledge led to access to consider a far wider range of strategies, tactics and options. This aspect was highlighted in the LXRP's Alliance Package 1 response to an adverse emerging problem being solved by a radical innovation developed through knowledge and insights from the design, contractor and road and rail operators. The alliance project team (APT) acted as a single united team. An effective system interface approach resulted in a highly successful 37-day blitz to work 24/7 to complete that project ahead of schedule and under cost (Walker, Matinheikki and Maqsood, 2018).

Rail crossing removals normally follow one of three construction methods (Victorian State Government, 2017, pp. 91–93) that present some complicatedness or complexity but are generally not inherently complex.

- Road over rail – The rail line remains at its existing level, and a new road bridge is constructed over the rail line.
- Road under rail – The rail line remains at its existing level, and the road is lowered to pass underneath the rail.
- Hybrid options – Hybrid options are variants of road over/under rail and rail over/under road options where both the road and rail are raised/lowered. The elevated rail over road with landscaped longitudinal park facilities replacing existing rail line ground is referred to as a 'skyrail' option.

Many technical aspects and system interface challenges contribute to complexity. The LXRP project design, delivery and facility operator team participants collectively have significant expertise and contextual knowledge about the technological and system interface between road and rail and signalling upgrade issues. The program also shares knowledge, innovations, and insights across the program of projects to reduce the level of complexity and unknowable unknowns.

Another identified complexity element is people-related (Remington and Pollack, 2007) presenting behavioural challenges that sometimes seem erratic, causing confusion and negativity. However, the Project Alliance Agreement (PAA) has strict APT behavioural requirements that limit human conflict complexity (see Chapter 8 and Chapter 9 relating to leadership). Other external people-related complexity issues are addressed through stakeholder and community engagement initiatives (see Chapter 12).

The third major complexity influence is caused by dynamic and turbulent external events. These include rare and unexpected PO-initiated issues, for example, the unwillingness or inability to facilitate access to parts of the site. Others relate to what may be considered force majeure when an extraordinary event or circumstance beyond the parties' control occurs. This complexity is managed through the Target Outturn Cost (TOC) and Target Adjustment Event (TAE) process. Essentially in Alliancing, the APT takes responsibility for all project risks and uncertainty that the team feels best to manage. Should such unexpected but foreseeable events covered within the TOC plan eventuate, the APT uses its resilience and innovative, proactive action to remedy them. TAEs that the APT cannot best manage are left to the PO to budget for, and these TAE claims are dealt with reasonably and fairly (Walker, Vaz Serra and Love, 2022). In this way, complexity in LXRP projects is managed coherently and effectively.

2.4   LXRP Historical and Political Influence Context

History and politics are entwined. Two major threads to this context's political complexity are the politics of road versus rail development and the emergence of special purpose quasi-government authorities to specifically take control of realising major initiatives. Murphy (2022) details the struggles in transport policy veering between the proponents of the perceived need for more freeway/tollways (expressing freedom of choice when, where how to travel) and those who proposed public transport improvements to rail and tram systems (expressing public good and fixed time schedule travel). The debate at roads versus rail has also taken on a stronger climate and physical environment focus. These struggles may be seen as common across many cultures and countries. The second thread involves the emergence over time of various forms of Government Authorities established to facilitate and manage infrastructure projects. See Section 2.5 for more on this aspect.

The VAGO (2018, p. 32) report provides the following historical insight:

> Victoria has a history of establishing administrative offices like [Regional Rail Link Authority] RRLA and the Level Crossing Removal Authority for specific time-bound projects. At the conclusion of the project, staff often move on to the next major project or to other roles in the private or public sector.

This Victoria Government experience extends back in time to authorities earlier than the RRLA and LXRA, as summarised below. The establishment of these authorities provided lessons to be learned, influencing the LXRP's development.

The Victorian Government understood the strategic importance of coordinating major infrastructure projects through a purpose-formed authority to manage programs of major projects. The Boston Consulting Group (BSC) identify several enablers that facilitate improved cost and schedule performance adherence.

- upfront investment in de-risking the project, particularly through early contractor engagement
- alignment of procurement approaches and contract models with the specific pressures and risks in each project

- clear legitimacy and support from government sponsors and community stakeholders
- adoption of new technology, tools, and innovation, bringing a focus on productivity improvement and efficiency throughout the project
- deep experience in relevant agencies in similar projects and programs
- setting expectations on cost and schedule which account for the degree of uncertainty.

(Boston Consulting Group, 2021, p. 7)

These are all characteristics of the approach to managing the LXRP as a program of alliance projects. The international benchmarking case studies discussed in their report focused on common drivers of cost and time overruns, and all stem from an inadequate understanding of the scope, scale and consequences of risk and uncertainty in major projects. This challenge was being addressed through experimenting with an Alliancing approach (for example, by the RRLA). This delivery form integrates the project design, delivery, POR and facility operator in an IPD form of collaborative Project Alliance arrangement. This approach has been shown to deliver a more reliable whole-team understanding of the project because of its multi-disciplinary perspective capacity. Walker et al. (2022) provide empirical evidence to demonstrate how an IPD team engaged in Alliancing can produce a more reliable cost/time and delivery plan due to the contractual nature of Alliancing and its focus on cross-team and multi-disciplinary dialogue that leads to a better understanding of a project's context, scope, challenges and likely risk and uncertainty consequences.

Experimenting with alternatives to conventional project delivery models was one set of lessons learned from Alliancing projects such as RRL but limitations of innovation adoption across projects may be due to staff changes and a 'start again from scratch' challenge. This challenge was tackled by developing an administrative/management authority program for a linked project.

2.4.1   *The Regional Rail Link Authority*

The LXRP had an important rail infrastructure precursor, the Regional Rail Link Authority (RRLA) program of projects. VAGO reveals interesting contextual content from its audit (2018) suggesting influences for the LXRA formation rationale. The report states that the DTP established the RRLA in August 2010. Its governance board was accountable for the (RRL) program of projects (work packages), with its board chairman reporting to the Secretary of DTP (2018, p. 20). The audit report also states that the RRL was delivered eight months ahead of schedule and $667 million less than its initial budget of A$4.32 billion. RRL used a range of project delivery approaches (2018, p. 20) to achieve a highly successful outcome that provided a model for future time-bound programs of projects costing billions of dollars over several/many years. Part of its success may be attributed to its program of projects (work packages) procurement strategy of using a delivery choice that best suited each work package's risk profile and the prevailing economic climate context.

Six RRL work packages comprised the program of work. According to the VAGO report (2018, p. 20) the program comprised three different procurement and delivery approaches.

- A work package delivered through a franchisee works model where the state signed a project agreement with the franchise operator Metro Trains Melbourne (MTM) to provide infrastructure works at the Southern Cross Station and its environs on behalf of the state.
- Two novated design and construct packages with the state undertaking limited reference design works, inviting potential suppliers to complete the design and construct works. These were for 'greenfield' environments with no existing operating rail or other rail infrastructure.

- Three other work packages were procured and delivered through competitive alliances (Victorian Auditor-General's Office, 2018, p. 21). These were for 'brownfield' sites, the first time this approach had been used in this context. The critical difference in this approach is that brownfield work involves continuing work while the facility remains operational. This presents considerable logistics challenges. One significant challenge is maintaining the safety of rail passenger and rail freight operations. Electrified 'live' rail areas pose potentially severe complex technical, human and system interface issues.

An organisational innovation that we can see in the evolution from the program management approach of the RRLA is the focus on an integrated and coherent program of projects so that attention could be concentrated on the overall rail link benefits identified for the RRL. This facilitated lessons learned on one project to be shared with others. However, this capacity was diminished by a lack of compulsion in the project contract forms. The more intensely deliverer-competitive Public Private Partnership (PPP) and design and construct (D&C) approaches did not encourage participants to 'give away' what they saw as their intellectual property. The project alliance packages had greater flexibility in doing so because collaborative behaviour is a core value of Alliancing, with or without key result areas (KRAs) and key performance indicators (KPIs), to reward that behaviour. One lesson learned from that aspect was that innovation sharing and diffusion could be operationalised into a KRA and KPIs that would ensure and reinforce collaborative innovation sharing behaviours.

Recent experience of improving and expanding the state's rail assets[8] suggests that a delivery approach of a program of alliance projects would provide the most effective project delivery model. Key personnel from the RRLA helped to develop the LXRP organisational design, drawing upon lessons learned and insights on how to adapt that model to suit the LXRP context.

This baton-passing of how to shape new project organisational forms, especially from an innovative IPD organisational design perspective, is seen in the United Kingdome (UK). Key senior project management people moved from the highly innovative London Heathrow Airport Terminal 5 (T5) project to Crossrail, then onto the Thames Tideway Tunnel and/or High Speed 2 (HS2). All these UK projects have an IPD organisational form as their modus operandi.

2.4.2   The Rail Projects Victoria (RPV) Authority

This government agency was formed in February 2015 and emerged from its previous form as the Melbourne Metro Rail Authority (MMRA), established to deliver the program of projects known as the Melbourne Metro Rail Project (MMRP). The RPV was later expanded in its responsibilities to include managing and planning several major infrastructure programs on the Victorian rail V/Line's regional rail services. The authority was renamed RPV in 2018 to reflect its expanded scope. It later became one of several project teams comprising the DTP's MTIA, with the MMRP as RPV's major responsibility. We now further explain the scope of this program of work.

MMRP's scope includes seven work packages:

1. Early works – Utility service relocation and works to prepare construction sites delivered through a management contractor approach.
2. Tunnel and stations – Twin 9km main tunnelling works, five underground stations, station fit-out, mechanical and electrical systems and specific maintenance services for the infrastructure delivered by the package and commercial opportunities at the new stations delivered through a PPP approach.

3. Rail infrastructure – Works at the eastern and western portals, including cut and cover tunnelling, decline structures, turn backs and local reconfiguration and realignment of existing lines delivered through a competitive alliance approach.
4. Construction power – Utility service provider, works delivered by a service supply contract.
5. Rail systems – Rail systems design (including conventional signalling, High Capacity Signalling (HCS), train and power control systems and operational control systems), installation works, rail systems integration and commissioning delivered through a competitive alliance approach.
6. Tram infrastructure works – Delivered by the tram franchisee through a service supply contract.
7. Wider network enhancements – Works required across the broader network, including track modifications, station upgrades and signalling system upgrades delivered through Market engagement developed on a case-by-case basis.[9]

It is a program of projects with an estimated completion cost of A$12.58 billion (Victorian Auditor-General's Office, 2022, p. 3). This is a highly complex endeavour in interface management with seven packages of mixed delivery approaches and different risk and uncertainty profiles.
The program summary information states that:

Successful delivery of the Metro Tunnel Project requires significant coordination and cooperation between MMRA, Project Co, the RIA and the RSA (each a Package Contractor), as well as with other Related State Project contractors (including the HCMT Project and LXRP) and third parties (such as the Train Franchisee).

Adding that

MMRA has undertaken significant work to understand the potential interfaces associated with the Work Packages and to develop strategies to manage them, including establishing a commercial framework that encourages Package Contractors to work together in a cooperative and collaborative manner with an appropriate level of risk transfer.
(Melbourne Metro Rail Authority, 2018, p. 28).

Considerable ongoing challenges are being met and lessons applied from knowledge either formally or informally recorded. Our interviews with senior members of LXRP Alliancing packages confirm that their experience on the LXRP has allowed them to bring insights to contribute to the MMRP.
This suggests that strategy has evolved in recent years to take the opportunity of the backlog of required infrastructure projects to provide enough job certainty and continuity to allow skilled senior project management staff to fuel greater diffusion of innovation and continuous improvement. This has provided a significant knowledge and capabilities infrastructure outcome. This phenomenon is not confined to senior professional employees of government agencies. As researchers, we observe people we interview say they move about in their employment across organisational roles on these projects, gaining cross-disciplinary competencies. One salient observation we make from numerous interviews with LXRP and RPV staff in Alliancing roles is that they have successfully moved between government organisations such as POR on the LXRP and RPV moving to design team consultant roles and/or with construction contractors engaged in alliances in LXRP and MMRP projects. Learning combines positive Alliancing experience through the way it provides new skill opportunity development by access to multiple

collegial perspectives and leading to critical thinking about challenges and their solutions. This has helped many individuals progress in their careers and competencies.

The same may well be true of the many hundreds, if not thousands, of tradespeople working on these projects and experiencing the collaborative and integrated work environment and ambience that alliancing may offer. Walker and Lloyd-Walker (2014) have documented how, in an alliance case study project, the workers on that project were positively motivated to learn, question and demonstrate critical thinking within a psychologically safe and cheerful workplace environment.

RPV provides another example of how establishing a system integrator and major project administration experiences can facilitate organisational learning in Victoria.

2.4.3   *The Victorian Infrastructure Delivery Authority (VIDA)*

The preceding sections explain how the concept of a single government infrastructure authority evolved first into the MTIA and on 2nd April 2024 into VIDA. In the closing days of 2018, The Victorian Government announced the formation of the MTIA, gazetted to come into effect on 1 January 2019.[10] In doing so, it abolished the Administrative Offices of the:

- LXRA
- Major Road Projects (focusing on road infrastructure projects as a dedicated government body charged with planning and delivering major road projects for Victoria)
- North East Link Authority (responsible for the A$10 Billion 26 kilometre link between The Melbourne Ring Road Greensborough and the Eastern Freeway (integrating road, busway, and bike networks)
- RPV and
- West Gate Tunnel Authority (delivering through a PPP a tunnel and elevated road linking the West Gate Freeway and East link as a second alternative to the West Gate Bridge to enter the docks and central business district)

This represented a key strategy – to draw together each of these existing separate infrastructure government administrative entities into one coherent, integrated authority with the MTIA – and was designed to better manage the program of project interfaces and facilitate collaboration between program teams across the road and rail infrastructure portfolio of transport infrastructure projects. With the formation of VIDA, the MTIA's organisational structure was further modified and restructured as illustrated in Figure 2-2, and its governance arrangements are explained in more detail in Chapter 3.

In April 2024 VIDA was created that incorporated the North East Link Authority and Westgate Tunnel Authority into Major Road Projects Victoria. The LXRP incorporated the Airport Rail Link along with additional rail projects previously administered under the Regional Rail Revival entity. The Director General position was filled by the LXRP CEO reflecting the status of its LXRP respect and origins.

Figure 2-1 illustrates key government authorities that shaped the LXRP concept's evolution showing traceable historical influences in dotted lines and structural influences of the LXRP and RPV in evolving into the MTIA organisational structure.

The Victorian DTP 2021–25 Strategic Plan outlines its philosophy through its vision, purpose, and core values. Its vision is stated as "An integrated and sustainable transport system that contributes to an inclusive, prosperous and environmentally responsible state," its purpose is "The department's purpose is to deliver simple, safe, connected journeys," and its cultural

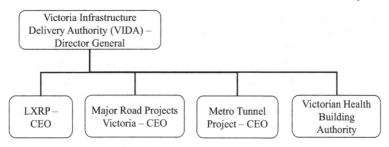

Figure 2-2 VIDA organisational chart

values as "to guide how staff work and enable the delivery of a simple, safe, connected transport network for everyone in Victoria. The Cultural Values support the department's staff to speak up, solve challenges and use shared language to work together" (2021, p. 15).

The language used is consistent with that used in a project alliance agreement. There is a focus on value, sustainability and safety. The driving philosophy appears to be about BV rather than VfM. MacDonald argued that BV should be viewed as a superior aspiration to VfM because BV is more nuanced to appreciate sustainability as a long-term goal that delivers better value.

Another underlying principle of the DTP strategy is the importance of system integration through sophisticated interface management. This has been widely accepted as 'best practice' in project management of complex inter-related programs of work. Morris (1983) recognised this decades ago. In his last book, he reflected on program management as coherently integrating projects within programs to deliver positive transformational change – particularly in relation to social and environmental benefits more effectively realised when the management of programs of projects are strategically thought through to better manage system interfaces (Morris, 2013, Chapter 21).

When we reflect on the evolution of the management of projects in Victoria across the authorities and programs discussed in this chapter and review Figures 2-1 and 2-2, we see a learning-by-experience journey in rail transport infrastructure administration and management. Organisations managing these programs have emerged as facilitating authorities. However, the RRLA's scope was never linked to the tram or bus public transport systems.

The RRLA began to take a more strategic and focussed approach to deliver megaprojects into individual but integrated project work packages that varied in their procurement approach depending on the risk profile and appetite for risk-taking by participants in their design and delivery. The use of project alliances, integrated and managed as a program alliance, for the brownfield project work was seen to be a successful project delivery strategy. Key to this success was the expertise and competencies of the senior leadership team of the RRL program, and this enabled an effective hands-on strategy with the RRLA as an active and productive participant.

Winch and Leiringer (2016) analysed the literature on PO capabilities and identified strategic capabilities as vital in shaping the brief of a project (or program of projects) to move beyond efficiency and effectiveness to commercial, governance, relational and adaptive dynamic capabilities. Dynamic capabilities are particularly useful during the delivery phase as they are demonstrated by resilience, coping with turbulence being open-minded and adaptable to take advantage of converting uncertainty and challenges into opportunities (Teece, Peteraf and Leih, 2016). Morris (2013) adds to this list the ability of sophisticated project leaders to achieve positive, sustainable transformational goals. Senior RRLA project managers developed their alliancing capabilities, and many of them moved on to the LXRP, MMT and RPV and were able to diffuse much of their tacit alliancing knowledge in doing so.

The formation of the MTIA and VIDA may be seen as a logical progression of aspiring to integrate related projects and manage system interfaces at a new level. They have many experienced senior staff to sustain this goal. The Director General of MTIA has a long history of managing large-scale rail infrastructure projects using an Alliancing approach through the RRL and LXRP. The VIDA Director-General was the CEO of LXRP from 2019.

2.4.4   *The role of organisational champion in the LXRA and MTIA formation*

Organisations such as the LXRA, RRLA or MTIA do not just appear out of thin air. Figure 2-1 indicates the evolutionary context, but we could ask how this trajectory drove the formation of the RRLA, initial LXRA and MTIA. One important motivating force is recognition of the need for a specific governance form to realise organisational goals and strategy.

These government authorities were specifically formed to manage the delivery of bespoke major rail infrastructure megaprojects. So, who were likely champions that promoted the emergence of these government authorities and how was that supported?

Understanding the development of these authorities as a learning project, a set of experimental projects that evolved, required a championing force that drove a logic that managed these projects as a time-bound government authority. Each learning project iteration brought with it lessons learned about each organisation's contextual logic, governance arrangement features, barriers and enablers and how to best develop the next iteration to suit the next iteration's context.

Project champion literature helps us understand championing mechanisms at work. Murphy's (2022) East-West Link project cancellation study, pursues three hypotheses explaining how a megaproject was initiated and finally killed off. These hypotheses assist understanding what powers and influences triggered the LXRP as follows (adapted from pp. 153–154):

- The Premiers Hypothesis: Australian state premiers dominate transport policy decision-making. Their political choices become the most important factor in advancing a 'pet' project.
- The Bureaucrats Hypothesis: State bureaucrats pushed Their 'pet' projects for their own reasons. They foisted the project on the government of the day and rushed it to achieve project lock-in. Their motives vary from personal ego to the protection of their departmental interests.
- The Pressure Politics Hypothesis: extra-parliamentary forces effectively influence government policy. Their efforts to initiate or stop a project are significantly important.

Murphy argues that the Premier's Hypothesis was not valid, but he cites global examples of political leaders who may have significant personal championing influence. However, this is insufficient to overrule countervailing political and other influences. He argues that where a political leader has accrued high levels of 'political capital' to push a pet project or policy, this capital is usually expended over time. Projects take a lot of time to gestate, and so political capital may be totally expended during this process.

The Bureaucrats Hypothesis explains a lot about pet projects or policy initiation being shepherded through a procurement process. However, the process involves many bureaucrats from diverse departments and competing interest groups. It is not easy for one group, or individual bureaucrat, to force through a policy or pet project because resistance builds if there is no 'public good' logic rationale or other reason why support should be continued.

The Pressure Politics Hypothesis was pivotal in the East-West Link case study in explaining both the project initiation and its demise through a different array of government-external pressure points. Murphey (2022) makes fascinating reading from a stakeholder engagement as well as political influencing force perspective. In the LXRP case, there was mounting lobbying pressure, as well as high levels of activism, to address the poor state of public transport due to high levels of population growth but also due to procrastination about delivering more effective rail infrastructure. Dissatisfaction with the status quo triggered internal pressures from external public feedback to identify and suggest solutions to a previously lacking integration strategy with and between public transport element interfaces such as train-bus-tram-taxi. Woodcock and Stone (2016, p. 3) noted, for example, that there were "opportunities for the fundamental re-organisation of Melbourne's bus system and its connection to the rail network." There was also considerable public pressure to 'fix' the many problems associated with disruptions and danger at many level crossings. Pressure politics, therefore, informed and legitimised the LXRP strategy as an election-winning policy and that an election promise to be kept legitimately 'trumped' the initially perceived inadequate business case argument – so long as a business case BV evaluation proved this policy decision to be correct.

This presented a 'perfect storm' confluence: Premier and Cabinet championing; bureaucratic support; and pressure politics of key voting stakeholders supporting the LXRP concept. The LXRP was a strongly supported and its strategy not only met the immediate need to 'fix' rail crossing disruption problems but also enhanced efficiency and effectiveness in the passenger rail and network and to conduct a transformational improvement to affected stakeholders through the urban planning sub-strategies and the program's legacy KRAs.

Pinto and Slevin (1989) also provide early ideas about project champions that Pinto later (Pinto and Patanakul, 2015) more rigorously investigated with the advantage of greater access to an authoritative and wider range of psychology and business social science literature.

Pinto and Slevin (1988) outlined a project champion's set of duties beyond non-traditional leader behaviours. They characterised them as:

- Cheerleader – providing the needed motivation (spiritual driving force) for the project team.
- Visionary – maintaining a clear sense of purpose and a firm idea of what is involved in creating the project.
- Politician – playing the necessary "political games" and maintaining important contacts to ensure broad-based support for the project.
- Risk-taker – being willing to take calculated personal and professional risks on behalf of the project.
- Ambassador – maintaining good contacts with all critical project stakeholders (top management, intended users, and the rest of the organisation) and representing the interests of the project.

(p. 17)

Pinto and Patanakul's study (2015) took a private sector perspective. Focussing on the champion as chief executive officer (CEO), they investigated instances of narcissistic behaviour. Pinto and Patanakul's paper cites famous CEOs that championed their pet projects—some of which were outstandingly successful, others were disastrous. Their paper provides a cautionary analysis of groupthink and weakness in governance through the leader's followers' inability to voice and sustain valid concerns, resulting in allowing a narcissistic champion to lead the organisation astray.

The LXRP's case study provides a public sector example. While some elements of the private sector examples provided by Pinto and Patanakul's study (2015) may have relevance in general, public sector governance generally guards against particular individuals demanding total loyalty to a pet cause. Government ministers and senior bureaucrats may wield considerable power, but the bureaucracy has various means to resist this power through delay and, as Murphy (2022) argues, even highly forceful entrepreneurial internal pressure has its limits. Championing change requires institutional acceptance.

Lehtonen and Martinsuo (2008) undertook a study of change management practices within a government public sector setting in Finland. Their study found that middle management ranks were central to the successful change taking place. Many of these individuals took boundary-spanning roles to bridge gaps in responsibility and accountability resulting from the emergent change. This not only facilitates transition but also provides many of the championing behaviours noted by Pinto and Slevin (1988). Lehtonen and Martinsuo state that boundary spanning refers to "the task of linking an organisation with its environment and coordinating that boundary" (2008, p. 22). In this chapter's instance, we are reflecting on how the LXRP emerged from various championing influences, one being whether a single powerful individual wields sufficient power to force through a 'pet' project or whether the impetus comes from an energy wave generated by middle-level bureaucrats that are committed and convinced that the initiative is worthy of their support and voluntary commitment.

The LXRP concept reflects lessons to be learned from the East-West link about stakeholder engagement and transparency and the successful use of Alliancing in the RRL. These lessons centre around planned stakeholder engagement strategies and identifying these megaprojects, in part, as transformational change management projects.

The long-term organisational form, taking a change management project approach, started with the RRLA and resulted in the LXRP, MTIA and VIDA. It was initiated by a group of public servants creating a specific bespoke organisation to initiate and delivery presented severe project complexity and interface challenges, including between departments that may present a danger of inertia. We can make sense of the evolution of the LXRP, MTIA and VIDA as the closing stages of this long-range change management program. We don't see any heroic or perhaps potentially narcissistic champion for the LXRP project delivery concept. We do see, however, evidence of institutional support for this kind of organisational authority and, in the case of RRLA and LXRP, support for the ideas of an integrated collaborative Alliancing team approach to infrastructure project delivery. This implies that support has been institutionalised (i.e., has become part of the project delivery option fabric).

IPD Institutionalisation was observed in the United States of America (USA) with the IPD form being adopted not only by the private sector health provider Sutter Health who initiated IPD in the USA but by other commercial enterprises adopting this project delivery approach (Hall and Scott, 2019).

Institutionalisation is a social process in which a way of working, workplace culture, becomes embedded into the fabric of an organisation. Scott (2014) identifies three 'pillars' of institutional acceptance of a workplace or organisational culture. First, there needs to be a clear set of regulatory arrangements that 'sets the rules of the game.' This sets the basis for shared agreement about 'how things should be done.' In the case of project delivery, this is largely about the contract form and its detailed requirements. Second, the institutionalisation process involves people that are part of the process and have cultural norms that shape their view of regulations and their legitimacy. People's norms and cultural predispositions affect how committed they may be to these regulations. They may support them or decide to ignore or undermine them. The third 'pillar' Scott identifies is cultural-cognitive. This is about how people interpret rules and regulations based on their norms and their cultural leanings. People may bend or manipulate the 'rules' to suit what they believe to be the 'right' way to do things. When we reflect upon the trajectory of organisational change and development of the LXRP/MTIA/VIDA, we argue that participants in this evolution of infrastructure delivery programs have been shaped by institutional forces. Individuals may stand out at various times to influence and 'pass the baton,' but overall, there is an inevitable wave or 'outbreak of common sense' driving the process. Institutional theory helps explain cultural drivers for changes in work practices leading to a particular workplace culture. The best-for-project mindset and swim-or-sink-together Alliancing culture became institutionalised as 'the way things are done here.'

2.5   LXRP Alliance Service Operator Inclusion Choice Context

Chapter 4 discusses the Alliancing choice but both RRL and LXRP cases were based upon delivering projects in a working rail line environment. The technical process of rail construction work is complicated but not intrinsically complex. Rail lines have been built for over a century. While much of the technology involved has not radically changed over that time, human and social standards of user safety and working safely has. The LXRP also dealt with the challenges of working within a live electrified rail network context. In interviews undertaken on alliance project Work Package 1, one interviewee explained the importance of the Rail Safety Act.[11] This act, as with any legal instrument, is difficult for a layperson to understand or interpret, associated with much uncertainty about responsibilities and accountability. Having the rail operator's team on the APT allowed its expertise to be applied to planning and decision-making to reduce risk and uncertainty. While compliance is important, understanding why certain rules have been made and how to interpret them intelligently can lead to innovative solutions being developed to solve difficult site rail interface problems.

The LXRP strategy to closely include the rail and road operator's team can be contrasted with Crossrail. Mark Wild, CEO of Crossrail, during its final stages stated, in a webinar,[12] that one failure of Crossrail was failing to take onboard the advice and technical input of the rail operator team during the critical scoping and problem-solving stages of developing the project delivery strategy. He mentioned that he saw how this could have been better achieved during a visit to Australia, meeting the LXRP senior team on his site visit itinerary.

An example of the value of dialogue that includes operational team participants in planning is the planning of the 37-day Blitz for LXRP Package 1. One of the senior planning team members had considerable rail operator experience. He contributed to proposing ideas about how to use an innovative sheet-piling rig and how it could meet safety compliance requirements. This enabled a radically different construction plan to be developed that enabled the APT to justify the rig's use and to gain appropriate regulatory approvals (see Walker and Matinheikki 2020, pp. 374–379).

A key responsibility of the APT was to ensure that the actions that alliance members took caused no harm to passengers or workers. The rail operator participant organisation was cognisant of safety risks leading to potentially severe repercussions. Safety was not only its paramount contextual issue, but it was also the APT's role to ensure that all alliance participants were aware of the complexity of working in this inherently hazardous environment that had major logistical construction methods implications. A brownfield work context results in the site being shared between live active rail operations and the construction works. This prompts potential worker electrocution risks in a space where trains are active. Also, workers had to contend with road traffic hazards because the work involved road and rail line intersections.

Conventionally procurement project strategies for this context expose the PO to potential claims from contractors for unscheduled disruptions. Brownfield rail infrastructure projects result in contractors having to schedule most of the work within a highly constrained time window – late night to early morning while accounting for late-night rail freight traffic. This context requires high levels of flexibility, grace, goodwill, and tight scheduling and planning logistics. Alliancing requires that all APT participants take responsibility for jointly managing the alliance. The task of scheduling is vastly improved with the operator as part of the APT. This is partly due to the operator's knowledge of train schedules and detailed knowledge of how the Rail Act should be applied. Being part of the APT means that the PO can facilitate rapid decisions being made in a collaborative and constructive manner when and where necessary and when required.

An added LXRP integrated team advantage was the road operator (VicRoads) being part of the APT facilitating extensive road traffic data being available for traffic modelling and logistics planning for soil removal, concrete and other materials/equipment delivery. This expert information helped deliver the highly successful 37-day Blitz on Package 1 works (Walker and Matinheikki 2020, pp. 374–379). Both the rail and road operators were also able to advise the APT on how to best apply for relevant permits for temporary road and rail closures/disruptions. None of this would have been as well-planned had it not been an effectively integrated collaborative alliance project.

Alliancing also has the advantage of binding alliance participants into an integrated collaborative best-for-project mindset, with performance assessment being based on the whole project delivery and not on the participants' individual contributions (Department of Infrastructure and Transport, 2015). This cements the motivation to collaborate as an integrated team because they all swim-or-sink together, and their gain-pain sharing arrangements are set on a range of KRAs, with cost and time being two important ones but others such as safety, innovation, environmental and social impact that determine the APT's commercial reward.

The initially developed TOC (including the time and delivery logistics plan) is more reliable than that for many conventionally project delivery forms because the facility operator is engaged and actively collaborating on the design and delivery strategy, usually for 6–9 months during the Alliance Development Agreement phase and so have a robust understanding of risks and uncertainties and how to cope with these (Walker, Vaz Serra and Love, 2022). The PAA contains workplace behavioural requirements such as APT participants respecting each other's expertise, exposing assumptions made by individuals to scrutiny and having strong perspective-taking abilities so that initial problem solutions may be solved through facilitating true dialogue (Walker, Love and Matthews, 2023).

Thus, when faced with 'brownfield' site conditions that present significant complexity, it is logical, from risk management and cost/time planning perspective, to use the Alliancing approach. Additionally, KRAs can be designed-in into the PAA (as with the LXRP) to encourage and reward innovation and its diffusion across projects in a program. This enhances the strategic links across integrated projects within a program, and under the MTIA then VIDA it now allows these advantages to flow across the divisions within that authority.

2.6   Conclusions

This chapter answered the question: *In what context did the Level Crossing Removal Program (LXRP) evolve as a program of integrated projects within that program?*

The chapter traced the evolution of single-purpose, time-based government authorities to effectively manage major projects and programs of projects. It placed the LXRP within this historical context. Figure 2-3 and Table 2-1 illustrate the logic of our argument to answer this chapter's question.

Figure 2-3 illustrates two prime influences on the choice of the program and its delivery form. The choice of the program was heavily influenced by strategic planning.

This was influenced by the logic of ensuring a PO's hands-on' engagement with the external stakeholder group – lessons learned from the impact of many stakeholders who had successfully opposed previous projects, the East-West Link in particular. There was a clear need for an integrated transport logic for all modes of transport that stakeholders required. The issue of road-rail interfaces at rail crossings had been long neglected and was seen as an urgent necessity to be strategically delivered rather than previous piecemeal approaches. There was a perceived transport project backlog, and a strategy was required to not only urgently remedy that deficiency but also to prepare for population growth over the coming decades. These informed and influenced the strategically targeted alliance program design and delivery approach.

VfM/BV was also a key driver of the decision to adopt a program of alliance projects approach to achieve an urban transformational legacy to areas affected by the level crossing removals. The complexity of working in a brownfield situation was successfully delivered using alliancing on the RRL with lessons learned applied to the LXRP.

Experience on the RRL supported the use of alliancing incentivisation to be deployed for continuous improvement and innovation diffusion not only within projects but across the program, both sequentially and dynamically.

The strategy to deliver the LXRP as a program of alliance projects could be described as an outbreak of common sense, given that the governing authority had the expertise, experience and competence to effectively engage in alliances.

Table 2-1 also helps answer this chapter's research question, summarising the argument features and rationale.

This chapter provides an important introductory step along the journey to understand how the LXRP came into being and how its culture, the rationale for its modus operandi and its contextual evolution was determined.

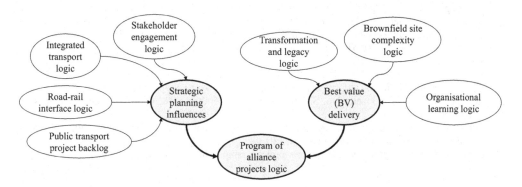

Figure 2-3 An outbreak of 'common sense'

Table 2-1 Chapter 2 question rationale summary

Argument features	Argument rationale
Contextual drivers to form a single-purpose, time-based government authority.	The identified strategic context for the LXRP, its complexity and historical evolutionary context shaped how the LXRP emerged in its present form.
The role of championing	The role of champions and institutional drivers led to LXRP adopting the program alliancing organisational form.
The operator inclusion in the APT choice context	The LXRP Alliancing choice for including the facilities operators in the APT improved addressing effective safety and logistical concerns, especially pertinent for brownfield site projects. Integrated and effective collaborative teams are most suitable for managing complicated and complex interface and system integration issues and engaging in effective dialogue to resolve problems.
The rationale for a program of project alliances.	Figure 2-3 illustrated the rationale for the program alliance form constituting an 'outbreak of common sense.'

Notes

1 The LXRP has gone through several name changes over its initial four years. It started out as the Level Crossing Removal Authority as a body established to deliver 50 level crossing removals but was absorbed into the Melbourne Transport Infrastructure Authority (MTIA) and renamed the Level Crossing Removal Program (LXRP) for some time before being referred to as the Level Crossing Removal Project.
2 https://population.gov.au/data-and-forecasts/projections/capital-city-and-rest-state-projections-2020–21–2031–32. Note that Covid-19 slowed down immigration growth but that growth is expected to recover to some extent, and resume approaching that projection.
3 www.melbourne.vic.gov.au/about-melbourne/melbourne-profile/Pages/facts-about-melbourne.aspx#:~:text=The%20entire%20Greater%20Melbourne%20area,170%2C000%20(as%20of%202021)
4 See Tomazin, 'Battle for the city,' *Sunday Age*, 15 September 2013, p. 19 cited in Murphy, J.C. (2022) *The Making and Unmaking of East-West Link,* Melbourne, Melbourne University Press, 149.
5 The Hong Kong and Singapore example refer to 'Transit Oriented Development' (TOD) rather than IDO and is more ambitious and involves linking rail stations to trigger extensive development growth precincts rather than smaller scale upscaling of land value and commercial development opportunities. IDO is more concerned with value capture around an existing precinct whereas TOD is more concerned with planning a precinct with a new rail station as its hub.
6 See https://navire.com/projects/level-crossing-removal-projects/ also Victorian Department of Premier and Cabinet. (2021). *Victoria's Value Creation and Capture Guidelines –Information for Precincts, Development of Public Land and Capital Investments* Melbourne. 33pp.
7 Department of Transport (DoT) was changed to Department of Transport and Planning (DTP) during 2021
8 https://en.wikipedia.org/wiki/Regional_Rail_Link_project
9 See https://bigbuild.vic.gov.au/projects/metro-tunnel/about/overview/project-delivery
10 Victorian Government Gazette No. S 580 Friday 21 December 2018, with the State Premier as the responsible Minister.
11 Rail Safety National Law Application Act 2013, www.legislation.vic.gov.au/in-force/acts/rail-safety-national-law-application-act-2013/013
12 University College London hosted webinar on leadership in Megaprojects, 11 August 2022.

References

Baccarini, D. (1996). "The concept of project complexity – A review." *International Journal of Project Management.* 14 (4): 201–204.

Boston Consulting Group. (2021). *International Major Infrastructure Projects Benchmarking Review*, Victoria O. o. P. Melbourne, Australia.

Department of Infrastructure and Transport (2015). *National Alliance Contracting Guidelines Policy Principles*, Department of Infrastructure and Transport A. C. G. Canberra, Commonwealth of Australia: www.infrastructure.gov.au/infrastructure/ngpd/files/NACG_Policy_Principles.pdf

Department of Transport. (2021). *Department of Transport Strategic Plan 2021–2025*, Government V. Melbourne.

Hall, D.M. and Scott, W.R. (2019). "Early stages in the institutionalization of integrated project delivery." *Project Management Journal*. 50 (2): 128–143.

Johansen, A., Eik-Andresen, P., Dypvik Landmark, A., Ekambaram, A. and Rolstadås, A. (2016). "Value of uncertainty: The lost opportunities in large projects." *Administrative Sciences*. 6 (3): 17.

Kurtz, C.F. and Snowden, D. J. (2003). "The new dynamics of strategy: Sense-making in a complex and complicated world." *IBM Systems Journal*. 42 (3): 462–483.

Lehtonen, P. and Martinsuo, M. (2008). "Change program initiation: Defining and managing the program-organization boundary." *International Journal of Project Management*. 26 (1): 21–29.

MacDonald, C.C. (2011). Value for Money in Project Alliances. DPM Thesis, *School of Property, Construction and Project Management*. Melbourne, RMIT University.

Melbourne Metro Rail Authority (2018). *Tunnel And Stations Public Private Partnership Project Summary*, Authority M. M. R.

Morris, P.W.G. (1983). "Managing project interfaces-key points for project success." In *Project Management Handbook*, Cleland, D.I. and W.R. King (eds), pp. 3–36. New York, Van Nostrand Rienhold.

Morris, P.W.G. (2013) *Reconstructing Project Management*, Oxford, Wiley-Blackwell.

Murphy, J.C. (2022) *The Making and Unmaking of East-West Link*, Melbourne, Melbourne University Press.

Pinto, J.K. and Patanakul, P. (2015). "When narcissism drives project champions: A review and research agenda." *International Journal of Project Management*. 33 (5): 1180–1190.

Pinto, J.K. and Slevin, D.P. (1989). "The project champion: Key to implementation success." *Project Management Journal*. 20 (4): 15–20.

Remington, K. and Pollack, J. (2007) *Tools for Complex Projects*, Aldershot, Gower.

Scott, W.R. (2014) *Institutions and Organizations*, Thousand Oaks, CA and London, Sage.

Snowden, D.J. and Boone, M. E. (2007). "A leader's framework for decision making." *Harvard Business Review*. 85 (11): 69–76.

Teece, D., Peteraf, M. and Leih, S. (2016). "Dynamic capabilities and organizational agility: Risk, uncertainty, and strategy in the innovation economy." *California Management Review*. 58 (4): 13–35.

VicRoads. (2014). *Strategic Framework for the Prioritisation of Road-Rail Level Crossings in Metropolitan Melbourne*, VicRoads Melbourne.

Victorian Auditor-General's Office (2015). *East West Link Project*, Government V. Melbourne.

Victorian Auditor-General's Office (2017). *Managing the Level Crossing Removal Program*, Melbourne.

Victorian Auditor-General's Office (2018). *Assessing Benefits from the Regional Rail Link Project*, Printer V. G. Melbourne.

Victorian Auditor-General's Office (2020). *Follow up of Managing the Level Crossing Removal Program*, Printer V. G. Melbourne.

Victorian Auditor-General's Office (2022). *Melbourne Metro Tunnel Phase 2: Main Works*, Melbourne.

Victorian Department of Premier and Cabinet (2021). *Victoria's Value Creation and Capture Guidelines – Information for precincts, Development of Public Land and Capital Investments*, Melbourne.

Victorian State Government (2017). *Level Crossing Removal Project – Program Business Case*, Government V. Melbourne.

Volden, G.H. (2019). "Assessing public projects' value for money: An empirical study of the usefulness of cost-benefit analyses in decision-making." *International Journal of Project Management*. 37 (4): 549–564.

Walker, D.H.T. and Lloyd-Walker, B.M. (2014). "The ambience of a project alliance in Australia." *Engineering Project Organization Journal*. 4 (1): 2–16.

Walker, D.H.T., Love, P.E.D. and Matthews, J. (2023). "Generating value in program alliances: The value of dialogue in large-scale infrastructure projects." *Production Planning & Control.* 1–16.

Walker, D.H.T. and Matinheikki, J. (2020). "IPD from a lean supply chain management perspective." In *The Routledge Handbook of Integrated Project Delivery*, Walker D.H.T. and S. Rowlinson (eds), pp. 365–392. Abingdon, Oxon, Routledge.

Walker, D.H.T., Matinheikki, J. and Maqsood, T. (2018). *Level crossing Removal Authority Package 1 Case Study*, RMIT University Melbourne, Australia.

Walker, D.H.T., Vaz Serra, P. and Love, P.E.D. (2022). "Improved reliability in planning large-scale infrastructure project delivery through Alliancing." *International Journal of Managing Projects in Business.* 15 (8): 721–741.

Winch, G. and Leiringer, R. (2016). "Owner project capabilities for infrastructure development: A review and development of the 'strong owner' concept." *International Journal of Project Management.* 34 (2): 271–281.

Woodcock, I. and Stone, J. (2016). *The Benefits of Lessons from Melbourne's Historical Experience*, University of Melbourne R. U. Melbourne.

Yang, P.P.-J. and Lew, A.S.H. (2009). "An Asian Model of TOD: The Planning Integration in Singapore." In *Transit Oriented Development: Making it Happen*, Curtis C. and J. L. Renne (eds), pp. 91–106. London, Routledge.

3   LXRP From Strategy to Governance

Derek H.T. Walker, Mark Betts and Peter E.D. Love

3.1   Introduction

How did the LXRP concept move from strategic intent to program delivery through its organi-sational structure design and governance?

This chapter's research question focuses on how the LXRP strategy shaped its organisational structure and governance arrangements. This chapter builds on Chapter 2, focusing on 'what' aspects by describing the initiating strategy and organisational form. It also explores the 'how and why' LXRP strategic rationale aspects in the ways it evolved, shaping its organisational pro-gram delivery, design and accountability governance arrangements. The governance approaches, designed into the contracts to meet the program's strategic objectives, are explained. The LXRP is a time-limited entity (established in 2015, expected completion in 2030). However, despite its temporary status, it may be considered an innovative, institutionalised organisational form.

How the LXRP has been designed to function in a specific way is central to this chapter's research question. LXRP avoided off-the-shelf solutions in designing its strategic aims. Lessons learned, specifically from the Regional Rail Link Authority (RRLA) experience (Chapter 2), informed and shaped the LXRP strategy. Figure 3-2 illustrates how four central questions influ-enced the LXRP's purpose, strategic approach, and chosen delivery process.

Strategic approach 1, project delivery strategy, was aimed at coping with inherent uncertainty found in brownfield infrastructure rail projects, recognises that best practice assumes that "one size does not fit all" (Shenhar, 2001) but is purposefully developed – based on project contextual contingencies.

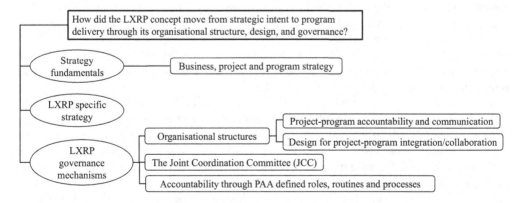

Figure 3-1 Chapter 3 content structure

DOI: 10.1201/9781003389170-3

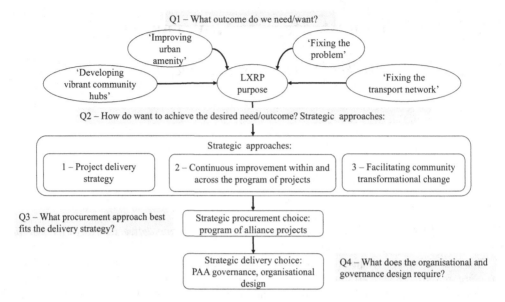

Q1 – What outcome do we need/want?

'Improving urban amenity'

'Fixing the problem'

'Developing vibrant community hubs'

LXRP purpose

'Fixing the transport network'

Q2 – How do want to achieve the desired need/outcome? Strategic approaches:

Strategic approaches:

1 – Project delivery strategy

2 – Continuous improvement within and across the program of projects

3 – Facilitating community transformational change

Q3 – What procurement approach best fits the delivery strategy?

Strategic procurement choice: program of alliance projects

Strategic delivery choice: PAA governance, organisational design

Q4 – What does the organisational and governance design require?

Figure 3-2 The LXRP strategy to deliver governance

Strategic approach 2, continuous improvement, took an emergent strategic path where plans and design for successive crossing removal works were consciously and systematically based upon previous as-built predecessor plans across the alliance projects. Each iteration presented improvement opportunities and grew from the intellectual property (IP) assets of previous experience through the program of 110 crossing removals. The project alliancing delivery choice (Chapter 4) allowed IP to be re-used across the program. We are unaware of this reuse of as-build documentation strategy being adopted anywhere else; it presents a unique strategy. The continuous improvement strategy and innovation is not about innovation for innovation's sake, often resulting in gold-plating (Schaefer and Hallonsten, 2023). Rather, this strategy was realistic and specifically designed to enhance program performance.

Strategic approach 3, facilitating community transformation, illustrates a program strategy to trigger community transformational change to improve the transport system and provide a richer community legacy. Community included rail station precinct residents, people using the infrastructure, and the construction industry community. The latter outcome being able to change the construction industry's culture, often cited as being toxic (Clegg et al., 2023).

The program's purpose, Q1, defines the outcome needed or wanted. A project's raison d'être should deliver a specific need or benefit. Question 2 addresses how to achieve this outcome-benefit. What strategic approach should achieve this? Q3's answer informs the procurement form, aligned to Q1 and Q2's logic. Question 4 addresses how to operationalise an appropriate organisational and governance design for Q3's procurement decision.

During September 2021 the LXRP (2023) strategy was refined and the Five Greats (5-Greats[1]) communicated through an internal document. This strategy was articulated as outlining 'how we intend to achieve our vision of Delivering Great Change – Transforming the way Victorians live, work and travel. We have five strategic objectives, the 5-Greats, and each one has associated success factors to guide our work' (Table 3-2). Central to this strategy is *transforming*. This program is firmly categorised as being a social infrastructure project and not just delivering a hardware product.

According to IV16, this strategic Blueprint (LXRP, 2023) is reviewed by the LXRP annually, to maintain relevance and ambition. For example, the 2023 version includes clearer linkage gained from Aboriginal inclusion insights, specifically the way that 'story telling' and communication framing is done to reinforce a sense of purpose. IV16 explained the role and importance of the LXRP. Big story, the LXRP team story and participant stories and how these are all connected to provide and define purpose. These stories form cultural artefacts that bond people and supports institutionalisation of the workplace culture (Chapter 7).

In understanding the importance of how strategy facilitates a satisfactory outcome-benefit delivery we need to focus on how the chosen strategic direction is shaped, why it is framed in a particular way, and how that is accomplished.

3.2   Business, Portfolio, Program and Project Strategy

A business strategy serves a purpose. The private sector delivers goods and services sustainably at a profit. The public sector's business supports a nation/state/community in its development and maintenance. Governments develop policies to prioritise these activities and establish administration mechanisms to coordinate and manage interfaces between its constituent delivery departments that implement its policies. The challenge is to effectively manage these interfaces so that each part supports and integrates an integrated whole government business. George et al. (2021) identify strategic purpose being beyond meeting a commercial interest vision, or being confined to efficiency. They argue that purpose also relates to service delivered, organisational and stakeholder values and a sense of stewardship.

Part of the business strategy challenge relates to disaggregating and integrating assigned activities to meet the organisation's mission and goals. The Project Management Institute (PMI) recommends that organisations be established in a hierarchy of organisational units comprising portfolios that are grouped into programs of projects (PMI, 2016). Government organisations establish departments with Government Ministers responsible for managing all resources for all projects and operational activities within those departments supported by its organisational structure. This may be viewed as a portfolio management role by balancing resources to maximise the value delivered (Meskendahl, 2010). For example, the Victorian Government formed the Melbourne Transport Infrastructure Authority (MTIA) in 2015 within the Department of Transport (now renamed Department of Transport and Planning (DTP)) to focus on major infrastructure projects (Chapter 2). The UK Transport for London entity was formed in response to major transport upgrades required for the 2012 London Olympics (Davies and Mackenzie, 2014).

Portfolios may be further disaggregated into programs of projects. Similar to portfolios, these include a strategically designed grouping of projects with a consistent, logically linked purpose to deliver benefits/value. Programs, such as the LXRP, may be smaller in scope than portfolios, but they share similar resources that need to be balanced, aligned, and prioritised (PMI, 2008; Artto et al., 2009; Näsholm and Blomquist, 2015).

At the individual project level, strategy is concerned with planning and applying resources to deliver a benefit/value output in a project form. Morris and Jamieson maintain that

> Project strategy management is widely recognised as a significant project management practice for ensuring that project definition and development are comprehensively considered and properly relate to corporate goals and strategies. Project strategy typically covers the entire project lifecycle, with review and optimisation occurring at specific points as the strategy is progressively developed.
>
> (2004, p. vii)

They distinguish outputs from outcomes, stating, "Project portfolio management is predominantly about choosing the right project, whereas project management is about doing the right project" (pp. 9–10). Thus, portfolio, program and project management strategies follow a corporate strategy to deliver detailed outputs through projects and more collectively (and more integrated strategically) through programs and portfolios of projects to deliver a beneficial outcome.

Strategy fundamentally links purpose and outcome. Integrated within a program, desired project outcomes reflect the corporate strategy while the project output contributes to the program outcome. This explains why a single project within a program may be considered as not delivering its planned output but may still critically contribute towards delivering the required program outcome. A project may be purposely conceptualised as an experiment or 'vanguard' project (Frederiksen and Davies, 2008). Similarly, a project may successfully deliver its output but diminish a program's beneficial outcome by starving resources for other vital projects within that program.

Each project's purpose should be clearly articulated, and, in program terms, maintaining focus on the program outcome is vital (George et al., 2021). One aspect of LXRP's project purpose includes continuous improvement through learning within and across the program. The no-blame and consensus decision making elements of LXRP collaboration, specified by the Project Alliance Agreement (PAA), supports an important organisational learning lesson. Initial lack of success in experiments should not be met with punishment or reproach but treated as learning opportunities that benefit the outcome (Anbari, Carayannis and Voetsch, 2008; Love et al., 2016).

Strategy has been perceived in competitive terms in finding the best way to deliver value/benefit at public/private business levels. But value or benefit to whom? Figure 3-2 illustrates this from a government agency perspective in delivering improved transport options while facilitating a healthy and vibrant community hub. The value/benefit is a public good. The aim is to provide the best value (BV) by providing sufficient integrated public and private transport efficiency and effectiveness value combined with improved community amenity while avoiding being myopic about 'the cheapest/fasted' option at one extreme and succumbing to special interest groups by 'gold-plating' at the other extreme (Henisz, Levitt and Scott, 2012). Clearly, a balance needs to be struck.

From a project owner (PO) suppliers' perspective, strategic value includes its public good competitive advantage. This indicates an ability to deliver value for money (VfM) as well as other BV aspects such as knowledge transfer skills, innovative capacity enhancements, rare expertise, reputational capability, and lower transaction costs (Haaskjold et al., 2019) through behavioural capabilities that lower dispute resolution and search/recruitment costs, and valuable dynamic capabilities availability (Davies, Dodgson and Gann, 2016; Teece, 2017). Hence, supply chain members' strategy is concerned with making themselves as attractive as possible to POs seeking collaboration and cross-team/discipline integration with participants to co-generate value. Vital behavioural assets include collaborative capabilities in developing project delivery plans and flexibility and resilience in responding to unexpected events (Walker, Love and Matthews, 2023).

General business strategy theory has evolved over decades. Mintzberg et al. (1998) identify ten strategic schools of thought from the extant business literature, summarised into three groups. The first group exhibits *prescriptive* ways of perceiving strategy, how processes facilitating planning and control should happen through strategy as fixed plans. The second relates to *descriptive* schools to understand and explain how strategy is developed and operationalised in practice. This focused on behaviours, power, and environmental strategy attributes to explain underlying assumptions about strategic development and how it emerged or was designed.

A third, *configuration*, group took elements of the other two groups. It perceived strategy as transforming things through purposeful design and implementation. This strategic perspective perceives a transformational flow between rational, constrained, and emergent action. It accepts that sometimes we are in a position of control and other times not. It assumes that we can influence, perhaps shape, a future path while understanding the environmental dynamics, technicalities, system interfaces and psychological factors in developing strategy. How we do this depends greatly on how we perceive the world and what should be an appropriate process to establish a legitimate and viable strategy in that context.

Figure 3-2 suggests the LXRP strategic approach comprising a conventional design-and-plan elements that conform to Mintzberg et al.'s (1998) prescriptive school of thought with elements of the descriptive schools of thought emphasising continuous improvement through becoming a learning organisation and strategic co-generation of value with its supply chain as described by Porter and Kramer (2011). Additionally, Figure 3-2 illustrates the LXRP's transformational aspirations that link with the Mintzberg et al. (1998) transformational school of thought.

The strategy adopted by LXRP's alliance participants was purposefully shaped using a consistent design thinking approach. Brown (2008) identifies personal traits of design thinkers with empathy, integrated thinking, optimism, experimentalism and collaboration. Leonard and Rayport (1997) frame it as the empathic design where designers work with end-users of a new idea with open minds to co-create value. Gharajedaghi sets out ten principles of design thinking. His ninth principle is particularly relevant to the LXRP strategy:

> Design is the instrument of innovation. Innovation starts by questioning the sacred assumptions and denying the commonly accepted constraints with playful reflections on technology and market opportunities. It was highly focussed with several key objectives in mind, and it recognised.
>
> (2011, p. 137)

A key LXRP assumption is that innovation and continuous improvement would deliver the means for VfM/BV through a collaborative, integrated team.

The LXRP's PAA form is a procurement process innovation that not only designs-in, but also seeks to change, the conventional project delivery mindset. It embraces collaboration and integrated team focus with governance mechanisms that support and reinforce value co-creation. All LXRP alliance participants purposefully and strategically positioned themselves as being competitive (from a VfM/BV perspective) to steer the program's projects through both prescriptive monitoring and control means and the alliance's dynamic capabilities expertise (Davies, Dodgson and Gann, 2016) being aware of early warning signs (Williams et al., 2012) and responding to them with resilience and ambidexterity (Turner, Kutsch and Leybourne, 2016; Pitsis et al., 2018).

3.3   The Specific LXRP Strategy

The Victorian Government's vision was to rectify an identified immediate problem (existing rail crossings disrupting traffic, being inherently dangerous and causing inefficiencies in Melbourne's transport network) and address projected future strains on the transport system while seizing an opportunity to deliver additional public benefits. The LXRP strategy, therefore, was primarily a transformational plan to configure a way to enhance road and rail network effectiveness, interface efficiently with other transport systems and leave a positive legacy to the communities affected by the removal of the level crossings. However, it was also a plan to deliver the immediate goal by focusing on the immediate symptoms (rail crossing disruption and danger).

The LXRP strategy was designed as a system with requisite resources to deliver the identified outputs and outcomes, as well as its transformational sense in terms of its intended interface management with the whole of Melbourne's intermodal (road, rail, tram, bus, bike and pedestrian) traffic network and its legacy of leaving impacted communities with a set of physical and socially beneficial assets. Thus, the LXRP strategy was an effective benefit value realisation tool.

Road-rail crossings throughout Greater Melbourne were, in many cases, highly dangerous and disruptive to road network traffic flow (VicRoads, 2014). The LXRP business case states that between "2003–2012, there have been more than 97 collisions between a train and a vehicle or pedestrian at level crossings, including 40 resulting in fatalities and serious injuries" (Victorian State Government, 2017, p. 12). This provides a pressing rationale for acting to reduce this toll dramatically. The 2014 VicRoads strategy plan identified ten level crossing removals, three completed and 7 funded for removal, with two under planning and preconstruction at the time of the report's publication (VicRoads, 2014, p. 10–11). The Labor Government pledged in their November 2014 election to remove 50 level crossing by 2022 and had achieved that goal ahead of time, pledging to remove 25 more at the 2018 election to be completed by 2025 and then announcing a further ten in 2021 followed by 15 more in 2022.

Initial LXRP strategic goals extend beyond the above. As its business case states, it supports four core goals (Victorian State Government, 2017, p. 6) as illustrated in Table 3-1 from Figure 3-2.

A change of government heralded the formation of the LXRP in May 2015 with a focused government authority, the Level Crossing Removal Authority (LXRA), given the mandate to take over administration and control of the crossing removal program from VicRoads. The LXRA had a VicRoads expert team and a Rail Operator's expert team on board to balance the perspectives of what constituted network operational effectiveness.

Table 3-1 LXRP Strategy goals and purpose

Strategic goal	Scope and purpose
Separating road and rail networks at critical junctions.	Using infrastructure solutions (including removal of the level crossing) designed for each level crossing site.
	The purpose being to remove dangerous crossings that have a tragic accident-death history and improve road and rail network traffic flow.
Implementing a Metropolitan Network Modernisation Program	Including new train stations, improved public transport access, and better pedestrian and cycling facilities.
	This aims to improve transport intermodal linkage effectiveness and rail passengers' travel experience, including access to trains and ending their rail journey to walk/tram/bicycle/taxi or another transport mode to their travel destination.
Improving the urban amenity and physical integration of activity precincts and communities along rail corridors	Using high-quality urban design to make public areas around train stations and level crossings more attractive, accessible, and secure.
	This aims to improve intangible but essential emotional value in the physical assets impacted by the rail station and rail corridors. For example, linear parks are being created for recreational, bicycle/scooter or pedestrian travel and exercise. Many improvements included table tennis, basketball, and exercise equipment fixture accessories.
Improving integrated land use along rail corridors to create vibrant community hubs	Exploring opportunities to undertake property development around stations to improve local amenities, make better use of currently underused land, encourage residential and commercial development around public transport networks, and contribute to more efficient growth patterns across the wider city.
	The purpose being to leave a positive and valuable legacy to those affected by the rail corridors and station refurbished precincts. It also aimed to lift property values in these community hubs by making the areas more attractive, thus lifting revenue for the state and local authorities through land tax and rate income.

On 1 January 2019,[2] with four years of LXRA and LXRP experience to draw upon, the Victorian Government announced the formation of the MTIA. It was gazetted to abolish the LXRA but incorporate the LXRP, with LXRA staff continuing their roles within the MTIA entity. The MTIA creation also abolished other infrastructure authorities, such as road link works in Melbourne's freeway and tollway network, to encompass them under one overarching entity. One strategic outcome was to leverage advantages in gaining expertise on continuous improvement and innovation diffusion processes developed on the LXRP across other critical MTIA-administered infrastructure programs and projects.

From 2019 onwards, the LXRA operations were recognised as the LXRP in the MTIA governance arrangements. This path followed a strategic plan to leverage expertise developed and gained through the successful LXRP to enhance the delivery capability and to upscale these capabilities more broadly within the MTIA.

Taking a road-rail network perspective and considering public and business stakeholders, the LXRP remit was broader and consistent with a systems-thinking approach. The above four supporting strategic goals demonstrate a sustainable and transformative problem solution being pursued rather than a quick fix to remove the symptoms. This broader ambition had organisational and immediate cost/time performance implications. When considering project success or achieving VfM/BV performance assessment becomes more complicated than a mere tick-the-box scenario (Barton, Aibinu and Oliveros, 2019). Chapter 13 discusses LXRP performance in depth.

We can conclude that the LXRP strategic plan, from its initiation to its delivery through its program of alliance projects, exhibited a comprehensive understanding of broader BV strategy rather than being a program to 'fix' the identified, the simpler, but highly important problem of level crossing removal. It appears to be closer to a transformational program.

Table 3-2 LXRP September 2021 articulated strategy

Greats	Description
Great network	Strive for a great transport network – Strive for a great transport network by reducing congestion, improving safety, and unlocking capacity to run more trains. • Remove 50 level crossings by 2022 • Remove 85 level crossings by 2025 • Complete Hurstbridge and Cranbourne line duplications by 2023 • Achieve practical completion for 58 level crossings by 2022 • Minimise disruption to communities, road and rail users • Improve accessibility and integration across all transport networks for a prosperous and sustainable future.
Great places	Create safe, vibrant, attractive, and connected places for communities that enhance the travelling experience. • Urban design solutions are elegant, enduring and take pride of place in the community. • Increase public transport patronage by creating inviting and activated spaces. • Activate under-used land and facilities. • Environmental sustainability integrated into design solutions.
Great partnerships	Leverage great partnerships to drive the safest and most efficient delivery for our works. • Have a shared vision for the project with our client. • Build strong, mutually beneficial relationships with our delivery partners. • Share knowledge to maximise skills, capability and technology and drive consistency, standardisation, and productivity. • Advocate the success of the project's delivery model. • Create enduring change in the way projects are delivered.

(Continued)

Table 3-2 Continued

Greats	Description
Great engagement	Harness the power of community feedback and collective wisdom to deliver great outcomes.
	• Take the community on a journey to inspire interest, create understanding and build support for our projects.
	• Listen, innovate, and share our knowledge to set new standards for community engagement.
	• Strive to innovate and use all available channels to reach more of the community.
	• Build trust by doing what we say we are going to do.
Great people	Unleash the potential of people by investing in their skills, capabilities and wellbeing.
	• Embrace hybrid working and place wellbeing at the heart of our practices and policies.
	• Support a culture of continuous learning and improvement.
	• Create meaningful career paths and opportunities for people to develop.
	• Grow industry capability and capacity.
	• Strive for gender equality and sustainable participation for First Nations' People, and influence industry to do the same.
	• Provide our people with integrated, accessible and intuitive systems.

3.4   Governance Mechanisms for LXRP Program Delivery

According to the Organisation for Economic Co-operation and Development

> Corporate governance involves a set of relationships between a company's management, its board, its shareholders, and other stakeholders. Corporate governance also provides the structure through which the objectives of the company are set, and the means of attaining those objectives and monitoring performance are determined.
>
> (OECD, 2004, p. 11)

This concept has evolved being adapted to project/program governance (Turner, 2006).

Müller (2017) identifies project governance as systems and controls, the processes by which organisations manage these in response to stakeholders' rights and reasonable expectations, and how to define relationships between the project organisation and external and internal stakeholders. Andersen et al. (2020) added to this conception for integrated project delivery (IPD) considerations for governance alliance team relationships and governmentality, the cultural norms and mindset relating to governance arrangements and how alliance project teams (APTs) perceive these. These processes apply once a project has been initiated. Broad objectives of project governance are both process and relationship based:

- *Transparency* to ensure that performance information is accurately presented in a timely manner consistent with the need to take necessary action.
- *Accountability* to ensure that the roles and responsibilities of all parties are clear and designed to support transparency. This is wider than developing an organisation chart to include identifying the ways and means that accountability is maintained and coordinated, for example, through an individual's direct reporting line and ad hoc and standing steering committees that probe and examine how the organisation delivers its strategies.
- *Responsibility* to ensure that all participants know what their role entails, how it fits within the project organisation and what standards, protocols, rules, laws, behaviour, and workplace norms are acceptable and valued.

- *Fairness* ensures that exploitation by anyone is not tolerated, behaviours and actions are ethical, and everyone understands what fairness means.

Program governance follows a similar path to project governance with the additional aim to maintain consistency across the projects to ensure that the program of projects remains coherent and aligned to the strategic goals set, or modified, for the program. Strategy is operationalised into project delivery through the above four objectives and its organisational structures, contracts, and obligation requirements, and means of ensuring collaboration and integration.

Konstantinou (2023) makes an interesting philosophical point about project governance and purpose. She argues that effective governance must be articulated to motivate those intended to be governed so that the governance arrangements are valid, purposeful and resonate with their values. Rather than being a control mechanism, governance should be seen as an enabler of creativity and collaboration (Chapter 6). Relying on governance as control leads to mere conformance rather than affective (want to) commitment and often frequent unintended consequences.

3.4.1   *Governance through organisational structures*

Program governance includes an organisation structure, managed by a steering committee group for individual projects' prioritisation and survival (PMI, 2008). The LXRP is a program of projects linked both by project type (similar in scope, technology, and aim) where program governance was also highly focussed on the diffusion of innovation and learning across the program of projects so that each project had the benefit of knowledge gained through other projects in the program.

Figure 3-3 illustrates a simplified form of the VIDA organisational chart. The VIDA can be seen as a form of network administrative organisation (NAO) as identified by Braun (2018). Its role is formal rather than informal and besides controlling and coordinating roles, it is also a support mechanism for the delivery of transport infrastructure projects. The VIDA Director

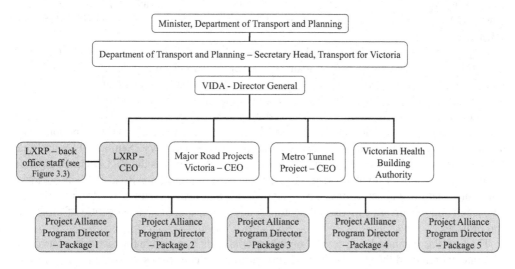

Figure 3-3 Organisation chart illustrating the LXRP role

General reports to the DTP Secretary Head of Transport for Victoria (TfV), who in turn reports directly to the Minister of DTP. The LXRP is one of four project/program entities that includes:

- Major Road Projects Victoria[3], established as a similar authority to the LXRP to oversee the delivery of major road projects around metropolitan Melbourne including the North East Link Project[4] links and integrates the M80 Ring Road in Greensborough to an upgraded Eastern Freeway and regional Victoria with a project scope spanning freeway upgrades, new bridges, roads and road widening and The West Gate Tunnel[5] project includes a 4km Public Private Partnership (PPP) toll road component linking the West Gate Freeway with the Melbourne Docklands and Melbourne Central Business District and associated amenities and works;
- The Melbourne Metro Rail Authority that oversees the development of the Melbourne Metropolitan Rail Tunnel link with associates with new rail stations, upgraded signalling and associated works. It also incorporates the former Rail Projects Victoria[6] (RPV) RPV that supervises a program of regional rail revival projects; and
- The Victorian Health Authority[7] that delivers crucial health services including several new hospitals.

The nature of these interface issues is complex. The programs have multiple project components with varying contract forms used for delivery depending on the project risk profile. Many have 'early works' components to prepare sites for the 'main works.' Services such as electricity, gas, water etc., often need to be relocated, upgraded, or installed from scratch. It is particularly complex from the interfacing organisation perspective because many local authorities (councils), private companies, utility providers, and associated organisations exist. These must be informed and coordinated to minimise public and business disruption while maintaining strict safety standards when managing temporary works.

The LXRP provides a vital position within VIDA. LXRP staff have intimate knowledge of working on highly complex brownfield site projects because all crossing removals involve interactions with a 'live' road and electrified rail system. The project delivery choice was made for Alliancing, which is an IPD form that features the close collaboration of an integrated team comprising the Project Owner Representative (POR) from (what was the LXRA, MTIA and now VIDA), the rail operator project team, the VicRoads project team, the design team, and construction team.

Project Alliancing is explained in Chapter 4. This alliance project team (APT) was integrated to perform as a united single project team that develops the design and delivery plan together with the agreed target outturn cost and time (TOC) that binds participating teams to a fixed time and cost with their profit margins held at stake with an incentivised gain-pain sharing system based on the final cost/time and other key result area (KRA) measures (Walker, Vaz Serra and Love, 2022). This Alliancing experience and knowledge allows the LXRP to function as an innovation, continuous productivity and quality improvement catalyst. Senior MTIA staff that have experienced Alliancing on complex brownfield projects such as the RRLA's three alliance projects and have been involved on LXRP as well as RPV and Major Road Projects Victoria. Therefore, there is a considerable pool of knowledge and experts with cross-disciplinary collaboration and Alliancing experience mentoring and supporting MTIA's portfolio of programs and projects.

The LXRP is organised with several organisational units responsible for each part of its overall strategy organised through the following roles and mechanisms, as illustrated in Figure 3-3. The organisational structure relationships are part matrix and part direct reporting. There were five alliance project work packages reporting to the Delivery Director, who reports to the Chief Operating Officer, who in turn reports to the CEO. Each of the five project alliance directors also has representatives of the functional units who report to their functional units and the specific alliance project director as summarised in Table 3-3.

While the organisational chart may appear complex and possibly prone to confused responsibilities and accountabilities, the Alliancing ethos and culture reinforce collaboration across the projects within the program and facilitates cross-project communication. Figure 3–4 illustrates how the LXRP organisation is designed to support and maintain responsibility and accountability governance measures across the program of projects. Project reporting strategically links key result areas (KRAs) and their key performance indicators (KPIs) through the functional units.

In 2019, KPIs were reviewed and updated to include at the alliance package level: safety, environment, disruption, stakeholder and community, and schedule. Additionally, each package had KPIs associated with continuous improvement and innovation, industry capability and inclusion, and effective engagement.

For example, Safety is a vital KRA with no gain-share incentives but pain-sharing for inadequate performance. The COVID-19 outbreak resulted in a need for cross-project dialogue to ensure that each project director could benefit from knowledge and lessons learned across the program.

Table 3-3 briefly explains LXRP key organisational people roles, further illustrating coherently designed-in strategy being transformed into action.

Each alliance project package has its own organisational structure, illustrated in Figure 3–5. For each project alliance, a project director reports to the LXRP Program Delivery Director and is reported to by the Alliance Leadership Team (ALT) chair and Alliance Manager (AM).

Figure 3–5 illustrates the ALT and Alliance Management Team (AMT) governance arrangements. The Alliancing form governance arrangements of integrated project delivery (IPD) differs from traditional project delivery forms by the inclusion of the Alliance Leadership Team (ALT) role (explained in the next section). This important governance element adds additional expertise value and oversight to enable alliance participating teams to effectively coordinate

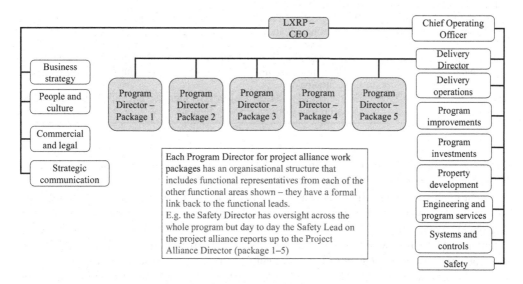

Figure 3-4 LXRP organisation chart illustrating activity roles across the program of projects

Table 3-3 LXRP program functional support roles

Functional role	Reports to	Role summary description
Business strategy	CEO	Supports CEO; integrating LXRP with whole of Government. Coordinates and manages LXRP strategic planning and overall Governance
People and culture	CEO	Manages LXRP resources including recruitment; retention and development via training.
Commercial and legal	CEO	Manages LXRP commercial and legal obligations including procurement; contract administration.
Strategic communications	CEO	Manages LXRP relationships with stakeholders – Government; Community and Internal.
Chief operating officer (COO)	CEO	Responsible for operational delivery of overall program of works and strategic objectives supported by Delivery Director and Program Directors.
Delivery director (DD)	COO	Responsible for Program delivery and coordination of resources across the program. Individual Program Directors report to Delivery Director.
Program Director (PD)	DD	Responsible for the Program of Work assigned from Definition; Planning; Implementation and Close out. The Program Director assembles a team drawn from Functional Groups to deliver Projects in conjunction with Alliance Partners; responsibilities of PD team include both Alliance Role and that of representative of PO under the PAA.
Delivery operations	COO	Managers Delivery support functions including specialist Rail Systems; Rail Occupation Coordination; Approved Rail Operator[8] (ARO) interface; Utilities; Asset Integration and Completions; Industry Capability and Inclusion.[9]
Program improvements	COO	Continuous Improvement across program with both internal and external focus; coordinates Joint Coordination Committee (JCC).
Program investments	COO	Management of Project Definition including coordination of Business Case development.
Property development	COO	Management of Property Development opportunities and maximisation of Property Value.
Engineering and program services	COO	Management of Project Requirements and Standards; Land Planning; Environment and Transactions; Digital Engineering Coordination; Managing Technical Advisor.
Systems and controls	COO	Management of Financial planning and accounting; Project Controls and Estimation including Risk Management, Benchmarking; Information Technology.
Safety	COO	Management of Safety incl Safety Program wide initiatives; Operations Safety and Corporate Safety Systems.

resources and activities as an integrated collaborative unified team as explained in detail by Walker and Lloyd-Walker (2020, pp. 178–182).

Traditionally, alliance participants each have a senior executive from their home organisation on the ALT with the ALT chair, usually the LXRP POR (Department of Infrastructure and Regional Development, 2015, see item 6, page 24–27). The LXRP PAA nominates the Program Director as ALT Chair with an additional senior representative as 2nd LXRP ALT representative; each non-owner participant (NOP) has a single senior ALT representative (usually a Divisional or Executive head of the NOP). The rail operator as NOP has an ALT representative on each alliance.

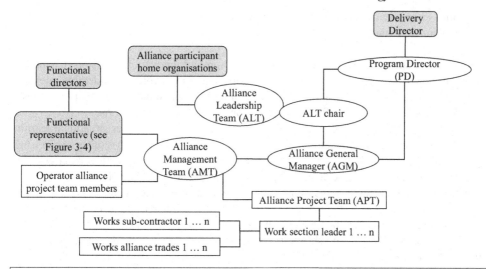

Alliance Manager referred to as Alliance General Manager – this is a bigger role than the traditional Alliance Manager as s/he is responsible for several alliance projects in all stages from development to closure. Typical concurrent work under management could be up to two projects in development, two in delivery and two in closeout comprising $2–3 billion of work.

Figure 3-5 LXRP organisation chart illustrating activity roles on each project alliance package

The AM chairs the AMT comprising participants from the design, delivery contractor(s), and road and rail operator teams. The AM reports to the ALT along with whoever the ALT need from the AMT to report on for specific issues. The LXRP functional directors also have representatives embedded in Project Alliance packages, these individuals report to the AM and their functional directors.

The project subcontractors and delivery team trades and team managers report to the program delivery team manager, who reports to the AM. The APT, AMT and ALT are all required by the PAA to operate collaboratively, making consensus decisions at the AMT and ALT meetings. This facilitates a no-blame workplace environment and full open discussion of issues raised because once consensus is reached, there is no logic in any party finger-pointing blame behaviours. Figure 3-5 illustrates each project alliance package organisational structure, but the LXRP is a program of these projects and is organisationally designed to be structurally linked. Several cross-project integration and collaboration mechanisms are deployed, including the JCC.

3.4.2   Governance through the JCC

The JCC plays two vital roles. First, it integrates the activities of each project alliance package (Figure 3-4) meeting formally each month. The JCC is a form of program steering committee. Its terms of reference (TOR) state its purpose as enhancing coordination, collaboration and sharing insights to achieve continuous improvement. LXRP's PO participation had gained significantly from the gain-pain sharing incentives enabling it to enhance the incentive pool to further facilitate innovation and continuous improvement. Its TOR states a JCC objective as harnessing opportunities for:

• identification of resourcing efficiencies and management of critical resources;
• coordination of track occupations and road closures;

- continuous improvement, learnings and innovations; and
- coordination and management of third-party utilities.

This mechanism significantly enables effective cross-project innovation diffusion (Chapter 9). Its monthly meetings highlight each project's challenges and devotes considerable energy to sharing insights and committing action to a range of emerging and unexpected opportunities. It uses its 5-Greats as a foundation for its coordination and collaboration efforts.

Second, the JCC acts as a community of practice (COP), that facilitates situational learning (Orr, 1990; Wenger, McDermott and Snyder, 2002). The JCC operates across the program where insights, lessons learned about problems and solutions are situated in practice experience and dialogue (Sense, 2007). The JCC learning mechanism is explained in Chapter 9.

Figure 3-6 illustrates the JCC's role in integrating project alliance members across the program. At the strategic level, each project is connected and integrated through a forum that brings together project directors, alliance general managers and alliance management teams to communicate, explore challenges and solutions and devise action plans and idea development and diffusion. At the Tactical level the JCC facilitates a series of sub-committees that are tasked with developing innovation and improvement initiatives, manage their communication through the LXRP intranet and extranet facilities and facilitate linking these initiatives to the KRA/KPI incentivisation processes. At the operational level four cross-project working groups, supported by 20 discipline group champions and JCC sub-committees, operate as a COP that both initiate and action various proposed improvements and innovations. The LXRP supports and manages this platform activity through its governance mechanisms – the JCC, KRA and KPI measurement and assessment, and project AMT and ALTs. This system engages with its partners both LXRP-internal (the NOPs and PO participants) and LXRP-external entities such as the MTIA and government relevant departments.

The JCC group determines the agenda for each meeting that discusses new ideas, innovations and insights that encourage continuous improvement and innovation diffusion (Chapter 9). The JCC is supported by a lessons-learned computer system (Chapter 10) to record ideas and assist organisational learning. VAGO (2020, p. 30) reported that 13 JCC groups integrate

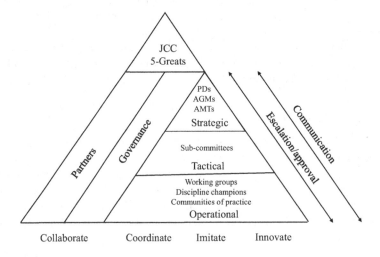

Figure 3-6 LXRP integration role of JCC Groups

(Source: LXRP internal document[10])

trade-discipline individual specialists across both trade and administration process levels to share their knowledge and insights and discuss experiments and innovations tried and tested. We explain the roles of the functional directors and their staff in more detail in Chapters 5 and 6, highlighting their team integration and collaboration across the projects in the program.

The above provides insights into how the integration of projects within the program has been designed-in to facilitate continuous improvement.

3.4.3   *Governance through PAA defined roles, routines and processes*

Ahola et al. (2014) identified two literature streams examining the nature of project governance and how the concept has developed. One stream focuses on economics, contract law, and organisational theory from a transactional cost efficiency perspective, while the other is more about the relationship between an owner and agent acting on behalf of an owner. The starting point of the cost implications of any transaction is considering market efficiency as an intermediary in producing goods and services – the classic 'make or buy decision.'

Williamson (1981) later developed a theory to explain how organisations are structured for optimal financial performance. He likened friction in mechanical devices to business transaction costs. His main idea was to find the most economical way to 'oil' the business machine. He termed this as transaction cost economics (TCE) and introduced concepts such as psychological motivation, trust between parties, and the role of information availability and exchange. These concepts help explain how people engaged in transactions and how they viewed these in fairness and reasonableness terms. Many researchers have since used TCE as a lens for investigating the effectiveness of contract forms in general construction and, more lately, in relation to collaborative, integrated project delivery (Sweeney, 2009; Bygballe, Håkansson and Jahre, 2012; Haaskjold, 2021). This economics and contract focus overlaps a second stream of governance literature that explains how and why TCE and governance are linked.

The relationship between an owner and agent in producing goods, services and projects has been identified as a hidden cost and value in TCE terms. This relationship-based perspective builds upon social and psychological theories to explain how people act and react, share, or hide information and knowledge or the degree to which they may trust or be sceptical of each other's intentions.

Governance, therefore, embraces oiling the business machine through arrangements that hold an economic logic (seeing TCE in cost-benefit efficiency terms) on the one hand with, on the other hand, considering social and behavioural logic (how to organise people and systems) that serve an effectiveness aspect (engaging in business transactions).

The LXRP adopted several program governance arrangements to ensure that the program of projects became a coherent collection of projects contributing to the program strategy. The PAA has been a key innovation in delivering projects that value and require integration and collaboration of the design, delivery, facility operations and owner-representative teams. The shift from these individual teams acting and being independently accountable to becoming a united team, accountable for the whole project performance, requires a designed-in governance changes in expectations in transparency, accountability, responsibility, and fairness that is oriented towards a collective and not individual project performance outcome.

Ross (2003) provided one of the first practitioner guides on how to structure a PAA and how the component parts should be understood. Australian government guidelines on project alliancing (Department of Infrastructure and Regional Development, 2015) explain the Alliancing concept and illustrates how the governance arrangements apply. We identify key governance processes and arrangements in the order they appear in its Alliancing guide.

The guidelines use the term VfM rather than BV, but it suggests that benefits that can be monetised and intangible and difficult-to-monetise benefits are assumed to be part of VfM.

> The Participants must work collaboratively to develop and agree a methodology or approach in designing, costing (as part of the development of the TOC) and constructing the Works which will produce a "value for money" outcome having regard to the PO's VfM Statement. The "value for money" outcome must be able to be demonstrated to the PO, the State and other stakeholders of the Project, and should be assessed and subject to a Value for Money Report at Practical Completion. The Participants must give a real commitment to ensuring that the PO is able to understand the methodology or approach used by the Participants in designing, costing and constructing the Works so that the "value for money" outcome can be demonstrated to the Project Owner.
>
> (p. 19)

This governance requirement clearly states that a VfM report must be delivered that links to the owner's VfM expectations as agreed through the TOC process. The LXRP TOC process is a lengthy (15–20 weeks) integrated team collaborative process of developing an agreed fixed cost and time for the works and the delivery method in detail. This process, explained in detail elsewhere (Walker and McCann, 2020; Walker, Vaz Serra and Love, 2022), highlights how this process differs substantially from other project delivery forms. This is because the integrated team and collaborative nature allow multiple perspectives to be taken when understanding the scope and nature of the work. This permits identifying and exploring how value may be generated through the project. In theory, this should make producing a suitable VfM report more achievable. Access to and re-use of as-built designs from previous LXRP works permits the TOC process being significantly shorter than the typical 6–9 months to 15–20 weeks. Additional transaction cost savings are significant – money, scarce resources and design and delivery intellectual property procurement costs.

The PAA also requires an alliance charter to be agreed upon, developed and published to facilitate all participants being aware of what is expected of them, and to regard this as a key driver to their workplace practices and attitudes. This leads to a best-for-project mindset that pervades the project organisation's culture (Department of Infrastructure and Regional Development, 2015, pp. 21–22). Open book commitment is also part of the PAA governance requirements, specifying that records management practice should ensure transparency and access be maintained. It should be remembered that the PO pays directly for all related costs and is entitled to this level of access to such records. Specifically:

> Participants must fully document their involvement in the Project (including all Reimbursable Costs reasonably and incurred by the Participants in performing the Works) and be transparent in all of their dealings with each other in respect of the Project. For this purpose, the Participants must agree to record-keeping and accounting practices and procedures that the Participants will implement.
>
> (p. 22)

Governance arrangements that support trust and commitment to the project are supported by a 'no litigation' clause in the PAA. The guide states

> Participants must use their best endeavours to avoid issues arising as between each other and, to the extent an issue arises, must resolve the issue internally and otherwise comply with the procedure for the resolution of issues set out in Schedule 14.
>
> (p. 23)

The guide also provides a provision for an exception to that rule where a participant has breached a statutory requirement or engaged in wilful default or deception.

Leadership governance is managed through the ALT that comprises a senior member of each relevant Participant's organisation not part of the AMT. Their role is similar to a corporate board of directors but with a best-for-project responsibility and accountability for the project. Apart from oversight of KPIs and the maintenance of the project's strategic direction, the ALT uses its business networks and influence with its parent organisation to ensure that the interests of the project take precedence. Should there be problems arising out of resource usage on an alliance project by an APT participating organisation, all APT members' ALT will use reasonable endeavours to overcome such problems with the advantage of having a strong voice in their home organisations to do so. The guide states that: "The chairperson must be a representative of the Owner Participant and a member (and not an alternative member) of the ALT" (p. 24).

ALT meetings are held regularly (often monthly), and decisions are made by consensus with a quorum of the Owner Participant and one representative of each non-owner APT organising team participant (NOPs). Consensus decision-making is consistent with the PAA no-blame behaviour and no-litigation requirements. This supports collaboration; logically, if the ALT and AMT operate on a consensus decision-making basis, then attributing blame for any misjudgement leading to problems for the project is illogical. Thus, the APT is incentivised to act as an integrated team in good faith. Incentivisation gain-pain PAA clauses reinforce collaboration because KPIs are based on project performance, not individual NOP team performance. This relationship-based IPD Alliance agreement contract form is markedly different from other relationship-based approaches, such as partnering, and significantly different to conventional procurement choices, such as design and construct (D&C) and design-bid-build (DBB).

Alliance managers are nominated as part of the TOC development process and lead the APT. The AMT comprise each NOP representative from the APT who is responsible and accountable for project delivery and performance. AMT decisions are also made by consensus. Issues not able to be resolved by the AMT may be referred to the ALT. There is a set and prescribed dispute resolution process based on fair and reasonable behaviours and respect for each APT member.

Performance is explicitly defined by KRAs and KPIs. Important TOC-related KPIs are the TOC cost and delivery time. These are fixed but subject to Target Adjustment Event (TAE) changes to the TOC for judicial reasons totally outside the control of the APT. This process follows a fair and reasonable approach. During TOC development, a series of risk identification meetings occur in which alliance bidding syndicates and the PO representative explore and discuss risk and uncertainty categories that may arise. Those that the APT agree is best managed by them and are assessed to ensure that the TOC will include allowance for those risk and uncertainty issues. The PO representative will take responsibility for risks and uncertainty categories that they can best manage. Thus, a fair and reasonable protocol is established for TAE treatment. It also allows the TOC to provide more accurate risk contingencies and for the PO's contingency to be more realistic (Walker, Vaz Serra and Love, 2022). This process also leads to the APT overcoming challenges for risk and uncertainty included within the TOC through innovation, resilience, and ambidexterity.

KRAs/KPI's have both a project and program focus with a series of Program Focused measures. The PAA provides for an Initial Works Package (IWP) as illustrated in Figure 3–4 with subsequent Additional Works Packages (AWP) – each AWP has its own TOC and is a unique project but fits into and is concurrently managed by its individual AMT/ALT. Separate AWP organisations are established – the lead of an AWP is more attuned to the traditional Alliance Manager. It is the IWP and AWP process with several concurrent projects at Development to Delivery to Closeout that makes the LXRP unique.

The overall project performance governs gain-pain incentivisation according to the PAA terms and conditions. Each APT participant agrees to their share of gain-pain based on the final project performance outcome. As adjusted fairly and ethically by the TAE process, the TOC provides the basis for gain-pain sharing through cost differences between the TOC and actual cost. If the actual cost is lower than the adjusted TOC, then a gain will be shared in proportion to the agreed PAA formula. Similarly, pain is shared if the actual cost exceeds the adjusted TOC.

The TOC and incentivisation protocol also provide gain-pain based on other KPIs. For example, the KPI associated with savings from innovation, if diffused to other alliance projects within the program that makes savings on their adoption or adaptation of the diffused innovation, will be allocated to the innovation instigator. Governance arrangements provide for registering such innovations and their recognition of them through the JCC when they deliver savings on other projects. This permits innovation diffusion to be a designed-in governance feature of the LXRP.

One interviewed LXRP AM stated (Love and Walker, 2020, p. 409) that:

> There is innovation, KRA1, and 1.1 is ideal, 1.2 is the execution of those ideas, and 2 is continuous improvement and efficiency. So, that one is about us getting more efficient in how we deliver them in demonstrating cost improvements from package to package to package. . . we're measured on idea generation, so are we generating new ideas and ways of doing things? And we're also measured on how well we implement those ideas and how well we share them across the program, and others benefit from them. For example, if we come up with a new way to pour concrete or something, and we reckon it saves us 2% per job if we share that with the other alliances and they also save 2% per job, we're rewarded for that.

LXRP Project Alliancing takes a radically different approach, compared to conventional D&C and DBB approaches, for direct works payment. The PAA requires the PO to reimburse the NOPs for all direct project-related costs. A rigorous process is put in place to avoid any intentional or unintentional breaches of trust by NOPs through an open-book audited governance arrangement. This gives the PO total and unfettered access to APT documentation to facilitate auditing and assurance of honest and fair NOP commercial behaviour. This is linked to sub-contract approval and other expenditure for materials, equipment, and services. The PAA is specific about how these items should be procured with due reference to compliance with statutory requirements (Clause 15.3) and Victorian Code and Industrial Relations guidelines (Clause 15.4).

The PAA is also clear on how the program's industrial relations plan operates and the expectations and requirements concerning health and safety. The LXRP has an established program-wide governance plan and policy for occupational health and safety. This is essential, given the potential hazards for workers involved in brownfield sites where electric train systems continue to operate and where there are potential road traffic risks for workers (and the public), offering greater certainty and consistency in maintaining workplace safety than for ad hoc, individually delivered projects.

Section 16 of the PAA relates to payment to the NOPs for reimbursement of direct project works undertaken by them and ensuring that the NOPs pay their sub-contractors and suppliers in accordance with the Building and Construction Industry Security of Payment Act 2002 (Vic). PAA governance rules are clear about the process to follow and the implications for breaches in those processes.

Section 17 of the PAA details the reports and records to be provided by NOPs to comply with the LXRP governance requirements. These include work status reports, KRA and KPI reports, and financial and VfM reports.

Section 18 deals with insurance that provides program-wide insurance for each project. This allows for insurability certainty and more economical insurance of the whole program and is often a forgotten benefit when considering the transaction cost savings that Alliancing brings, such as having a consistent insurance process and assurance that required aspects of the work are effectively and efficiently covered. The governance arrangements set out in Schedule 9 of the PAA identify responsibility for insurance cover by the PO and NOPs.

The strategically designed program approach also initiates and maintains adherence to governance through the process of tendering and award for individual program projects. Section 5.5 discusses the recruitment process for project alliance syndicates from the personnel recruitment and the process undertaken for each alliance syndicate to be considered to undertake projects. There is a pool of alliance participants engaged in projects that have a life of three to five years each over the entire decade-long program life. These participant organisations have a steady and reliable pipeline of project work available through the program. It is in their interest to ensure that they fully comply with, and are committed to, the governance arrangements so that as one project ends, they can demonstrate their capacity and capability to be engaged in the 'next' AWP. It is in the LXRP's interest for participating organisations to fully support and understand the governance arrangements. Transaction costs involved in learning, understanding, applying and policing these arrangements impose unnecessary costs (Williamson, 1981) and are reduced by the PAA design.

3.5   Conclusions

This chapter links the LXRP strategy with action. We first discussed strategy from the general to specific project-program perspective and explained the LXRP governance strategy. Our goal was to answer the question: *How did the LXRP concept move from strategic intent to program delivery through its organisational structure design and governance?*

Figure 3-2 illustrates the approach to answering this chapter's core question. Table 3-1 outlines the four strategic goals of the LXRP. We explained how the LXRP could be considered a novel form of a strategic integrated program with innovative and well-developed governance arrangements.

The LXRP strategy of a program of project alliances responded to the context of the program's evolution, as discussed in Chapter 2, and from lessons learned from the RRLA program. The PAA adopted by the LXRP program is consistent with the Department of Infrastructure and Regional Development (2015) that was, in turn developed from experimentation and continuous improvement from PAAs explained by Ross (2003) and earlier PAA developed by the Victorian government's Department of Finance and Treasury (2006). We argue that the LXRP PAA is a process innovation designed to most effectively deliver a highly complex brownfield site program of projects not dissimilar to components of the RRLA program.

The organisational structure was discussed and illustrated. Figure 3-3 shows the LXRP within the MTIA context and Figure 3–4 the program organisational structure, Figure 3–5 the project organisational structure and Figure 3–6 how the projects are integrated into the program through the JCC. The PAA governance requirements were explained in detail, relevant to how the program delivered its intended results.

This program illustrates and explains how the LXRP demonstrates, through its designed-in and institutionalised features, that the program is integrated and collaborative with well-defined (through its KRAs) strategic cross-project and intra-project learning and innovation to deliver BV and the transformational benefits explained in Table 3-1. The LXRP's strategic and governance novelty is associated with it being an effectively integrated program of projects reinforced through its PAA mechanisms and routines and the JCC.

Notes

1 Internal LXRP document LXRP Blueprint 2023 updated from 2021 original version.Devlin September 2021
2 Victorian Government Gazette No. S 580 Friday 21 December 2018 with the State Premier as the responsible Minister.
3 See https://bigbuild.vic.gov.au/projects/major-road-projects-victoria/about
4 See https://bigbuild.vic.gov.au/projects/north-east-link-program
5 See https://bigbuild.vic.gov.au/projects/west-gate-tunnel-project
6 See https://bigbuild.vic.gov.au/about/mtia/rail-projects-victoria
7 See https://www.vhba.vic.gov.au
8 ARO, the approved rail operator, has responsibility for both rail operations and ensuring that legislative requirements (safety etc.) is adhered to.
9 Industry Capability and Inclusion – the policy for expanding the range of employee diversity to address gender, first nations people etc. industry imbalances. It also addresses more general upskilling and industry cultural transformation strategy.
10 PowerPoint presentation LX PRESENTATION 20190710 JCC AGM PD Forum.pptx 23 July 2019, pp. 8–13

References

Ahola, T., Ruuska, I., Artto, K. and Kujala, J. (2014). "What is project governance and what are its origins?" *International Journal of Project Management*. 32 (8): 1321–1332.
Anbari, F.T., Carayannis, E.G. and Voetsch, R.J. (2008). "Post-project reviews as a key project management competence." *Technovation*. 28 (10): 633–643.
Andersen, B., Klakegg, O.J. and Walker, D.H.T. (2020). "IPD Governance Implications." In *The Routledge Handbook of Integrated Project Delivery*, Walker, D.H.T. and S. Rowlinson (eds), pp. 417–438. Abingdon, Routledge.
Artto, K., Martinsuo, M., Gemünden, H.G. and Murtoaro, J. (2009). "Foundations of program management: A bibliometric view." *International Journal of Project Management*. 27 (1): 1–18.
Barton, R., Aibinu, A.A. and Oliveros, J. (2019). "The value for money concept in investment evaluation: Deconstructing its meaning for better decision making." *Project Management Journal*. 50 (2): 210–225.
Braun, T. (2018). "Configurations for interorganizational project networks: The interplay of the pmo and network administrative organization." *Project Management Journal*. 49 (4): 53–61.
Brown, T. (2008). "Design thinking." *Harvard Business Review*. 86 (6): 84–92.
Bygballe, L.E., Håkansson, H. and Jahre, M. (2012). "A critical discussion of models for conceptualizing the economic logic of construction." *Construction Management and Economics*. 31 (2): 104–118.
Clegg, S.R., Loosemore, M., Walker, D.H.T., van Marrewijk, A.H. and Sankaran, S. (2023). "Construction Cultures: Sources, Signs and Solutions of Toxicity." In *Construction Project Organising*, Addyman, S. and H. Smyth (eds), pp. 3–16. Chichester, John Wiley & Sons.
Davies, A., Dodgson, M. and Gann, D. (2016). "Dynamic capabilities in complex projects: the case of London Heathrow Terminal 5." *Project Management Journal*. 47 (2): 26–46.
Davies, A. and Mackenzie, I. (2014). "Project complexity and systems integration: Constructing the London 2012 Olympics and Paralympics Games." *International Journal of Project Management*. 32 (5): 773–790.
Department of Finance and Treasury. (2006). *Project Alliancing Practitioners' Guide*, Department of Treasury and Finance G. o. V. Melbourne, Australia.
Department of Infrastructure and Regional Development (2015). *National Alliance Contracting Guidelines Template 1 Project Alliance Agreement – Template 1 Project Alliance Agreement*, Department of Infrastructure and Regional Development A. C. G. Canberra, Commonwealth of Australia: 165. www.infrastructure.gov.au/infrastructure/ngpd/files/Template_1_PAA.pdf
Frederiksen, L. and Davies, A. (2008). "Vanguards and ventures: Projects as vehicles for corporate entrepreneurship." *International Journal of Project Management*. 26 (5): 487–496.
George, G., Haas, M.R., McGahan, A.M., Schillebeeckx, S.J.D. and Tracey, P. (2021). "Purpose in the for-profit firm: A review and framework for management research." *Journal of Management*. 0 (0): 01492063211006450.

Gharajedaghi, J. (2011). "Design thinking." In *Systems Thinking*, 3rd Edition, Gharajedaghi, J. (ed.), pp. 133–157. Boston, Morgan Kaufmann.

Haaskjold, H. (2021). *The Puzzle of Project Transaction Costs. Optimising Project Transaction Costs through Client-Contractor Collaboration*, Trondheim, Norway, Norwegian University of Science and Technology.

Haaskjold, H., Andersen, B., Lædre, O. and Aarseth, W. (2019). "Factors affecting transaction costs and collaboration in projects." *International Journal of Managing Projects in Business*. 13 (1): 197–230.

Henisz, W.J., Levitt, R.E. and Scott, W.R. (2012). "Toward a unified theory of project governance: economic, sociological and psychological supports for relational contracting." *Engineering Project Organization Journal*. 2 (1–2): 37–55.

Konstantinou, E. (2023). "A philosophy of governance." In *Research Handbook on the governance of projects*, Müller, R., S. Sankaran and N. Drouin (eds), pp. 9–20. Cheltenham, UK, Edward Elgar Publishing.

Leonard, D. and Rayport, J.F. (1997). "Spark innovation through empathic design." *Harvard Business Review*. 75 (6): 102–113.

Love, P.E.D., Teo, P., Davidson, M., Cumming, S. and Morrison, J. (2016). "Building absorptive capacity in an alliance: Process improvement through lessons learned." *International Journal of Project Management*. 34 (7): 1123–1137.

Love, P.E.D. and Walker, D.H.T. (2020). "IPD Facilitating Innovation Diffusion." In *The Routledge Handbook of Integrated Project Delivery*, Walker, D.H.T. and S. Rowlinson (eds), pp. 393–416. Abingdon, Oxon, Routledge.

LXRP (2023). The LXRP Blueprint. Melbourne, Level Crossing Removal Project: 1

Meskendahl, S. (2010). "The influence of business strategy on project portfolio management and its success: A conceptual framework." *International Journal of Project Management*. 28 (8): 807–817.

Mintzberg, H., Ahlstrand, B.W. and Lampel, J. (1998) *Strategy Safari : The Complete Guide Through the Wilds of Strategic Management*, London, Financial Times/Prentice Hall.

Morris, P.W.G. and Jamieson, A. (2004) *Translating Corporate Strategy into Project Strategy: Realizing Corporate Strategy Through Project Management*, Newtown Square, PA, PMI.

Müller, R. (2017). "Organizational Project Governance." In *Governance and Governmentality for Projects : Enablers, Practices, and Consequences*, Müller, R. (ed.), pp. 24–37. Abingdon, Oxon, Routledge.

Näsholm, M.H. and Blomquist, T. (2015). "Co-creation as a strategy for program management." *International Journal of Managing Projects in Business*. 8 (1): 58–73.

OECD (2004) *OECD Principles of Corporate Governance*, Paris, Organisation for Economic Co-operation and Development.

Orr, J. (1990). Talking About Machines: An Ethnography of a Modern Job. PhD Thesis. Ithaca, NY, Cornell University.

Pitsis, A., Clegg, S., Freeder, D., Sankaran, S. and Burdon, S. (2018). "Megaprojects redefined – complexity vs cost and social imperatives." *International Journal of Managing Projects in Business*. 11 (1): 7–34.

PMI (2008) *The Standard for Program Management*, Newtown Square, PA, Project Management Institute.

PMI (2016) *Governance of Portfolios, Programs, and Projects : A Practice Guide*, Newtown Square, PA, Project Management Institute.

Porter, M.E. and Kramer, M.R. (2011). "Creating shared value." *Harvard Business Review*. 89 (1/2): 62–77.

Ross, J. (2003). *Introduction to Project Alliancing*. Alliance Contracting Conference, Sydney, 30 April 2003, Project Control International Pty Ltd.

Schaefer, S.M. and Hallonsten, O. (2023). "What's wrong with creativity?" *Organization*. 0 (0): 13505084231179383.

Sense, A.J. (2007) *Cultivating Learning Within Projects*, New York, Palgrave MacMillan.

Shenhar, A.J. (2001). "One Size does not fit all projects: Exploring classical contingency domains." *Management Science*. 47 (3): 391–414.

Sweeney, S.M. (2009). Addressing Market Failure: Using Transaction Cost Economics to Improve the Construction Industry's Performance. PhD Thesis, Dept. of Civil and Environmental Engineering. Melbourne, University of Melbourne.

Teece, D.J. (2017). "Towards a capability theory of (innovating) firms: Implications for management and policy." *Cambridge Journal of Economics*. 41 (3): 693–720.

Turner, J.R. (2006). "Towards a theory of project management: The nature of the project governance and project management." *International Journal of Project Management*. 24 (2): 93–95.

Turner, N., Kutsch, E. and Leybourne, S.A. (2016). "Rethinking project reliability using the ambidexterity and mindfulness perspectives." *International Journal of Managing Projects in Business*. 9 (4): 845–864.

VicRoads. (2014). *Strategic Framework for the Prioritisation of Road-Rail Level Crossings in Metropolitan Melbourne*, VicRoads Melbourne.

Victorian Auditor-General's Office. (2020). *Follow up of Managing the Level Crossing Removal Program*, Printer V. G. Melbourne.

Victorian State Government. (2017). *Level Crossing Removal Project – Program Business Case*, Government V. Melbourne.

Walker, D.H.T. and Lloyd-Walker, B.M. (2020). "Foundational elements of the IPD Collaboration Framework." In *The Routledge Handbook of Integrated Project Delivery*, Walker, D.H.T. and S. Rowlinson (eds), pp. 168–193. Abingdon, Oxon, Routledge.

Walker, D.H.T., Love, P.E.D. and Matthews, J. (2023). "Generating value in program alliances: The value of dialogue in large-scale infrastructure projects." *Production Planning & Control*.

Walker, D.H.T. and McCann, A. (2020). "IPD and TOC Development." In *The Routledge Handbook of Integrated Project Delivery*, Walker, D.H.T. and S. Rowlinson (eds), pp. 581–604. Abingdon, Routledge.

Walker, D.H.T., Vaz Serra, P. and Love, P.E.D. (2022). "Improved reliability in planning large-scale infrastructure project delivery through Alliancing." *International Journal of Managing Projects in Business*. 15 (8): 721–741.

Wenger, E.C., McDermott, R. and Snyder, W.M. (2002) *Cultivating Communities of Practice*, Boston, Harvard Business School Press.

Williams, T., Klakegg, O.J., Walker, D.H.T., Andersen, B. and Magnussen, O.M. (2012). "Identifying and acting on early warning signs in complex projects." *Project Management Journal*. 43 (2): 37–53.

Williamson, O.E. (1981). "The economics of organization: The transaction cost approach." *American Journal of Sociology*. 87 (3): 548–577.

4 LXRP Delivery Choices

*Derek H.T. Walker, Peter E.D. Love, Mark Betts
and Mattias Jacobsson*

4.1 Introduction

What project delivery choices did the LXRP consider and why did they choose Project Alliancing for the program of projects?

The purpose of this chapter is to explain the process of alliance projects choice and to compare and contrast the motivation and context of this choice by relating it to studies of other similar program contexts. This provides the *how* and *why* part of this chapter's research question. We focus on the LXRP project/program procurement choice as our unit of analysis.

This chapter expands upon Chapter 3 of this book and builds on Chapter 2 which explained the LXRP context in detail.

4.2 General Project Procurement and Delivery Choices

Most of the project procurement and delivery literature focuses on *conventional* approaches that are project rather than program centric, and assume a clear briefing, design, delivery, and

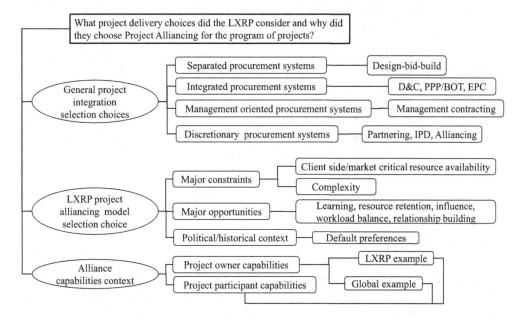

Figure 4-1 Chapter 4 structure

DOI: 10.1201/9781003389170-4

handover sequence. As such, these approaches are generally ill-equipped to deliver the project owner's (POs) expected outcomes. According to a critical review of project procurement methods, Naoum and Egbu (2015) identify several barriers. Specifically, they (2015, p. 6) refer to the impact of "(1) separation of design from construction; (2) lack of integration; (3) lack of effective communication; (4) uncertainty; (5) changing environment, (6) changing clients' priorities and expectations, and (7) increasing project complexity" presenting project procurement process problems. They draw upon many reports highlighting these issues identified over decades (Latham, 1994; Egan, 1998; Murray and Langford, 2003). So, what are these conventional approaches and why are they so entrenched?

Conventional project delivery systems have evolved over the past five decades to encompass numerous systems. Masterman (2002, p. 18) categorises these into four broad areas:

1. Separated procurement systems – where the main elements of the project-implementation process, i.e., design and construction, are the responsibility of separate organisations, e.g., design consultants, quantity surveyors, and contractors. The client deals separately with all members of the project team and is responsible for the funding and eventual operation of the facility. This category contains the *conventional system* known as design-bid-build (DBB).
2. Integrated procurement systems – where one organisation, usually but not exclusively a contractor, takes responsibility for the design and construction of the project and, in theory at least, the client only deals with one organisation. *Design and build, novated design and build, develop and construct,* the *package deal* method and the *turnkey* approach are the main procurement systems in this group. Turkey contractors often provide or arrange project funding and the client is responsible for the subsequent operation of the facility.
3. Management-orientated procurement systems – where the management of the project is carried out by an organisation working with the designer and other consultants to produce the design and manage the physical operations which are carried out by works, or package, contractors. When using these systems, the client will need to have a greater involvement with the project than when employing any of the other methods in the previous two categories. The main systems contained in this group are *management contracting, construction management* and *design and manage*. Integrated project delivery (IPD) forms are included here because they are formalised management collaborative project delivery models (CPDMs) distinct in the anglosphere partnering modes from forms of partnering as understood in Nordic countries (Nevstad et al., 2022). Framework agreements may arguably be positioned in this category.
4. Discretionary relationship-based systems – where the client lays down a framework for the overall administration of the project within which s/he has the discretion to use the most appropriate of all the procurement systems contained within the other three categories. *Partnering* and the little-used *British Property Federation system* are the two main frameworks included within this grouping.

These four categories of procurement systems are discussed below.

4.2.1   *Separated procurement systems*

The construction sector tends to be very traditional compared to, for example, the automotive and aircraft industry as observed many years ago in industry reports (Latham, 1994; Egan, 1998; Murray and Langford, 2003). Partly, this is because many clients (POs) have been conservative and traditional in their view of what constitutes efficiency and value for money (VfM), often opting for the cheapest (contract price) or quickest quoted delivery time.

Government agencies and utilities have for decades used traditional and conventional delivery forms for much of their construction contracting, mainly because their critics have expressed a need for competitive tendering to ensure value for money (VfM), while implicitly assuming that VfM equates to best value (BV) or ignoring non-monetised value. A more recent focus on sustainability (economic, social and environmental) prompted a paradigm shift towards BV terms. In his PhD thesis, MacDonald (2011) explored the concepts of BV and VfM for Alliancing projects in order to determine the optimum configuration to assist in the achievement thereof. Based on his study, it can be argued that BV is a more appropriate term than VfM for Alliancing The United Nations' Sustainable Development Goals Report 2021 also identify a range of environmental and social sustainability measures (United Nations, 2016). Highly conventional contract forms recognise 'iron triangle' cost, time and quality (fitness for purpose) performance measures, but largely ignore environmental and social measures such as improving the social conditions of stakeholders impacted by a project, or environmental aspects beyond legislated requirements through avoiding pollution.

For decades many project procurement textbooks highlight the same general conventional project delivery choices (Franks, 1984; Masterman, 1992; Rowlinson and McDermott, 1999; Masterman, 2002; Moreledge, Smith and Kashiwagi, 2006) as dominating the project procurement landscape. These focus on separated teams and how they are coordinated, not how they relate to one another. The narrative concentrates on hierarchical organisational structures and mechanistic transactional rather than on inter-team behavioural issues. For example, there was little discussion in these early texts about how POs and their teams may benefit from organisational learning project performance measures, and how their contracts do not facilitate conditions for reflection, collaboration to enhance learning lessons learned, or even learning from mistakes and errors, in what has been termed the 'praxis of stupidity' (Love, Ackermann and Irani, 2018).

The most highly traditional/conventional form has been DBB. This approach is used when the PO identifies a need to develop civil engineering infrastructure or commercial, industrial, institutional or residential buildings, and proceeds sequentially. The PO starts working with a design team to develop a brief that outlines the end product characteristics and functions it will perform. Thereafter, then to varying extents, the PO engages with a design team to develop a set of design drawings and specifications to facilitate a tendering/bidding process for delivery entities to be awarded a contract to deliver the infrastructure or building, as described in the design and specification. The bid and contracted cost, time, design and specified products are just the starting point on a long journey over the delivery phase where tendered scope and details are amended and adjusted through negotiated cost, time and outcome quality contract changes. This designed process separates all teams in that process over time – as a sequential 'waterfall' process flowing like water downstream over rapids and obstacles, experiencing serene periods (if lucky), but heading towards its conclusion – often ending up in a somewhat, or very different, shape/form from what was intended in its initial stages.

The DBB form is perhaps the most *fragmented* team project procurement form of all four broad categories. It is often used by inexperienced POs who are once-off or infrequent clients to the construction industry (McGeorge and Palmer, 1997) and who may be novices to the construction industry and how its various players interact and how the whole process of DBB works in practice (Masterman, 2002, p. 7). Before outsourcing of government public works services became popular from the late 1970s to early 1980s, many government agencies, globally, also adopted DBB but often with design being undertaken internally.

In contrast to the above described, the following three broad categories of project delivery form attempt to widen the scope for delivering to include more *reasonable* and thoughtful

approaches to performance beyond iron triangle constraints. POs generally expect to pay a reasonable price for what they get which reflects the need for all participants to be compensated
fairly for their inputs without resorting to opportunistic behaviour. However, as with DBB, the
contracted cost and time is a starting point when projects are subject to client-initiated changes
due to, for example, realising the implications of various design brief decisions made, or when
other circumstances demand design or specification changes.

4.2.2   *Integrated procurement systems*

Before the term *outsourcing* became commonplace, construction clients chose to essentially
outsource the effort of taking a design brief through the DBB stages. They did so for a variety
of reasons, some related to their confidence in participating in the traditional DBB approach
and not being taken advantage of when dealing with contractors, consultants and other project
participants and governing authorities. Others may have felt that while internal staff resources
were sufficiently knowledgeable, they may be unavailable. A further advantage of having an
integrated design and construct (D&C) entity (known as design and build (D&B) in the UK)
to deliver a project, is the simplification of the relationship between the PO and those charged
with delivering it. The PO often appoints a briefing consultant (internally or externally) to work
with the PO representative to scope and define the project output product and the intended
outcome. The project's strategic need led to the brief. For example, a health facility to serve X
patients in Y and Z patient care areas. The PO would then either negotiate with a single D&C
organisation based on the brief or else call for expressions of interest with a shortlisted group of
often two or three organisations bidding on their design based on the brief for a stipulated (often
fixed) contracted price and delivery time. This form emerged in the United Kingdom (UK) and
United States of America (USA) in the 1960s and 1970s and has remained a popular procurement choice globally. The PO pays for the construction delivery in stages or periods (usually
monthly) based on assessment and advice from the PO's (internal or external) construction cost
consultant. The contractor often subcontracts much of the work and holds and manages contracts between it and the sub-contractors and suppliers.

Main variations on this theme included *novated design and construct (or build),* the *package
deal, develop and construct,* and *turnkey.* In the novated D&C approach, the PO organisation
develops the brief and a preliminary design, sometimes referred to as a reference design. Several
D&C organisations then bid based on that design, but these organisations have an integrated
design and construction team who interpret and finalise the design so that it becomes a novated
(new version) with full specifications, contract cost and time plan. The PO then selects which of
the D&C bids to accept and proceed with. The package deal is essentially the same as D&C as
it is an approach where the client deals with the delivery entity to provide a package solution.

A variant of the D&C package deal is where the delivery entity provides the initial capital to
design and construct the project. The client pays for it at handover when they figuratively are
given the key. The term Engineering Procurement and Construction (EPC) is used in engineering and infrastructure projects, particularly in North America for large industrial engineering
projects such as oil and gas plants, chemical processing plants, and transport infrastructure
projects (Merrow, 2011), and is often also referred to as turnkey projects.

Further variants that became popular during the 1980s onwards were the Build Own Operate
Transfer (BOOT) forms. In these forms of project delivery, the PO may choose a turnkey-like
option but instead of paying on completion and handover, a service fee may be negotiated.
For example, a road or bridge may have a toll payable by the user, or a government agency
shadow toll based on traffic numbers. This BOOT arrangement may last in perpetuity with the

construction operation including asset maintenance as well as toll-taking arrangements. Frequently, a concession period for the asset is arranged for a set time period after which the asset is handed over to the PO in working order as specified. Alternatively, an extension of the concession period is negotiated (Walker and Smith, 1995). The BOOT forms became re-badged during the late 1980s as a Public Private Initiative (PPI) or Public Private Partnership (PPP). This whole group of variants has been widely written about in the literature from a contract form explanation (Grimsey and Graham, 1997; Akintoye et al., 2003; Infrastructure Australia, 2008; Jefferies and Rowlinson, 2016; Hodge and Greve, 2017) to financing perspectives (Regan, 2009; Regan, Love and Smith, 2013) with many critiques of these forms (Hodge, 2004; Hodge and Duffield, 2010; Roehrich, Lewis and George, 2014).

4.2.3   *Management-orientated procurement systems*

During the 1970s there was also a rise in the popularity of several *managed-oriented* approaches to project procurement following experimentation of the approach in the USA during the 1960s. These came under the management contracting banner (Sidwell, 1983). These forms often had a consultant team as project coordinators and team integrators on a fee-for-service basis to cover overhead and profit. Generally, contracts remained between the design team, contractor and subcontractors and the PO with the management contractor taking on the role of coordinator-integrator agent. Alternatively, the management contractor could take on the project management role for a guaranteed stipulated sum (and delivery time), taking on the risk of managing *to* that agreed sum/time by wearing any losses or receiving any surplus as a super-profit.

The process of recruiting and contracting a management contractor has varied over time and location. The approach is contingent upon the context and the PO's preferences, expertise, and resource base. The management contractor may be from an integrated professional practice with cost and time consulting expertise, possibly with engineering and architectural design capabilities such as companies like AECOM or JACOB or a construction contractor organisation may take on that role.

However, this continues a fragmented team approach, partially integrated through a coordinating entity, either on a fee basis or a price/time guarantee with the entity taking the risk. Either way, the agreed contract price is the starting point, similar to DBB. Any change from the tender documents is subject to a contract variation order. An advantage of the management contracting approach is that the project can be organised as a series of work packages often, as with an experienced construction delivery company taking the main role, there are opportunities for early works to proceed and fast-track construction where design and delivery are concurrent with various work packages tendered as and when sufficiently design and specification detailed to speed up the whole process (Masterman, 2002, Section 5.3).

IPD integrates teams, systems and business routines and practices to facilitate high-level collaboration that increases the value and optimises project output results. Key features include integration of client-side PO representatives, designers, and the delivery team which includes the general contractor and project first-tier services and other key sub-contractors. IPD took root in the USA after the 1994 Northridge Earthquake in Southern California which triggered legislators to pass the Hospital Facilities Seismic Safety Act that required all hospital buildings to be retrofitted with earthquake impact devices. Sutter Heath, a large health services provider concluded that IPD was the only feasible way to undertake the required construction works and this led to the Integrated Form of Agreement (IFOA) contract, binding participants into a single integrated team (Lichtig, 2005). IPD in the USA evolved from a sense of urgency to deliver projects and to overcome impending severe project delivery resources.

Project alliancing evolved over the past two to three decades as a seminal IPD from across the globe (Lahdenperä, 2012). According to the Australian Commonwealth Government Department of Infrastructure and Transport, it is a highly collaborative form of an integrated form of project delivery in which the PO's representative, design and construction contractual parties engage in a Project Alliance Agreement (PAA). They further state that (2011d)

> All Participants are required to work together in good faith, acting with integrity and making best-for-project decisions. Working as an integrated, collaborative team, they make unanimous decisions on all key project delivery issues. The alliance structure capitalises on the relationships between the Participants, removes organisational barriers and encourages effective integration with the Owner.
>
> (p. 9)

UK New Engineering Contract (NEC) NEC3/4 and PAA projects are further examples of megaprojects delivered through IPD-Alliancing forms of contract.

Central to the general success of Alliancing and other similar IPD models is the common logic illustrated by a focus on enabling relational attitudes such as high trust and commitment, levelling power and authority asymmetry, teamworking quality supported by shared leadership, and incentives based on project performance rather than individual/team performance (Suprapto, 2016; Nevstad et al., 2022). These project delivery models generally lead to better project performance than lump-sum and reimbursable contracts, however, as Suprapto's study found, the contract form does not guarantee such success with the distinguishing factors being people-related. Chapter 6 discusses collaboration in IPD/Alliancing more fully. Interestingly, recent Netherlands infrastructure quantitative research concluded that alliancing-like procurement choices, and their required behaviours and workplace environment and culture supported and influenced good teamwork that leads to high levels of project delivery performance (Koolwijk, van Oel and Moreno, 2020).

A system of framework agreements has been used for many decades where a program of work is outsourced to a series of pre-qualified and approved providers based on a negotiated agreed time, materials and profit margin reimbursement (Walker, Bourne and Rowlinson, 2008). These agreements are made by large, often public sector, clients or utilities with small-medium sized specialist service organisations to allow outsourcing with agreed protocols for the PO to effectively manage the providers and gain access to emergency or specialised resources that would otherwise be underutilised if 'owned' by the PO (Manchester Business School, 2009; Walker and Lloyd-Walker, 2015, pp. 50–51). Local governments use this as a supply chain management tool (Khalfan, McDermott and Kyng, 2006) and utility companies have also been active in this space (Smalley, Lado-Byrnes and Howe, 2004; Department of Health, 2012). The British Airport Authority (BAA) evolved its strategy from framework agreements it had been using for 15 years (Tennant and Fernie, 2010) into what became its Heathrow Terminal Five (T5) agreement that also mirrored many aspects that evolved into the NEC4/5 form (Doherty, 2008; Davies, Gann and Douglas, 2009). Framework agreements tend to be with a group of project content providers rather than a single agreement form on a specific project as is the case with IPD IFOA contracts and Alliancing PAAs. A recent example of this evolution can be seen in the UK with the emergence of the UK Institute of Civil Engineers (ICE) *Project 19* initiative. The UK's ICE web site (see www.project13. info/) contains several case study documents about government infrastructure authorities and agencies adoption of Project 19 principles and offers access to it framework and a raft of other useful reference materials.

4.2.4   *Discretionary relationship-based systems*

Masterman (2002, Chapter 6) describes a series of systems where the client decides upon the overall administration framework for the project but uses and adapts elements from the other three broad categories. He suggests *Partnering* and the little-used *British Property Federation system*. During the period since that time, several relationship-based systems have evolved. There has been a strong trend over the past decades towards what some (for example: Bygballe and Swärd, 2019; Engebø et al., 2020) term Collaborative Project Delivery Models (CPDMs). These evolved from partnering arrangements between the PO, the design team, and the project delivery supply chain.

The concept of partnering became popular during the late 1980s and early 1990s with the USA Construction Industry Institute study (CII, 1996) being highly cited in the project procurement literature. Partnering is an informal or semi-formal contractual arrangement where project participants across the design, main contractor and sub-contractors agree to form a partnership on the project to deliver a superior outcome rather than a business-as-usual, as per the project specifications and design outcome. It can produce a project with less risk of conflict when functioning as intended, in a collaborative team atmosphere with shared/common project goals (Eriksson, 2010). However, as Green (1999) argues, often the partnering arrangements are a cover for exploitation by the PO or main contractor. Larson's (1995) study of 280 partnering projects found that intended outcomes are often not achieved because participants feel that they have been coerced into agreement.

Eriksson (2010) observed that the literature on partnering, at that time, failed to provide a common definition of partnering and that it seemed to mean different things in different countries and cultures. He noted that many definitions appearing in the literature mixed partnering mixed procedures with outcomes making it an ambiguous term. Eriksson's work was inspired by research undertaken around the same time on the nature of partnering and how best to understand it. The research idea that Nyström (2005) offered resonated so effectively with others researching partnering and later Alliancing (for example: Yeung, Chan and Chan, 2007). Nyström distinguished partnering's "general prerequisites, components and goals when discussing the concept" (2005, p. 473) and then used the concept attributed to the philosopher Ludwig Wittgenstein in playing the family-resemblance game where family members are identified through their forbears' physical or other features, to describe a family of partnering forms. Through this concept, Nyström matched partnering-like attributes or component features to form a family resemblance matrix (2007). The central attribute was an inter-team mutual understanding of project goals, and how best to achieve them founded upon inter-team and inter-individual (benevolence, competence, integrity) trust. Often a partnering 'charter' was developed that participants agreed upon that included agreed team relationship behaviour guidelines. Surrounding this core element are grouped other essential ingredients: an effective partnering participant choice protocol; relationship-building activities; a pre-determined and agreed dispute resolution protocol; economic incentive contracts; a partnering facilitator; participant openness; and continuous structured meetings. Building on the work of Nyström (2005), it has been suggested that the core elements of partnering are interrelated in nature and develop sequentially as the project evolves (Jacobsson and Wilson, 2014). For example, mutual understanding and trust are the results of well-working team relationship behaviour guidelines and are thus dependent on actors' engagement. Also, it has been shown that the enactment of partnering elements has the potential to promote a shift towards a more service-oriented mindset among involved actors (Jacobsson and Roth, 2014).

Individual partnering forms cited in case studies or offered in various partnering frameworks could be explained by characteristics into of their general prerequisites, their components, and the outcomes or goals that the arrangement sought to achieve through partnering. This advance unbundled the partnering definition into something that became more understandable because it better contextualises the form deployed and so enables benchmarking, case study comparison and analysis of specific partnering instances.

Other types of agreements used for smaller projects within a programme of work are often referred to as framework agreements. These have been used widely in the UK utility sector for example, where consultants or contractors may sign contracts with a PO to be on call to respond to emergencies or to undertake smaller projects (Walker and Hampson, 2003a, pp. 21–23).

Recently, literature has emerged from Norway on the partnering approach adopted by the oil and gas industry and numerous government agencies for infrastructure projects where a form of partnering is used that more deeply entrenches the key collaborative and closer team integration elements (Aarseth et al., 2012; Børve et al., 2017; Børve, 2019; Nevstad et al., 2021), and in Sweden for electrical distribution infrastructure projects (Jacobsson, 2011b). These studies indicate a closer focus on achieving genuine partnering principles, which has been the case with findings such as Larson's (1995) study in the USA, or in the UK as found by Green (1999).

A further approach that emerged through the European Commission was what became known as the competitive dialogue (CD) process (Hoezen, Voordijk and Dewulf, 2013). This process presents an additional layer in the procurement process where, usually at the early stages of the project design development, formal and informal negotiation takes place over the project design and delivery mechanism, so that the PO and the delivery entity can improve project objective sensemaking through a dialogue process that exposes means, ends, needs, and wants that can provide superior outcomes through partial and temporary integration of teams and their collaboration.

CD takes relationship-based project delivery towards an IPD approach and more specifically alliancing, but it has been applied to other integrated forms such as PPP and D&C as part of the briefing and design clarification and front-end project definition processes.

4.3   LXRP Project Alliancing Selection Model Choice

A defining feature of this book is the rationale explanation of the LXRP's infrastructure mega-program adoption of a program alliance delivery form for this.

Alliancing has been increasingly used in Australia over the past two or more decades when a project has been perceived as complex and where the client (government PO representative) needs to take a hands-on involvement in the entire project design and delivery process (Ross, 2003). We see this logic applied in early landmark projects such as the Australian National Museum (Auditor-General of the Australian National Audit Office, 2000; Walker and Hampson, 2003b) and other infrastructure projects where motivations may include speedy delivery, emergency recovery, relation building, best value, and coping with risk and uncertainty (Walker and Lloyd-Walker, 2016).

Alliancing has been championed and supported by government departments for almost two decades. This is evident from the publication of Alliancing guidelines and explanatory documents with the Victorian State Government's initially developing guidelines (Department of Finance and Treasury, 2006) which was updated in 2011 by the federal government to apply across all states (Department of Infrastructure and Transport, 2011d). This is complemented by a publication of best practice examples (Department of Infrastructure and Transport, 2010) and several guidance notes for Alliance participants (Department of Infrastructure and Transport, 2011a; 2011c; 2011b). This institutional support suggests that Alliancing projects are welcomed by state and federal governments in Australia. It also suggests that IPD forms are broadly championed by infrastructure delivery agencies for specific risk profile projects where the PO (government department or agency) has the requisite skills and competencies to take a hands-on approach in forming alliance syndicates.

The above-cited government departments do not champion Alliancing as the sole/dominant procurement option to be adopted. The Regional Rail Link (RRL) program's experience (Chapter 2), however, with its brownfield project alliancing provided promising performance results. The LXRP uses Project Alliancing for all its brownfield projects. Therefore, championing was not characterised by blind support for Alliancing but by being supported *where appropriate*. A major LXRP motivation for Alliancing was institutional support for integration to best manage complex system interfaces and to facilitate productivity improvements through learning and innovation diffusion. The logic of having a managing government authority had gained significant support from influential individuals with a highly positive experience of this governance arrangement. More importantly, support was more generally institutional, similar to that found by Lehtonen and Martinsuo (2008) in their case study.

The LXRP's rationale for this preference is now discussed by investigating the major constraints and opportunities facing the LXRP, and the political and historical context that guided this preference.

4.3.1   *Major constraints*

One important constraint facing POs is their ability to demonstrate that the procurement choice delivers VfM/BV. Many adherents to traditional/conventional, D&C or BOOT/PPP contract forms appear to believe that competitive market forces will deliver the cheapest/quickest price/time offer to deliver a project. However, as criticised in numerous government and construction industry reports (Latham, 1994; Egan, 1998; 2002; Murray and Langford, 2003; Wolstenholme, 2009), the bid/contract price is often the starting price with these systems open to substantial claims for contract extras. Few POs may have allowed sufficient PO contingency to cover these contract variations resulting in many nasty surprises. This unpleasant truth seems to blind many supporters of these procurement choices and by the time real/actual VfM is known, it is too late for POs to extract themselves from this awkward situation.

The consideration of BV, in contrast to enthusiastic support for VfM, is based on the project's *purpose* (Palaneeswaran, Kumaraswamy and Ng, 2003; Staples and Dalrymple, 2006; MacDonald, Walker and Moussa, 2013). This purpose may include many benefits that are difficult to monetise for a cost-benefit analysis to support VfM. The context of the LXRP (Chapter 2) and the program benefit objectives (Chapter 3 Figure 3-1) confirms expected benefits including safety improvements at the level crossing locations and urban legacy transformational improvement. It is difficult to comprehensibly monetise *value* in a cost-benefit analysis (Barton, Aibinu and Oliveros, 2019; Volden, 2019), for example for life or injury value or for many LXRP environmental and social legacy project objectives. The key to understanding the project constraints that influence a project delivery choice is to first and foremost understand the purpose of the project.

Another constraint is project complexity. Chapter 2 identifies complexity in technical, political and system interface terms. The LXRP is almost totally engaged in brownfield working conditions, meaning that workers are exposed to risks from both road traffic and high voltage rails, i.e., highly complicated and complex hazards. Snowden's Cynefin framework (Kurtz and Snowden, 2003; Snowden and Boone, 2007; Snowden and Rancati, 2021) identifies complex and chaotic situations as being characterised by many unknown unknowns and unknowable unknowns that require special expert tacit knowledge that can only be accessed by subject matter experts collaborating with project delivery teams in a knowledge-power symmetry environment. Critical response requires cross-team/discipline interaction and knowledge sharing. This shared leadership approach (Scott-Young, Georgy and Grisinger, 2019) requires low power-information asymmetry to allow dialogue to tease out finer points needed to understand the context and find problem solutions based on a best-knowledge and evidence basis (Womack, 2011; Walker, Love and Matthews, 2022).

Traditional/conventional project delivery approaches fail to provide a supportive knowledge-sharing ecology to achieve this because knowledge transfer becomes sticky and difficult (Szulanski, 2003). Figure 4-2 illustrates constraints needing consideration when choosing a project delivery form.

LXRP project technical complexity-complicatedness issues, primarily driving the delivery choice, include a consistent continuous improvement and innovation, focus on safety, a history of successive governments procrastinating over investments in critical rail/road interface infrastructure, and the need for access to rail and road expertise. Chapter 2 discusses the LXRP complexity context in detail and one of the key strategic aims, illustrated in Chapter 3 Figure 3-1, is continuous improvement within and across projects within the program. This needs collaborative access to the system operators, rail and road so that their knowledge can be accessed to make effective and beneficial technical and operational decisions. Alliance project 1 LXRP study results (Walker, Matinheikki and Maqsood, 2018) revealed that VicRoads, the road operator on the alliance team, was able to provide travel volume and pattern data essential for optimising truck and equipment logistics as well as the rail operator's critical knowledge of how to best operate construction work within a hazardous train-construction site context. Thus, gaining collaborative access favoured an IPD approach. Previous Victorian Government Alliancing PAAs experience informed the decision to seriously consider a program of project alliances to lock-in the necessary rail and road technical expertise access. This strategy helped to address the 'fixing the problem' of removing rail crossings and 'fixing the transport network' to improve train services through clearer and less dangerous rail/road grade separations and improving the rail station facilities, updating aging and outmoded rail signalling, and improving intermodal transport transfers for passengers.

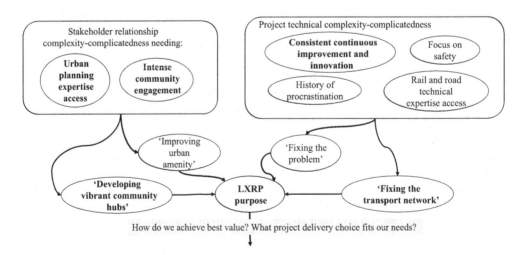

Figure 4-2 Project objectives constraint considerations for project delivery choice

The history of major infrastructure projects is replete with examples of community resistance and project-perceived failure by large sections of the public and other stakeholders. Thus, effective community engagement becomes a critical component of project delivery strategy and may be as important as technical/complexity issues. Similarly, stakeholder positive legacy value propositions for local businesses, community and 'passenger travel experience' requires careful consideration of intermodal interfaces and creating a vibrant community hub around rejuvenated stations are prime considerations.

4.3.2   *Major opportunities*

Creating value often arises from opportunities to overcome constraints. Conceptualising value is central, therefore, innovation and continuous improvement become vital value enablers. LXRP's value proposition was valued in technical, economic, social and environmental terms. Reverse thinking about constraints was reframed. For example, for decades governments failed to upgrade rail stations, intermodal interfaces, signalling and removing level crossings while Melbourne's population grew from about 2.8 million in 1980 to 5.1 million in 2020.[1] This offered opportunities to tackle shelved technical and network problems in a more holistic and eco-socio-friendly manner.

By removing a series of rail crossings within an integrated collaborative program of projects, it is possible to *design* a delivery model that requires continuous improvement outcomes, ensures this through designed-in incentivised governance arrangements combined with an innovation and knowledge diffusion process to minimise 'corporate amnesia.' Critically, innovation/continuous improvement and knowledge was recognised as an important valuable asset (Beckett, 2000).

Delivering projects through an integrated collaborative model provided opportunities to gain access to knowledge, skills and data assets owned by end-product rail and road facility operators. Their knowledge was vital in planning and executing the project so that uncertainty about legislative and regulatory details embedded in manuals, guidelines and rules could be effectively interpreted to maintain compliance but allow innovation and efficiency to prevail and be optimised and understood in context. This allowed both workforce and public safety to be achieved through intelligent planning and operational delivery. Innovations were made through the design, production, and administration stages because teams were integrated and effectively collaborating (Love and Walker, 2020).

Figure 4-2 illustrates the stakeholder relationship impact on project delivery choice, providing constraints but also offered opportunities. Community stakeholders had valuable insights into what mattered most to them and how they perceived value. Engaging effectively with them provides a channel for designing-in facilities and features that best suited their needs and how they could best cope with inevitable construction disruptions. This provided an impetus for extending the project delivery collaborative mindset to include stakeholder engagement (Chapter 12). Urban planning aspects also provided opportunities with an appointed board to help craft station precinct design and environmental and social value assets into the project brief.

According to IV16, significant opportunity arose through this program to overcome severe resource and skill constraints through designing-in a strategy to overcome constraints to the program and deliver a learning and skills legacy for the state that could benefit future projects. The LXRP had important skills and human resources development units structured into the LXRP organisational framework (Chapters 8 and 9). Having an integrated program of projects locked in for more than a decade meant that participating organisations had a steady and largely predictable volume of work to plan their recruitment and people development processes to weave into the LXRP. The timeline and its implication for cash flow and profitability became an attractive value proposition for participating organisations. This meant that workforce quantity-quality skills and knowledge base at all levels could be raised through the LXRP and that local supply chain participants had a stable environment with which to contribute, participate, and grow. Often delayed, or neglected, diversity participation aspirations were also able to be addressed.

The LXRP had a designed-in PAA requirement for the inclusion of under-represented groups in the workforce. This provided benefits in reducing hidden costs of un/(under)employment, potential improvements to redress the recognised toxic workplace culture associated with the construction industry generally (Clegg et al., 2023), and making available a stream of workers that are often ignored. Strict workplace safety and environmental standards that were build-into the PAA also ensured that it minimised the potential for short-cut blind-eye attempts to get away with harming the environment. KPIs designed into the governance system provided an incentive that complimented a best-for-program mindset to prevail. Additionally, the LXRP designed its program of works as a highly integrated series of alliance projects let as a series of alliance initially as alliance projects and then as a pipeline of alliances additional work packages (AWPs). This program mindset with its continuity of the pipeline of AWPs has advantages in motivating alliance syndicates and their participants by providing a career path (Chapter 8) and promoted a coherent and positive program and project workplace culture (Chapter 7).

The above opportunities led towards an integrated collaborative project delivery choice to achieve long-term economic, environmental, and social sustainability rather than opt for short-term objectives.

4.3.3   *Political and historical context*

Governments are long-term and economically significant project sponsors and owners. They also generally have corporate knowledge and experience to draw upon. Chapter 2 discusses the political context in detail. Suffice to say that the LXRP project delivery choice was informed by past RRL experience. The RRL had a mixed D&C, PPP and Alliancing program of integrated projects. That experience suggested that Alliancing was most suitable for brownfield work with its inherent uncertainty and interface complexity but knowledge and innovation diffusion across projects was hindered by mixing alliancing with more conventional and commercially driven project delivery forms because competing motivations inhibited open knowledge sharing (Szulanski, 2003).

The Victorian Government recommends considering Alliancing for complex large-scale projects (Department of Infrastructure and Transport, 2011d) recognising the constraints and opportunities that this delivery form offers. The LXRP project was born out of popular demand for fixing the problem, institutional demand for improving the rail and road networks around the level crossing location, and nascent but real demands by local businesses to improve tired rail station precincts to attract local communities.

Logically, while competitive conventional and even D&C and PPP options may appear superficially attractive, they fail to credibly deliver continuous improvement across the program of works or are likely to sufficiently gain sufficient community support and value from a stakeholder group that might easily become exhausted by a prolonged period of rail crossing removals and ancillary works. Given past negligence and procrastination by successive governments, a long-term *strategic* solution was needed to maintain community support that translates into political support of several government terms. The sheer scale of resources required, skills and knowledge of professional and sub-contractors and suppliers demanded a more integrated, collaborative and highly professional program management approach. Thus, a program of alliance projects appears to be the most appropriate and feasible sustainable project delivery choice to make.

4.4 Alliance Capabilities Context

Chapter 3 discusses how the strategy was translated into governance to deliver the LXRP. Chapters 8 and 9 discuss how the people-side of staffing, developing them and guiding an appropriate leadership context was developed. Deciding on a project delivery choice is one important issue to face but effectively enacting it is a different but linked issue. To deliver a program of this scale and scope required a distinct set of project owner and project participant capabilities.

4.4.1 Project owner capabilities

The LXRP owner capabilities are special in numerous ways but foremost, program *and* project capabilities are necessary. The LXRP is effectively, a corporate entity as Figures 3-2 and 3-3 in Chapter 3 suggest. Managing complex interfaces is a project management skill but, far more intense in a program setting (or where a 'project' includes complex commissioning activities). Failure to appreciate this has a potentially calamitous impact on project delivery success. Mark Wild, CEO of Crossrail in his (2022) webinar explained how Crossrail was a massive success from an engineering perspective, yet has lost its gloss, receiving criticism for the megaproject's final phase (integrating legacy and new signalling and communication systems and commissioning of new trains). The Heathrow Terminal Five (T5) project hit a similar snag by failing to fully test and integrate the baggage handling system at its opening (Brady and Davies, 2010), similar problems occurred on the Denver International Airport project (de Neufville, 1994).

At the core of program management capabilities lies interface management, adequately segregating project tasks while integrating corporate functions that support production activities. Commissioning services (as experienced in Crossrail, T5 and Denver International Airport) are vital with complex interfaces between various operational systems, not being obviously linked. Physical (or electronic) communication systems may need extensive training and upskilling that is not readily considered. Signage and wayfinding systems may need to be considered such as that for T5 where Brady and Davies (2010) note that baggage handling staff, flight crew and passengers were confused about how to get from A to B which contributed to the T5 fiasco. The interesting aspect about Figure 3-3 in Chapter 3 is that LXRP painstakingly developed a matrix organisational structure with functional specialist groups having participants embedded in AWPs that reported to both their project and corporate office leaders. This potentially leads to confused chains of command but can be managed through effective communication, integration, and collaboration practices (Chapters 5 and 6). Having fine-tuned interface management competencies is vital to achieving this goal.

Required program competencies include *soft skills* and high levels of *systems thinking* (Chapter 8). Soft skills refer to the ability to collaborate with individuals and groups who are disbursed organisationally upwards and sideways so that function (what the person's role is, what they do and what skills and expertise they possess) is more important than hierarchy so a shared leadership approach is necessary (Drescher et al., 2014; Scott-Young, Georgy and Grisinger, 2019; Love et al., 2021). Systems thinking is being aware of, and curious about, how various systems interact dynamically (Browning and Ramasesh, 2015; McCuen, 2023), in a program such as LXRP this is essential where continuous improvement is sought through cross-project learning which makes the context particularly dynamic.

The unexpected frequently occurs across projects in a program despite the best laid plans, sophisticated strategising, and skilled experienced staff. No two projects behave identically and so while modularisation is sought in the Skyrail component design, for example, there is often a need for customising and resilient reaction to adverse trends and events. This requires staff high levels of program management ambidexterity to be able to think quickly and effectively (Turner,

Kutsch and Leybourne, 2016; Turner et al., 2020) and to be able to rapidly move from 'slow to quick thinking' by reacting using mixed heuristics and analytical thinking when appropriate (Kahneman, 2011).

Finally, the continuous improvement goal across projects requires high levels of curiosity, critical thinking and organisational learning facilitation skills. Lessons learned need to be mentally processed in context and applied using the above levels of thought and consideration. Systems need to be developed to facilitate, encourage, reward and nurture the exchange of knowledge and insights across the program of projects (Chapter 7). Figure 3-2 illustrates how the LXRP links into the Melbourne Transport Infrastructure Authority (MTIA) through the LXRP CEO. The MTIA also engages in the management of a program of other infrastructure projects and programs of projects with a goal of achieving cross-learning and continuous improvement.

In general, at the program management level, the question posed to the corporate level staff illustrated in Figure 3-3 is 'how can we facilitate and enable project knowledge/insights and innovation initiatives across the projects, what systems, ways of thinking and working are required?'

4.4.2   *Project participant capabilities*

Similar capabilities to all the above need to be made at the project level. However, at the project level, the question in mind should be about addressing the dissemination of knowledge assets to other projects and within the participant's project.

Each participant needs to have high-level collaboration capabilities (Chapters 6, 7 and 8) and soft skills, such as perspective-taking, to permit them to appreciate the value of alternative views (Parker, Atkins and Axtell, 2008) and engage in dialogue to achieve optimal problem solutions (Floris and Cuganesan, 2019). The workplace culture needs to be one where challenging assumptions are expected and it is psychologically safe to do so within a no-blame environment (Lloyd-Walker, Walker and Mills, 2014). Participants also need to be trusting to gauge the trustworthiness of their within-team, cross-team or cross-project colleagues when sharing knowledge and insights (Walker and Lloyd-Walker, 2015). Wild, in his webinar reflecting on his time as CEO of Crossrail (Wild, 2022), specifically mentioned curiosity of project participants as being a key capability, he wanted participants to avoid blame when encountering unexpected events. He valued their ability to question how and why an adverse event may occur and what may be done to mitigate consequences or use the crisis as a trigger for seeing hidden opportunities (Wolstenholme, 2009), suggesting resilience.

In general, at the project level, the question that is posed to the staff project participants is 'How can we facilitate and enable project knowledge/insights and innovation initiatives *within* and across other projects, what systems, ways of thinking and working are required?'

4.4.3   *LXRP and global project capabilities comparison*

This section summarises the evidence on PO and participant capabilities to compare and contrast the LXRP with other similarly complex global megaprojects.

For years, British Airports Authority (BAA) had gained experience in an IPD form of delivery starting with Terminal 4 (T4) and later Terminal 2 (T2) redevelopment. As noted earlier, BAA designed a new relational contract form for T4 based on its experiences with framework agreements and elements of the New Engineering Contract (NEC) forms evolving at that time (Doherty, 2008; Davies, Gann and Douglas, 2009). Throughout the development of T5 and T2, the BAA PO team developed a deep understanding of the value of collaboration, team integration and both PO and team participant new-knowledge transfer across parts of these projects to benefit its overall outcome. Doherty's (2008) book is particularly enlightening on how T5 senior project delivery managers and key subcontractors experimented, modelled and innovated on numerous technical and process aspects of the project, but what is clear from that book, and numerous other journal papers (e.g., Davies, Dodgson and Gann, 2016) is how both the PO and participants developed dynamic capabilities by dealing with unexpected events in a collaborative manner that produced win-win outcomes. This experience led them to learn how to learn. Evidence supporting the claim about PO capability transfer is demonstrated by many similar learning and transfer examples that the PO team transferred to other UK megaprojects. Senior staff moved from T5 to Crossrail, Thames Tideway Tunnel, T2, the 2012 Olympics and High Speed 2 (HS2) with their clear understanding of the importance of innovation and continuous improvement and how to structure systems that encourage, acknowledge and reward improvement initiatives (Davies et al., 2014). These highly complex and resource-demanding megaprojects were viewed as opportunity-laden for developing and sharing knowledge and insights. Organisational and individual learning through action could be argued as the key capabilities developed by PO staff and project participants in that stream of projects. Central to that development was their understanding that team integration *and* effective collaboration through dialogue, action learning, and constant questioning context and history to effectively reconfigure and adapt knowledge. This resulted in continuous improvement (Zerjav, Edkins and Davies, 2018). In extensive research on this cohort of megaprojects, Denicol and Davies (2022) identified program management capabilities development with integration and learning passing across the above projects through those people involved.

How does the LXRP compare with this? First, the LXRP is a program of alliance projects stretching for over a decade with a stable set of participants at the PO team and other alliance project team (APT) levels. While some may move from one alliance project to another in the program, their knowledge, insights and default workplace cultural remain within-program. These people also learn from dialogue-based decision-making and problem-solving to facilitate action learning. Second, the foundational learning and experience from the RRLA alliancing example on brownfield sites presents a clear *flow* of learning and a culture of continuous improvement. The main difference between the LXRP and the UK experience is that the LXRP maintained a highly strategic program plan to link project experiences into the program that seems more consistent than BAA for example that was acquired by Ferrovial in 2006. This disrupted strategic continuity of BAA's organisational learning as evidenced by the T5 baggage handling fiasco (Brady and Davies, 2010).

From a North American perspective, we can compare the US Health care provider Sutter Health's approach to conducting a program of IPD projects. While Christian and Bredbury discuss how IPD has helped realise much value capture (2014), our discussions with one of the authors suggest that Sutter Health is yet to effectively formalise its PO or project participant learning. Literature about the organisation's program of projects is scant and it appears that the LXRP's more formally strategised approach is more advanced.

A third useful comparison can be made within a Nordic context. Several papers have detailed how the Norwegian Oil and Gas production industry has adopted what they term partnering, though with additional contract conditions similar to project alliancing (Børve et al., 2017; Nevstad et al., 2022). However, there is no strong evidence of a systematic program of projects approach that has been pursued by these POs and continuity of participants on successive projects has been disrupted when compared with the LXRP. A Swedish electrical distribution infrastructure project called Destination 2011 (Jacobsson, 2011b) also provides a useful comparison. The project was set up as a partnering agreement between Jämtkraft, a Swedish power producer and supplier, and Skanska but also included several subcontractors. The overall aim of the project was to secure the region's power distribution from severe weather conditions by switching all Jämtkrafts's overhead power lines to ground-based cables (more than 1,500km of new cables, through partially inaccessible and remote terrain). The contract, which lasted from 2005 until 2011 and was strategic in its nature, included a gain-share-pain-share clause and was divided into more than 300 sub-projects. In theory, the undertaking might sound simple but was highly complex in practice which was also the rationale for the partnering setup. An important ingredient of the project's success was that the PO and the main contractor decided to co-locate almost all functions which not only expedited the concurrent design process for each subproject but also facilitated transparency, learning between projects, mutual understanding and the development of trust (Jacobsson, 2011a; Jacobsson and Wilson, 2014). Despite being called a partnering agreement, thus resembling the alliance LXRP setup in both its rationale and implementation.

4.5   Conclusions

This chapter answers the question: *What project delivery choices did the LXRP consider and why did they choose Project Alliancing for the program of projects?*

We started with describing the possible options available for the LXRP to deliver its program of projects and found that complex brownfield sited projects where there are likely to be disrupted availability of requisite resources, knowledge, skills and an experienced workforce to effectively deliver the projects present critical challenges. Such challenges and constraints have been identified as requiring an IPD/Alliance form of delivery (Department of Infrastructure and Transport, 2011d; Walker and Lloyd-Walker, 2016). Moreover, we outline opportunities that alliancing offers, particularly in continuous improvement, innovation and knowledge diffusion and upskilling of the PO and project participants. Additionally, we discussed opportunities for the program's precinct transformation and stakeholder positive legacy purpose and motivation to be realised.

IPD/Alliancing literature notes, that merely having a contract, plan or intent for this depth of relationship-based project delivery is necessary but insufficient (Suprapto, 2016). What is needed is a well-delivered program strategy to effectively link projects and their learnings and synergistic outcomes together within and across projects in the program. This was the stated aim and purpose of the LXRP and it was pursued through its governance arrangements as discussed in Chapter 3.

While managing a program of projects is far from new, managing an integrated program of IPD projects to effectively harvest lessons learned, continuous improvement and deliver a positive legacy is relatively rare globally. The LXRP stands amongst some of the world's best-practice examples such as Sutter Health in the USA, the stream of megaprojects flowing on from T5 in the UK and other examples from Nordic countries. It could be argued that the LXRP may be an exemplar in this field.

Note

1 Data Source: United Nations – World Population Prospects (https://population.un.org/wpp/)

References

Aarseth, W., Andersen, B., Ahola, T. and Jergeas, G. (2012). "Practical difficulties encountered in attempting to implement a partnering approach." *International Journal of Managing Projects in Business.* 5 (2): 266–284.

Akintoye, A., Hardcastle, C., Beck, M., Chinyio, E. and Asenova, D. (2003). "Achieving best value in private finance initiative project procurement." *Construction Management and Economics*. 21 (5): 461–470.

Auditor-General of the Australian National Audit Office (2000). *Construction of the National Museum of Australia and the Australian Institute of Aboriginal and Torres Strait Islander Studies*, Canberra, Australia.

Barton, R., Aibinu, A.A. and Oliveros, J. (2019). "The value for money concept in investment evaluation: Deconstructing its meaning for better decision making." *Project Management Journal*. 50 (2): 210–225.

Beckett, R.C. (2000). "A characterisation of corporate as a knowledge management system." *Journal of Knowledge Management*. 4 (4): 311–319.

Børve, S. (2019). Project partnering defined and implications thereof. PhD by publication, *Faculty of Engineering, Department of Mechanical and Industrial Engineering*. Trondheim, Norway, Norwegian University of Science and Technology.

Børve, S., Ahola, T., Andersen, B. and Aarseth, W. (2017). "Partnering in offshore drilling projects." *International Journal of Managing Projects in Business*. 10 (1): 84–108.

Brady, T. and Davies, A. (2010). "From hero to hubris: Reconsidering the project management of Heathrow's Terminal 5." *International Journal of Project Management*. 28 (2): 151–157.

Browning, T.R. and Ramasesh, R.V. (2015). "Reducing unwelcome surprises in project management." *MIT Sloan Management Review*. 56 (3): 53–62.

Bygballe, L.E. and Swärd, A. (2019). "Collaborative project delivery models and the role of routines in institutionalizing partnering." *Project Management Journal*. 50 (2): 161–176.

Christian, D. and Bredbury, J. (2014). "Four-phase project delivery and the pathway to perfection." In *22nd Annual Conference of the International Group for Lean Construction*, Kalsaas, B.T., L. Koskela and T.A. Saurin (eds), pp. 269–280. Oslo, Norway, International Group for Lean Construction. http://iglc.net/Papers/Details/1013

CII (1996). *The Partnering Process – Its Benefits, Implementation, and Measurement*, Austin, Texas.

Clegg, S.R., Loosemore, M., Walker, D.H.T., van Marrewijk, A.H. and Sankaran, S. (2023). "Construction cultures: Sources, Signs and solutions of toxicity." In *Construction Project Organising*, Addyman, S. and H. Smyth (eds), pp. 3–16. Chichester, West Sussex, John Wiley & Sons.

Davies, A., Dodgson, M. and Gann, D. (2016). "Dynamic capabilities in complex projects: The case of London Heathrow Terminal 5." *Project Management Journal*. 47 (2): 26–46.

Davies, A., Gann, D. and Douglas, T. (2009). "Innovation in megaprojects: Systems integration at London Heathrow Terminal 5." *California Management Review*. 51 (2): 101–125.

Davies, A., MacAulay, S., DeBarro, T. and Thurston, M. (2014). "Making innovation happen in a megaproject: London's Crossrail suburban railway system." *Project Management Journal*. 45 (6): 25–37.

de Neufville, R. (1994). "The baggage system at Denver: Prospects and lessons." *Journal of Air Transport Management*. 1 (4): 229–236.

Denicol, J. and Davies, A. (2022). "The megaproject-based firm: Building programme management capability to deliver megaprojects." *International Journal of Project Management*. 40 (5): 505–516.

Department of Finance and Treasury (2006). *Project Alliancing Practitioners' Guide*, Department of Treasury and Finance G. o. V. Melbourne, Australia.

Department of Health (2012). *The ProCure21+ Guide – Achieving Excellence in NHS Construction*, Service N.H. Leeds, UK, Department of Health.

Department of Infrastructure and Transport (2010). *Infrastructure Planning and Delivery: Best Practice Case Studies*, Department of Infrastructure and Transport A.C.G. Canberra, Commonwealth of Australia. www.infrastructure.gov.au/infrastructure/nacg/files/National_Alliance_Contracting_WEB.pdf

Department of Infrastructure and Transport (2011a). *National Alliance Contracting Guidelines Guidance Note 3: Key Risk Areas and Trade-Offs*, Department of Infrastructure and Transport A.C.G. Canberra, Commonwealth of Australia. www.infrastructure.gov.au/infrastructure/nacg/files/NACG_GN3.pdf

Department of Infrastructure and Transport (2011b). *National Alliance Contracting Guidelines Guidance Note No 5: Developing the Target Outturn Cost in Alliance Contracting*, Department of Infrastructure and Transport A.C.G. Canberra, Commonwealth of Australia. www.infrastructure.gov.au/infrastructure/nacg/files/NACG_GN5.pdf

Department of Infrastructure and Transport (2011c). *National Alliance Contracting Guidelines Guidance Note No. 4: Reporting Value-for-Money Outcomes*, Department of Infrastructure and Transport A.C.G. Canberra, Commonwealth of Australia. www.infrastructure.gov.au/infrastructure/nacg/files/NACG_GN4.pdf

Department of Infrastructure and Transport (2011d). *National Alliance Contracting Guidelines Guide to Alliance Contracting*, Department of Infrastructure and Transport A.C.G. Canberra, Commonwealth of Australia. www.infrastructure.gov.au/infrastructure/nacg/files/National_Guide_to_Alliance_Contracting04July.pdf

Doherty, S. (2008) *Heathrow's T5 History in the Making*, Chichester, John Wiley & Sons Ltd.

Drescher, M.A., Korsgaard, M.A., Welpe, I.M., Picot, A. and Wigand, R.T. (2014). "The dynamics of shared leadership: Building trust and enhancing performance." *Journal of Applied Psychology*. 99 (5): 771–783.

Egan, J. (1998). *Rethinking Construction – The Report of Construction Task Force*, London.

Egan, J. (2002). *Accelerating Change. Rethinking Construction*, London, Construction Industry Council.

Engebø, A., Klakegg Ole, J., Lohne, J. and Lædre, O. (2020). "A collaborative project delivery method for design of a high-performance building." *International Journal of Managing Projects in Business*. 13 (6): 1141–1165.

Eriksson, P.E. (2010). "Partnering: what is it, when should it be used, and how should it be implemented?" *Construction Management and Economics*. 28 (9): 905–917.

Floris, M. and Cuganesan, S. (2019). "Project leaders in transition: Manifestations of cognitive and emotional capacity." *International Journal of Project Management*. 37 (3): 517–532.

Franks, J. (1984) *Building Procurement Systems: A Guide to Building Project Management,* Ascot, UK, Chartered Institute of Building.

Green, S.D. (1999). "Partnering: The Propaganda of Corporatism?" In *Profitable Partnering in Construction Procurement*, Ogulana, S.O. (ed.), p. 735. London, E & FN Spon.

Grimsey, D. and Graham, R. (1997). "PFI in the NHS." *Engineering Construction and Architectural Management*. 4 (3): 215–231.

Hodge, G.A. (2004). "The risky business of public-private partnerships." *Australian Journal of Public Administration*. 63 (4): 37–49.

Hodge, G.A. and Duffield, C. (2010). "The Australian PPP experience: Observations and reflections." In *International Handbook on Public–Private Partnerships*, Hodge, G.A., C. Greve and A. Boardman (eds), pp. 399–455. Cheltenham, UK, Edward Elgar Publishing Limited.

Hodge, G.A. and Greve, C. (2017). "On public–private partnership performance: A contemporary review." *Public Works Management & Policy*. 22 (1): 55–78.

Hoezen, M., Voordijk, H. and Dewulf, G. (2013). "Formal bargaining and informal sense making in the competitive dialogue procedure: An event-driven explanation." *International Journal of Managing Projects in Business*. 6 (4): 674–694.

Infrastructure Australia (2008) *National Public Private Partnership Policy Framework* Canberra, ACT, Government of Victoria.

Jacobsson, M. (2011a). "On the importance of liaisons for coordination of projects." *International Journal of Managing Projects in Business*. 4 (1): 64–81.

Jacobsson, M. (2011b). Samordningens dynamik: om samordningens samspel och förändring i ett interorganisatoriskt anläggningsprojekt. PhD Thesis, Umeå University, Faculty of Social Sciences, Umeå School of Business and Economics (USBE). Umeå, Sweden, Umeå University.

Jacobsson, M. and Roth, P. (2014). "Towards a shift in mindset: Partnering projects as engagement platforms." *Construction Management and Economics*. 32 (5): 419–432.

Jacobsson, M. and Wilson, T.L. (2014). "Partnering hierarchy of needs." *Management Decision*. 52 (10): 1907–1927.

Jefferies, M. and Rowlinson, S. (eds) (2016). *New Forms of Procurement: PPP and relational contracting in the 21st century*. Series New Forms of Procurement: PPP and relational contracting in the 21st century. Abingdon, Taylor & Frances.

Kahneman, D. (2011) *Thinking Fast and Slow*, New York, Ferrar, Strauss and Giroux.

Khalfan, M.M.A., McDermott, P. and Kyng, E. (2006). *Procurement Impacts on Construction Supply Chains: UK Experiences*. Symposium On Sustainability And Value Through Construction Procurement, CIB W092 – Procurement Systems, CIB Revaluing Construction Theme, The Digital World Centre, Salford, UK, 29 November – 2 December, McDermott, P. and M.M.A. Khalfan: 449–458.

Koolwijk, J.S.J., van Oel, C.J. and Moreno, J.C.G. (2020). "No-blame culture and the effectiveness of project-based design teams in the construction industry: The mediating role of teamwork." *Journal of Management in Engineering*. 36 (4): 04020033.

Kurtz, C.F. and Snowden, D.J. (2003). "The new dynamics of strategy: Sense-making in a complex and complicated world." *IBM Systems Journal*. 42 (3): 462–483.

Lahdenperä, P. (2012). "Making sense of the multi-party contractual arrangements of project partnering, project alliancing and integrated project delivery." *Construction Management and Economics*. 30 (1): 57–79.

Larson, E. (1995). "Project partnering: Results of study of 280 construction projects." *Journal of Management in Engineering – American Society of Civil Engineers/ Engineering Management Division*. 11 (2): 30–35.

Latham, M. (1994). *Constructing the Team*, London.

Lehtonen, P. and Martinsuo, M. (2008). "Change program initiation: Defining and managing the program-organization boundary." *International Journal of Project Management*. 26 (1): 21–29.

Lichtig, W.A. (2005). "Sutter Health: Developing a contracting model to support lean project delivery." *Lean Construction Journal*. 2 (1): 105–112.

Lloyd-Walker, B.M., Walker, D.H.T. and Mills, A. (2014). "Enabling construction innovation: The role of a no-blame culture as a collaboration behavioural driver in project alliances." *Construction Management and Economics*. 32 (3): 229–245.

Love, P., Ackermann, F. and Irani, Z. (2018). "The praxis of stupidity: An explanation to understand the barriers mitigating rework in construction." *Production Planning & Control*. 29 (13): 1112–1125.

Love, P.E.D., Ika, L., Matthews, J. and Fang, W. (2021). "Shared leadership, value and risks in large scale transport projects: Re-calibrating procurement policy for post COVID-19." *Research in Transportation Economics*. 90 (100999).

Love, P.E.D. and Walker, D.H.T. (2020). "IPD Facilitating innovation diffusion." In *The Routledge Handbook of Integrated Project Delivery*, Walker, D.H.T. and S. Rowlinson (eds), pp. 393–416. Abingdon, Oxon, Routledge.

MacDonald, C.C. (2011). Value for Money in Project Alliances. DPM Thesis, *School of Property, Construction and Project Management*. Melbourne, RMIT University.

MacDonald, C.C., Walker, D.H.T. and Moussa, N. (2013). "Towards a project alliance value for money framework." *Facilities*. 31 (5/6): 279–309.

Manchester Business School. (2009). *Study on Voluntary Arrangements for Collaborative Working in the Field of Construction Services – Main Report Part 2: Best Practice Guide and Case Studies*, Manchester.

Masterman, J.W.E. (1992) *An Introduction to Building Procurement Systems*, London, E & FN SPON.

Masterman, J.W.E. (2002) *An Introduction to Building Procurement Systems*, London, Spon.

McCuen, R.H. (2023). "A new perspective on critical thinking." In *Critical Thinking, Idea Innovation, and Creativity*, pp. 105–149. Milton, Taylor & Francis Group.

McGeorge, W.D. and Palmer, A. (1997) *Construction Management New Directions*, London, Blackwell Science.

Merrow, E.W. (2011) *Industrial Megaprojects – Concepts, Strategies, and Practices for Success*, London, John Wiley & Son.

Moreledge, R., Smith, A. and Kashiwagi, D. (2006) *Building Procurement*, Oxford, Blackwell.

Murray, M. and Langford, D.A. (2003) *Construction Reports 1944–98*, Oxford, Blackwell Science Ltd.

Naoum, S. and Egbu, C. (2015). "Critical review of procurement method research in construction journals." *Procedia Economics and Finance*. 21: 6–13.

Nevstad, K., Karlsen, A.S.T., Aarseth, W.K. and Andersen, B. (2022). "How a project alliance influences project performance compared to traditional project practice: Findings from a case study in the Norwegian oil and gas industry." *The Journal of Modern Project Management*. 9 (3): 139–153.

Nevstad, K., Madsen, T.K., Eskerod, P., Aarseth, W.K., Karlsen, A.S.T. and Andersen, B. (2021). "Linking partnering success factors to project performance: Findings from two nation-wide surveys." *Project Leadership and Society*. 2: 100009.

Nyström, J. (2005). "The definition of partnering as a Wittgenstein family-resemblance concept." *Construction Management and Economics*. 23 (5): 473–481.

Nyström, J. (2007). Partnering: definition, theory and evaluation. PhD, School of Architecture and the Built Environment. Stockholm, Royal Institute of Technology (KTH).

Palaneeswaran, E., Kumaraswamy, M. and Ng, T. (2003). "Targeting optimum value in public sector projects through 'best value'-focused contractor selection." *Engineering Construction and Architectural Management*. 10 (6): 418–431.

Parker, S.K., Atkins, P. and Axtell, C. (2008). "Building better work places through individual perspective taking: A fresh look at a fundamental human process." In *International Review of Industrial and Organizational Psychology*, Hodgkinson, G.P. and J.K. Ford (eds), pp. 149–196. Chichester, John Wiley & Sons Inc.

Regan, M. (2009). *A Survey of Alternative Financing Mechanisms for Public Private Partnerships – Research Report 110*, Mirvac School of Sustainable Development B.U. Gold Coast, Queensland.

Regan, M., Love, P. and Smith, J. (2013). "Public-private partnerships: Capital market conditions and alternative finance mechanisms for Australian infrastructure projects." *Journal of Infrastructure Systems*. 19 (3): 335–342.

Roehrich, J.K., Lewis, M.A. and George, G. (2014). "Are public–private partnerships a healthy option? A systematic literature review." *Social Science & Medicine*. 113 (0): 110–119.

Ross, J. (2003). *Introduction to Project Alliancing*, Alliance Contracting Conference, Sydney, 30 April 2003, Project Control International Pty Ltd.

Rowlinson, S. and McDermott, P. (1999) *Procurement Systems. A Guide to Best Practice in Construction*, London, E&FN Spon.

Scott-Young, C.M., Georgy, M. and Grisinger, A. (2019). "Shared leadership in project teams: An integrative multi-level conceptual model and research agenda." *International Journal of Project Management*. 37 (4): 565–581.

Sidwell, A.C. (1983). "An evaluation of management contracting." *Construction Management and Economics*. 1 (1): 47–55.

Smalley, M., Lado-Byrnes, L. and Howe, M. (2004). *NHS Wales: Construction Procurement Review – Selection of a Preferred Option for the NHS in Wales*, Ashford T.C.W.C. a. B. Cardiff, Wales.

Snowden, D. and Rancati, A. (2021). *Managing Complexity (and Chaos) in Times of Crisis. A Field Guide for Decision Makers inspired by the Cynefin Framework*. Luxembourg, Publications Office of the European Union.

Snowden, D.J. and Boone, M.E. (2007). "A leader's framework for decision making." *Harvard Business Review*. 85 (11): 69–76.

Staples, W. and Dalrymple, J. (2006). "Best value public sector construction procurement." *International Journal of Public Sector Management*. (forthcoming).

Suprapto, M. (2016). Collaborative Contracting in Projects. PhD, Faculty of Civil Engineering and Geosciences. Delft, the Netherlands, TU Delft University of Technology.

Szulanski, G. (2003) *Sticky Knowledge Barriers to Knowing in the Firm*, Thousand Oaks, CA., Sage Publications.

Tennant, A. and Fernie, S. (2010). "A contemporary examination of framework agreements." In *26th Annual ARCOM Conference*. Egbu, C. (ed.), p. 685–694. Leeds, UK, Association of Researchers in Construction Management. www.arcom.ac.uk/-docs/proceedings/ar2010–0685–0694_Tennant_and_Fernie.pdf

Turner, N., Kutsch, E. and Leybourne, S.A. (2016). "Rethinking project reliability using the ambidexterity and mindfulness perspectives." *International Journal of Managing Projects in Business*. 9 (4): 845–864.

Turner, N., Kutsch, E., Maylor, H. and Swart, J. (2020). "Hits and (near) misses. Exploring managers' actions and their effects on localised resilience." *Long Range Planning*. 53 (3): 101944.

United Nations (2016). Global Sustainability Goals www.un.org/sustainabledevelopment/sustainable-development-goals/

Volden, G.H. (2019). "Assessing public projects' value for money: An empirical study of the usefulness of cost–benefit analyses in decision-making." *International Journal of Project Management*. 37 (4): 549–564.

Walker, C. and Smith, A.J. (1995) *Privatised Infrastructure – The BOT Approach*, London, Thomas Telford.

Walker, D.H.T., Bourne, L. and Rowlinson, S. (2008). "Stakeholders and the supply chain." In *Procurement Systems – A Cross Industry Project Management Perspective*, Walker, D.H.T. and S. Rowlinson (eds), pp. 70–100. Abingdon, Taylor & Francis.

Walker, D.H.T. and Hampson, K.D. (2003a). "Procurement choices." In *Procurement Strategies: A Relationship Based Approach*, Walker, D.H.T. and K.D. Hampson (eds), pp. 13–29. Oxford, Blackwell Publishing.

Walker, D.H.T. and Hampson, K.D. (2003b) *Procurement Strategies: A Relationship Based Approach*, Oxford, Blackwell Publishing.

Walker, D.H.T. and Lloyd-Walker, B.M. (2015) *Collaborative Project Procurement Arrangements*, Newtown Square, PA, Project Management Institute.

Walker, D.H.T. and Lloyd-Walker, B.M. (2016). "Understanding the Motivation and Context for Alliancing in the Australian Construction Industry." *International Journal of Managing Projects in Business*. 9 (1): 74–93.

Walker, D.H.T., Love, P.E.D. and Matthews, J. (2022). "The Value of Dialogue in Alliancing Projects." In *The 9th International Conference on Innovative Production and Construction*, Melbourne, Australia, CIB. https://virtual.oxfordabstracts.com/#/event/public/1970/program?session=35425&s=600

Walker, D.H.T., Matinheikki, J. and Maqsood, T. (2018). *Level Crossing Removal Authority Package 1 Case Study*, RMIT University Melbourne, Australia.

Wild, M. (2022). Delivering the Elizabeth Line, www.youtube.com/watch?v=p9QFy1HnFwA

Wolstenholme, A. (2009). *Never Waste a Good Crisis – A Review of Progress since Rethinking Construction and Thoughts for Our Future*, London.

Womack, P. (2011) *Dialogue*, London, Taylor & Francis.

Yeung, J.F., Chan, A.P.C. and Chan, D.W.M. (2007). "The definition of alliancing in construction as a Wittgenstein family-resemblance concept." *International Journal of Project Management*. 25 (3): 219–231.

Zerjav, V., Edkins, A. and Davies, A. (2018). "Project capabilities for operational outcomes in inter-organisational settings: The case of London Heathrow Terminal 2." *International Journal of Project Management*. 36 (3): 444–459.

5 LXRP Project-Program Integration

Derek H.T. Walker, Peter E.D. Love and Mark Betts

5.1 Introduction

How did the LXRP integrate its program of projects, and its organisational structure to administrate each of its projects within the program?

This chapter explores team integration in both the project alliances and the program of projects. It explains how and why project teams chose and actioned their formation of a single united team to deliver each project and the team tasked with coherently integrating the stream of projects.

It fits into the book by extending our understanding of how integrated LXRP teams collaborate. Chapter 2 outlines the LXRP context and Chapter 3 explained the LXRP purpose and strategy and how that led to developing a suitable governance framework. Chapter 4 explained the rational for the IPD/Alliancing delivery decision. This chapter links directly to Chapter 6 about effective collaboration that is largely dependent on effective team integration.

We next discuss general alliance project integration principles, framing these against LXRP case study research findings, followed by a discussion on how project alliance teams were

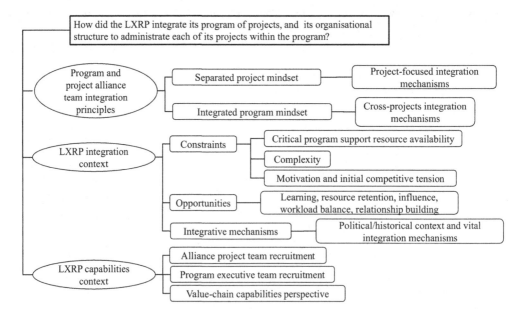

Figure 5-1 Chapter 5 structure

DOI: 10.1201/9781003389170-5

recruited and maintained and how the program management team were recruited to effectively integrate the program of alliance projects.

5.2   Program and Project Alliance Team Integration Principles

So, *what does effective IPD/Alliancing integration look like*?

Fischer et al.(2017, p. 148) define an integrated organisation as "a collection of people organized in an integrated structure that is aligned to the project.... A group of individual organizations – and their employees – that embrace a common set of values and goals and act as if they were one company." This guiding principle has framed IPD's form but omits to mention its aim, to drive united inter-team collaboration to fulfil the project's purpose. Critically, effective team integration is essential for delivering the LXRP purpose as illustrated by Figure 3-2 in Chapter 3.

Why should team integration be recommended for projects such as the LXRP? Why shouldn't 'market competition' be relied upon to achieve a best-for-project outcome through competition on price and time delivery and the contest of ideas? Competition may well drive superior outcomes when the intended project outcome is clear, and ways and means are known and essentially simple, perhaps slightly complicated, with readily available expertise easy to source, apply and integrate (Snowden and Boone, 2007). However, many projects are far from simple/standard, exhibiting varying levels of complexity. As Snowden argues through the Cynefin Framework (Kurtz and Snowden, 2003; Snowden and Rancati, 2021), the best way to deal with complexity is to access high levels of expertise, critical thinking, and resilient action learning. This requires, as detailed in Chapter 6, integrated teams to collaborate through effective dialogue. The reason why team integration is necessary, therefore, is to motivate teams to pursue a best-for-project mindset and to engage in effective collaborative dialogue.

Early scientific management writers (e.g., F.W. Taylor and F.B. Gilbreth), and organisational studies literature (e.g., H. Fayol, C. Barnard, and M.P. Folett), highlight advantages of task specialisation allowing consistent, repeatable and efficient production of goods by specialised teams grouped into an organisational structure to maximise efficiency. Seminal project management (PM) studies followed the scientific management theme and early theory tended to focus on operational aspects such as planning and scheduling, and organisational aspects such as work breakdown structures and team integration based on a project life cycle (PMI, 1996; 2017). Over the past 50 or so years, PM has substantially developed with breakthroughs – e.g., the re-thinking project management literature (Winter and Smith, 2006) and expansion of PM thinking in terms of strategy, social benefit, and mega-major projects by influential authors including thought leaders such as Morris (2013). Fundamental requirements for efficiency, and more importantly effectiveness, addresses *how* to split tasks and work into specialised areas undertaken by expert teams followed by *how* to effectively coherently integrate these to deliver the desired product-outcome.

IPD/Alliancing takes the differentiation-integration concept to a higher level, and organisations such as the LXRP raises the program of projects management concept to new hights. Figure 5-1 illustrates two mindsets exhibited by the LXRP, an integrated project and an integrated program mindset. Both mindsets are essential to ensure project and program delivery coherence. Teams need first to be integrated to collaborate at the project level as their primary responsibility and accountability with this *separate* integrated within-project mindset characteristic being their core concern. However, teams in a program of projects also need to adopt an integrated program mindset to collaborate across projects within the program with an integrated *best-for-program* mindset.

At the program level, the LXRP needs to integrate all alliance projects effectively to achieve program performance success – to fulfil the program's purpose. The LXRP CEO and its executive team (Chapter 3, Figure 3–4) also need to integrate the project alliance packages and develop mechanisms that promote separate Alliance Project Teams (APTs) to behave collaboratively in an integrated way. Critical to effective integration is development of the requisite level of trust and commitment to effectively collaborate so that the whole is greater than its parts (Chapter 6).

5.2.1   *Separate project mindset*

Traditional contractual forms segregate teams, ensuring each participating organisation is accountable for *their* portion of the work. This discourages integration through teams' fiduciary duties of care, loyalty, and obedience to their home-based organisation while ostensible not harming the organisations' beneficiaries/shareholders (Clark, 2011). Fiduciary duty complicates integrated separate project teams due to its balance between reasonableness and rationality, long and short term views, and distributive justice (Eccles, 2018). Consequently, narrow interpretation of this duty often inhibits prioritising a best-for-project mindset and encourages opportunistic behaviours. This presents an organising challenge to develop a contract form that enshrines workplace behaviours to create a united, integrated and highly collaborative team culture that prioritises the *project* outcome over *individual* team performance. The key focus is on project performance.

Resolving this challenge developed through IPD/alliancing practices. A project alliance agreement (PAA) specifically ties all participants to a *single* agreement, tied to project performance key result areas (KRAs), rather than individual participating team performance measures. This united team is incentivised by being accountable for the *entire* project outcome, not just their *part* of the project.

Key elements of an IPD/Alliancing integration framework are a united project team comprising: (1) project owner's representative; (2) design team; (3) project delivery team; and (4) facility operations management organisation's representatives. This integrated project team collaborates (based on their shared vision of the project objectives) and engages in shared accountability and responsibility to ensure a project's success. A gain-pain sharing arrangement is accepted, ensuring that the united APT is motivated to form an integrated collaborative team (Chapter 3). Fiduciary duty is then interpreted as the project being the main beneficiary. Therefore, team participants are relieved of a potential dilemma of which 'master' they serve (their home-based organisation or the alliance project) and what loyalty behaviour is expected of them by the PAA.

Team integration is supported beyond signing a PAA agreement. IPD/Alliancing facilitates a united one-team culture and mindset. Walker (2020) developed an alliancing Integration Framework (Figure 5-2), re-analysing data used to produce the Collaboration Framework developed by Walker and Lloyd-Walker (2015) and identified five platform facilities supporting project alliance and IPD integration: the motivation and context to integrate teams; integrating governance arrangements through the PAA; integrated risk mitigation through a PO funded common insurance policy; integrated joint communication facilities such as a common communication platform that may include building information modelling (BIM); and substantial co-location. These tools were identified from relevant literature such as co-location in a 'Big Room' used frequently on IPD projects in North America (Fischer et al., 2017) and Finland (Alhava, Laine and Kiviniemi, 2015) where all five elements were present. Platform facilities focus attention on project planning and action through a united team collaboratively responding to dynamic issues with dexterity.

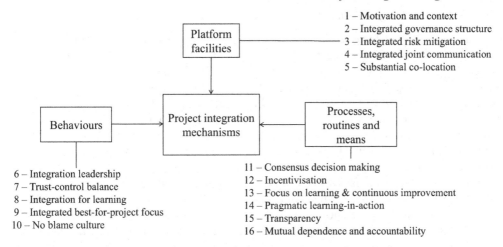

1 – Motivation and context
2 – Integrated governance structure
3 – Integrated risk mitigation
4 – Integrated joint communication
5 – Substantial co-location

Platform facilities

Behaviours

Project integration mechanisms

Processes, routines and means

6 – Integration leadership
7 – Trust-control balance
8 – Integration for learning
9 – Integrated best-for-project focus
10 – No blame culture

11 – Consensus decision making
12 – Incentivisation
13 – Focus on learning & continuous improvement
14 – Pragmatic learning-in-action
15 – Transparency
16 – Mutual dependence and accountability

Figure 5-2 Integrated project delivery elements required to support collaboration

Platform facilities are necessary but insufficient. People collaborate but require established integration of teams and collaborative behaviours supporting this. Walker (2020) identifies: supportive proactive leadership to sustain team integration practices; a balance between trust and control in systems and teams; integrated team learning with open sharing of knowledge, information and insights; a focus on a best-for-project priority; and an open workplace culture supporting dialogue, questioning assumptions and critical thinking supported by a no-blame workplace culture.

Platform and behavioural components require processes, routines and means that reinforce integration through: consensus decision making ensures that no-blame is logically reinforced; incentives are built into the PAA to support team joint and common integration; organisation focussed whole-team learning mechanisms for knowledge and innovation diffusion and continuous improvement; a pragmatic approach to action-learning through experimentation and modelling; organisational and cross-team transparency; and PAA requirements for whole-team responsibility and accountability.

Figure 5-2 identified elements and three components are supported by an extensive systematic literature review revealing 105 relevant articles on project integration of multi-organisational teams. This formed the basis for a "questionnaire tool assessing different integration capability dimensions and through five evaluation workshops" (Saukko, Aaltonen and Haapasalo, 2022, p. 81) undertaken with 49 Finnish project alliance design and delivery team members. Saukko et al. identified four main capability components, each with a range of identified elements. Their main elements were: administrative, organisational, and contractual integration capability; behavioural and relationship-based integration capability; technological and process integration capability; and capability for continuous management and the adaptation of integration and collaboration.

This results in collaboration through integration in general that promotes and supports a best-for project mindset. We see these features commonly present in IPD case study findings (Fischer et al., 2017) and in Alliancing (Department of Finance and Treasury, 2006). A best-for-project mindset prioritises and maximises the impact of the project in delivering value and benefits. However, the focus on a particular project alliance within a stream of projects may deflect vital attention away from the overall goal of the program of works. Thus, an integrated program mindset is necessary when managing a program of alliance projects.

5.2.2   *Integrated program mindset*

Program management focusses on how an overall program of work (projects) can be effectively delivered to fulfil a strategic aim that realises identified benefits to a range of impacted stakeholders (PMI, 2008; Artto et al., 2009; PMI, 2016; Patanakul and Pinto, 2017). A key program-integrating driver is the shared strategic purpose of engaged project managers and the program management team to understand how each project *fits* with the program strategy to deliver the anticipated benefits and value (Martinsuo and Killen, 2014), however while being aware of and sharing understanding is necessary it is insufficient. As Drouin and Turner ague (2022, p. 27), value and benefits are *context-dependent* relating to the type of infrastructure, *time-sensitive* relating to situations when either the benefit is no longer important or if the long-term benefit turns out far greater than initially envisaged, and *stakeholder-dependent* relating to the benefit value-proposition focused on by the beneficiaries. The challenge of developing benefit measures and performance indices is compounded by each program of projects and the projects within a program having slightly different strategically nuanced aims and objectives to meet its purpose. Therefore, developing an integrated mindset for program participants to focus on a consistent vision, presents a difficult test to meet. Chapter 12, performance, attempts to explain how this may be measured and Chapter 11 discusses stakeholder engagement.

An integrated program mindset reflects understanding the interaction and interfaces between projects within a program. The aim is to optimise the whole program's effectiveness. This affects timing of projects, resource allocation, monitoring and control and creation of synergy between project elements to *integrate* these, making the whole greater than its parts. Framework agreements go partway to integrating projects within an integrated program and have been used for decades in the UK (Smalley, Lado-Byrnes and Howe, 2004; Department of Health, 2012) and these have been evolving as happened for example in the Heathrow Terminal Five T5 Agreement (Doherty, 2008) and more recently in the UK Project 13 initiative (Institute of Civil Engineers, 2021). These focus on the skills, competencies and attributes needed to gain value through innovation diffusion from one project to others in a program as is the case for program alliances (Walker and Rowlinson, 2020, pp. 50–51).

Integrating projects within a program requires corporate-level skills and expertise. Using a similar logic to the argument for IPD/Alliancing being superior to DBB and D&C for complex projects, effective integration of projects' outputs and outcomes can be seen as essential. This has interesting implications. For example, in the LXRP case, which rail crossings should be removed first? The most dangerous (benefit being saved lives)? The most efficient (improving either/both the road and rail network performance)? Also, from a stakeholder and urban legacy perspective, should station precinct renewal/development take priority? Clearly, priority must be guided by the program's purpose and strategic aims. Competing perspectives on what really counts and takes priority makes program management highly complex and challenging.

An example of this dilemma can be found in early LXRP performance evaluations. The Victorian Auditor General's Office (VAGO) first report (2017) raised concerns about the way that work was packaged and undertaken when compared to the approved business case. The business case had stipulated a set of criteria that had not been fully met in practice during the initial stages of the program at the time of the 2017 audit. However, a follow-up VAGO audit (2020) noted that initial concerns had been addressed and the audit presents an overall favourable picture of progress. The program for the initial 50 level crossing removals has expanded several times (Chapter 2) to its current 110 removals and each expansion has been accompanied by better than anticipated time and cost program outcomes with the scope-creep being accompanies by greater value-creep. Barton et al. (2019) note that cost benefit analysis is notoriously difficult to accurately assess 'true' costs and benefits because many benefits are difficult to monetise and, as seen in Figure 3-2, strategic transformational aims are a case in point, legacy is best assessed after many years/decades and 'developing vibrant urban hubs' is impossible to accurately monetise. Moreover, at a program level, how can road-rail network improved efficiency be assessed with continued population growth and shifts in demographics triggered by project outcomes?

One key strategic LXRP aim was to achieve continuous improvement. While it was impossible to reward innovation and knowledge diffusion across the program for integrated APTs as a final program outcome incentive, it was possible to reward this diffusion to them for innovations that were diffused from one project to another that demonstrated additional cost saving and/or value capture. The program had an important mechanism to facilitate this (Chapters 3 and 9), the Joint Coordination Committee (JCC), set up by LXRP to integrate projects within the program and administer and enable *innovations* to be registered, assessed and savings within projects rewarded at the project level and, when applied to other projects in the program, would apply as rewards to the instigating APT.

Integration of projects mechanisms were also instigated through the LXRP corporate organisational structure (Chapter 3, Figure 3–4). While that matrix organisation appears complex, resources can be effectively shared across APTs while maintaining program leadership. For example, safety was a top priority, so the safety directorate could liaise with the embedded safety specialists within APTs to ensure that safety was a paramount consideration and that emerging best practice on one APT could be diffused to other APTs. This was particularly evident with the constraints imposed by COVID-19 management requirements. As Interviewee IV10 noted, COVID-19 experiences changed the mindset about the essential nature of team co-location. The LXRP found that while co-location had many advantages for cross-fertilisation of ideas and tacit knowledge, it was not essential to the extent previously thought. It was found that people could work from home (WFH) and "when COVID sent the Designers to WFH. Collaboration didn't suffer and in fact anecdotally, productivity increased."

In a similar manner, stakeholder engagement could be addressed at both a program level (Chapter 11) and at a local project alliance package level. Special events were frequently held to showcase various removals with open days, that invited residents and others to view project progress and techniques. Some of these were held during holiday periods allowing family outing days where APT staff explained to visitors, details of the project and responded to frequently asked questions. Other initiatives included website stories communicating how knowledge and benefits from one APT is shared with others.[1] This adopted a program perspective, later diffused to the Victoria's Big Build website for other projects under the Victorian Infrastructure Delivery Authority (VIDA). Other parts of the corporate program to project organisational matrix share similar characteristics and features. Importantly, this strategy did not simply evolve but was designed-in from the program start.

5.3 LXRP Integration Context

We now turn to *how* the program addressed its project integration by discussing constraints, opportunities and integrative mechanisms.

5.3.1 Constraints to integration

To integrate projects at the program level requires highly specialised knowledge, skills, and expertise and other necessary resources such as machinery and equipment and information communication technology (ICT). Machinery and equipment vary from the more standard items, concrete making and placement, mobile and specialised cranes, to more specific equipment to make precast concrete components. Constraints included both acquisition (timing and volume) as well as skilled operators of the equipment. ICT advancement such as BIM and other model-ling tools uptake required at scale for the LXRP, presents serious challenges that could not be ignored. As discussed in Chapter 4, these constraints influenced the Alliancing procurement choice. Integrating project alliances teams through the 15-year program strategy made invest-ments in training and development viable. Tension between a team integration PO intervention and relying on market forces and competition is often articulated through value for money (VfM) arguments. High market-oriented inclination assumes free flow of resources and suf-ficient supply to meet demand. High alliancing-orientation inclination assumes a resources im-balance and resolving that by managing that constraint.

Project-focussed staff were needed that held a best-for-project mindset to ensure that each project performed efficiently. Required skills range from technical and engineering to administrative and managerial, complimented by 'soft' relationship skills to engage with sub-contractors, suppliers, and many stakeholder groups (because this was a highly visible state project). It became clear that the number and type of skilled employees could not be obtained 'off the shelf' and many needed existing skills enhanced. An effective way to do this is through providing a collegial workplace culture featuring mentoring, knowledge diffusion support over time to create rich and fertile learning and experimentation (Szulanski, 2000). This suggests team integration with a united-team mindset to encourage collaboration. Participants need to identify with the expected shared values expected of their project and the program (Chapter 11).

Detailed discussion of required skills appears in Chapter 7 and leadership skills in Chapter 8. Addressing these required a program level integration response to avoid competitive 'bidding' for scarce human resources. The early establishment of the LXRP in 2015 as a stand-alone entity evolved into the VIDA structure (Chapter 3, Figure 3-2). Not only was there dire need for suitable talent at the LXRP project level but also similar demands for talented staff for the other VIDA pro-jects valued over 100 billion dollars. The VIDA response through the Victoria's Big Build program concept was to pool recruitment and development so that hiring of graduates, and training and development programs could be centralised. The website that most graduates could first access[2] il-lustrates how it carefully explains what graduates may expect in terms of salary and workplace con-ditions, what is expected of them in terms of qualifications and training and importantly, links to the projects that they may be assigned to. The web site lists a diverse range of graduate discipline areas that reflects the project and program support fields. Starting salary and prospects are attractive.

Program complexity poses a leadership-management constraint because there are numerous elements (Chapter 3, Figure 3-3) that require high levels of liaison and integration of special-ists within APTs that also report to their corporate program-level directorate. The alliancing behaviour mindset helps balance this complexity. Reducing power-information asymmetry to allow the levels of integration and collaboration across projects presents a shared leadership capabilities challenge (Drescher et al., 2014) of all participants to enable their confidence and independent critical thinking capability to challenge assumptions and also contribute insights that are valued by APTs and the program directors. This quality is often difficult to develop where people have been mainly exposed to a high face-saving workplace culture where poten-tial mistakes or failure of argument-decision logic is not flagged and potential improved ideas proposed (Clegg et al., 2023).

5.3.2   *Opportunities offered by team and project-program integration*

Many opportunities offered through team integration relate to scale-complexity management and reflexivity-ambidexterity. The duration, scale and complexity of the LXRP afforded oppor-tunities for upskilling on an industrial scale. Closer integration of the APTs with key suppliers such as the pre-cast concrete Skyrail components led to greater collaboration in design and erection to permit greater use of sustainable materials, modularisation and erection processes and efficiencies gained through knowledge and data available from VicRoads historical traffic flow data. For example, interviewee IV05's role in APT packages was extensive. VicRoads had 3D and geotechnical data on the crossings that provided value at the design stage, they also have deep corporate memory about requirements and information needed for utility services associated with level crossing construction and they were able to advise the APT on traffic flow patterns and truck logistic permit requirements for planning and target outturn cost (TOC) development. Other interviewees highlighted examples of as-build design details being shared; facilitated through an integrated cross-project digital platform (Chapter 10).

IV07's role, for example, was with the rail operator in an early program alliance. He had sufficient tacit knowledge to be able to lead a study into using an alternative sheet piling system to that originally proposed in the TOC that saved considerable time and cost and was also approved for safety in working close to live rail lines. It was this participant's team integration and detailed knowledge that provided an opportunity to find innovative solutions to challenges in project alliances that was then shared across the program.

Project director-level integration across projects within the program though the JCC provided a steering committee integration opportunity for continuous improvement and innovation diffusion. As discussed in Chapter 3 (Figure 3–6) in depth, the JCC was a critical LXRP integrative mechanism.

Both at the APT and program level, the alliancing structure allows flexibility facilitating resource sharing across APT participants, when required, because these resources are directly paid for by the PO. Reflexivity and ambidexterity are promoted by the alliancing united-team integrated form through united common purpose, whole-project performance view and specified KRAs. If any alliance participant flags looming problems e.g., resource shortages, then it is in everyone's interest that all other participants to help recover from setbacks. Thus, team integration provides valuable social and knowledge capital to demonstrate reflexivity and ambidexterity.

5.3.3   *Integrative mechanisms*

Two main integrative mechanism are discussed here. First, there was a political and historical context (Chapter 2) that provided corporate memory that shaped perceptions about the desirability and need for such mechanisms. Second, there were the mechanism themselves (Figure 5-2).

Chapter 2 discusses the historical and political LXRP context in depth. The most recent historical influence on the LXRP team integration strategy was its experience on the RRL program and many of those working on the LXRP had previously worked on the RRL program. One key lesson learned from that megaproject that was transferable to the LXRP was that mixing competitive delivery forms such as the RRL's public-private partnership (PPP) and D&C with alliance projects posed difficult integration issues because while the fiduciary duty for alliances state specifically that the *project* is the beneficiary, while for PPP and D&C it is mixed – mainly being participant *home-based organisations* in conjunction with the *project*. Thus, sharing sticky knowledge, innovation and intellectual property was not easy for the RRL program across projects due to its lack of integration with a common unity of purpose (Szulanski, 2000).

Once the LXRP's initial project alliance work packages got underway, APTs became more easily integrated as an integrated program and through the JCC shared ideas and knowledge freely. A sense of project priority remained because of the project incentivisation mechanisms but added KRAs for innovation diffusion across projects greatly assisted program integration.

In January 2019, the LXRP as a stand-alone entity ceased, not because it had failed but because it was merged into the Melbourne Transport Infrastructure Authority (MTIA), later restructured into VIDA, which modelled itself substantially on the LXRP as it had operated from May 2015. The number of rail crossings to be removed was increased from the initial 50 to 110 (Chapter 2) with the integration strategy maintained and supported by its governance arrangements (Chapter 3).

Figure 5-2 illustrates the integration framework with several important mechanisms. Effective integration requires unfettered communication between parties to facilitate the integration, collaboration, and joint action purpose. In many IPD/Alliancing contexts this is achieved through a common communication platform to minimise technical interoperability problems and to facilitate human communication interaction.

This was achieved at the technical-system level by the LXRP through the PO maintaining a common information and data transfer platform using BIM, file sharing facilities, common email and document processing software and associated technologies. There was also a database managing JCC communication to encourage innovation diffusion.

At the human level, interaction was maintained through supporting technology interfaces by the strategic communications program group as well as through daily toolbox meetings. Many of the IPD and some Alliancing literature discusses the operation of a 'Big Room' where APT participants are accommodated in mainly open-plan office space with breakout rooms, large information board and digital projection spaces for posting and assembling ideas into plans and action details (Fischer et al., 2017) although Aaltonen and Turkulainen (2023) add that creating such spaces has important, psychological implications about group and subculture dynamics with formation of in and out group tendencies that need to me carefully managed. They note that people in Big Rooms may form numerous cliques, dependent on their activities and task responsibilities. These cliques need to be psychologically safe to effectively integrate and be members of numerous sub-group task teams and this may require briefing and training.

The LXRP used its strategic communications entity to also inform project teams about important events and discussions through its internal internet facility as well as through access via its open web presence for external stakeholders. Chapter 11 discusses *internal* stakeholder engagement and Chapters 3 and 9 discuss the JCC more fully.

An important feature of the LXRP was its application of its continuous improvement strategy through experimental explorative co-learning across program packages, most notably the use of digital engineering solutions such as BIM within a common data environment (CDE) context (Chapters 9 and 10). The cross-project/within-program aspect was crucial to enhance effective alliancing *united single team* performance. This supported program-integrated processes such as focus on learning and improvement and pragmatic learning-in-action.

An example of how the LXRP applied integrate cross-team and cross project collaboration was observed through CDE initiatives undertaken on several alliance work packages. The LXRP has initiated an innovative practice of re-use of as-built designs and modularisation documentation whereby alliance projects are incentivised to share final design BIM data and insights about what works well and how to overcome challenges in adapting as-built common data. Several contractor and designer non-owner participants (NOPs) are industry-recognised global leaders in BIM, digital technology application and developers of an effective CDE. Interviewee IV15 explained the process of sharing of as-built design to the LXRP for other alliances' use.

we have to hand over 3D model as well as our survey data. So that's our drawing and point-cloud [data] capture as well as our drawings and documents. So, both in PDF [Portable data file] format and native [file format].

He explained that his NOP organisation collaborated and co-learned with other LXRP project alliance NOPs that were recognised global leaders in this infrastructure field. The JCC digital engineering community of practice (COP) facilitates participants comparing insights and learning from each other. This demonstrates an example of how emerging technological application integration was enabled by JCC COPs across the program. Chapters 9 and 10 explain how innovation diffusion and digitisation has had a positive impact on integrating teams via real-time communication technologies that enable within-project integrated collaboration as well as cross-project integration.

Cross-team, and cross-project communication integration through a CDE has been a rapidly evolving LXRP project-program integration mechanism. However, it is far from ubiquitous, partially explained by innovation knowledge transfer 'stickiness.' New skills and routines need to be learned and the output from the CDE is often not compatible with the organisations that the LXRP submit this valuable resource to. The Department of Transport and Planning, for example, require 'traditional' as-built documents as PDF drawings. This results in duplication, and potential inaccuracy between submitted PDF and CDE format. This poses a barrier to opportunities for more productive integration of as-built information to be used later for operations and maintenance by rail operators or for the road authorities. IV14 noted that the lack of integrated or interoperative standards across the LXRP clients for accepting digital format data, and the skills to work with these formats by those agencies, remain a challenge from a program-to-client interface integration efficiency perspective.

The above scratches the surface of LXRP mechanisms used to integrate its program of alliance project teams and the data, information and knowledge that has been co-created. Clearly, mechanisms must include governance-institutional elements i.e., providing a facilitating and leadership structure to require, encourage and manage how integration of people, information, and knowledge is addressed through the 'rules of the game.' Also, the human dimension must be addressed so that there is adequate enthusiasm and motivation for people to accept being integrated in project alliance teams and identify themselves as being active contributors to their project and team program. That requires behavioural and psychological mechanisms discussed above. Third, the technology used to enable and facilitate integration needs to be developed and refined to suit the dynamic program delivery context. Figure 5-2 illustrates how the integration elements logically fit, however, we must remember that the *purpose* of integration is to facilitate and enable collaboration leading to positive outcomes.

5.4   LXRP Integration Capability Context

How did the LXRP ensure that it had the necessary effective within-between project team integration capabilities? As explained in Chapter 4 the program delivery strategy was an integrated program of alliance projects. This was based on trust and mutual commitment focussed on a best for project-program basis.

5.4.1   Alliance project team recruitment

Project alliance syndicate team recruitment and section processes for potential has been well established for several decades. The National Museum of Australia project alliance (Auditor-General of the Australian National Audit Office, 2000) provides a model that provided a benchmark for many alliances. This approach led to the 'single TOC' approach (Walker, Hampson and Peters, 2002; Walker and Hampson, 2003). In summary, a general expression of interest call is made by the PO representative for alliance proponents to form syndicates of design and delivery participants that usually includes not only managing contractors but also major services sub-contractors that will join with the facility operators and PO representative to form an alliance team.

The expression of interest requires hard evidence of the proponents' expertise and capacity to work effectively within a project alliance. In the National Museum of Australia project alliance case (Auditor-General of the Australian National Audit Office, 2000, p. 120), they had to demonstrate with credible evidence demonstrating their:

1. Ability to complete the full scope of works illustrating their complex projects performance.
2. Ability to minimise project capital and operating costs without sacrificing quality.
3. Ability to achieve outstanding quality results.
4. Ability to provide the necessary project resources.
5. Ability to add value and bring project innovation.
6. Ability to achieve outstanding safety performance.
7. Ability to achieve outstanding workplace relations.
8. Successful public relations and industry recognition.
9. Practical experience and philosophical approach in the areas of developing ecological sustainability and environmental management.
10. Understanding and affinity for operating as an alliance member.
11. Substantial acceptance of the draft alliance documented for the project including related codes of practice, proposals for support of local industry development, employment opportunities for Australian indigenous peoples; and
12. Commitment to exceed project objectives.

The above required evidence of past project performance of each criterion. Clearly the benchmark standards are very high and while the above requirements were framed in the late 1990s, many of the present LXRP participating organisations have had two decades alliancing experience to be able to cite sufficient evidence of their capacity. These general minimum requirements have since been embedded in Australian government guidelines for procurement (Department of Finance and Treasury, 2006; Department of Infrastructure and Transport, 2011). Examples of project alliance agreement insights is accessible at PCI (www.pcigroup.com.au/our-publications).

The 'single TOC' approach, where alliance proponents were selected purely on demonstrated team excellence, was questioned by various state government Treasury and Finance Departments. A 'competitive TOC' approach was adopted on many projects from the late 2000s, where two competing alliance proponent teams, shortlisted using similar criteria to the above, worked on their TOC plan and the most attractive one (not necessarily the cheapest/quickest) proposal was selected with compensation paid to the unsuccessful proponent team. However, there is still mixed views on whether a 'competitive TOC' provides better VfM or best value (BV) than the single TOC (Ross, 2008; Walker and McCann, 2020). The core principle is that the best integrated team be recruited that can collaborate and deliver a BV outcome. TOC proposals were subject to an external VfM and viability scrutiny. External probity consultants ensured due process being consistently followed.

The project alliance proponent team must, therefore, offer highly skilled and experienced staff to be able to mount a credible proposal. The TOC process normally takes 6–9 months of intense dialogue between participants from its initial formation under an Alliance Development Agreement until the TOC plan is accepted by the PO signing the PAA. At this stage the facility operator team and PO participant team formally unite into one APT (Walker, Vaz Serra and

Love, 2022). These teams need to be nurtured and supported with participant skills development and mentoring-coaching a key feature to maintain the united APT spirit (Chapters 7 and 8).

The LXRP started with a mixture of Single and Competitive TOC approaches but now that it has delivered many crossing removals and station precinct revival projects in the program it has sound benchmarking reference data to be able ensure that VfM/BV is maintained on each new project and re-use as-built designs as a starting point to adapt for any new project's site conditions.

The LXRP moved from an initial competitive TOC basis for project alliance Work Packages 1–5. However, after the initial alliance project packages had been awarded the recruitment process was modified to allow for modified supplementary additional work package (AWP) project alliances to be awarded through its initial work package (IWP) mechanism. Project Alliance Work packages 1–5 were aligned with works associated to one of five rail network lines with existing APTs bidding on these in a similar way to previously, however, essentially on a single TOC basis.

LXRP's procurement strategy of packaging alliance AWPs in a pipeline of ongoing work had a dual impact. First, it provided motivation for each alliance syndicate to retain key staff, several interviewees highlighted the motivation to see the LXRP as a genuine career path supported by training and development and undertaking shadowing roles. Second, the benchmarking data and innovation diffusion through the shared knowledge/experience JCC mechanisms engendered both highly competitive (against benchmarked reference data) TOCs with a *program* mindset.

Interviewee IV10 explained the rationale was primarily twofold. First, the experience of resource intensity requirements from potential market syndicate players indicated such an availability constraint that the value of conducting dual TOC developments (with two teams competing and requiring to tie up rare human expert resources during that TOC period) proved unfeasible and unstainable. The TOC process under the evolved AWP approach took 15–20 weeks and required one TOC development team. Second, with scores of level crossing removals and initial project alliance projects completed, there was not only a considerable body of valid and useful benchmarking data available, but also there was (through the LXRP PAA) an ability to adapt existing as-build designs and delivery TOC plans to ensure greater efficiency and effectiveness and credible and reasonable TOC plans for the expanded program from the initial 50 to 110 removals and additional stations and precinct works. Thus, greater VfM/BV was possible than adhering to a dual TOC process.

Interviewee IV18 also explained how the combination of the AWP approach using a single TOC selection method, and the impact of the KRA1.2 (innovation adoption/adaptation) was enlightening. Having a pipeline of work through the AWPs incentivised the five alliance syndicates to maintain their interest in keenly bidding a TOC that met or beat the benchmarks set by previous LXRP AWPs. Innovation sharing, encouraged by the KRA 1.1 and 1.2 governance measure and performance based on KRA 1.2, proved a powerful incentive. When presenting an AWP TOC for PO acceptance, the question 'what innovations from other LXRP alliance works have you adopted or adapted?' is posed. Failure to consider continuous improvement, with the program's now over seven years' experience and detailed benchmark cost and design data, meant invited alliance syndicates must consider a highly competitive benchmark. IV18 said of the implications of the program and AWP procurement strategy, and the requirement to adopt/adapt innovation and continuous improvement that:

> they're [the invited syndicate] coming back proactively with all of the wins that the other teams have already had so that they can be as successful, you know as possible. And that's of course, because all of those wins, if they've driven cost improvements that's playing out in our owner's benchmark costings as well.

Thus, a single TOC approach through existing concept and idea adaptation is more feasible than relying on a competitive TOC. Additionally, by providing a more certain workflow of projects through the program, current APT members can integrate more easily through having worked together over time to build trust (benevolence, ability and integrity) as explained by Davis and Walker (2020).

5.4.2   Program executive team recruitment

Figure 3–4 in Chapter 3 illustrates the LXRP program organisation chart with its integrated corporate functions and embedded staff from these functions within APTs. The matrix organisation also needs close attention to maintain its integration integrity. Specialists in areas such as business strategy, people and culture, commercial and legal and strategic communication may not necessarily have worked within either a project alliance or an infrastructure project delivery context. These and other support function staff illustrated in Figure 3–4 also need to be recruited with many of the above 12 criteria prominent in their appointment.

5.4.3   Value-chain capabilities perspective

Organisational integration and differentiation have been studied for many decades following seminal work by Lawrence and Lorsch (1967) that concluded that organisational form followed contingency and context where each element of the organisation took responsibility for coping with its external environment. The high levels of integration with LXRP projects (the alliance form) and program (high levels of cross-project integration) is noticeably prominent when compared with global megaprojects. Many comparable megaprojects have been delivered as projects and not as integrated program of projects, though they have been perceived as requiring intensive systems integration (Whyte, Davies and Sexton, 2022) for example Crossrail, Heathrow's T5 (Davies, Gann and Douglas, 2009), the Panama Canal widening (van Marrewijk et al., 2016) and a host of others. The London 2012 Olympics was delivered as a series of projects integrated by the Olympic Delivery Authority (ODA), linking the separate Olympic Games facilities and support infrastructure to ensure the smooth delivery of that highly complex array of integrated individual events (Davies and Mackenzie, 2014).

What stands out when considering delivering a megaproject as a single 'project' or as a program of projects, is the contingent nature and context of the challenges being faced and how best to integrate participants (companies engaged in project delivery or projects within an integrated program) and any advantages of creating a pipeline of work to motivate program participants to remain in the program (Chapters 7 and 8). As discussed earlier, and in Chapter 3, project purpose and strategy is a defining factor in how best to integrate teams. Two aspects of the program strategy, purpose and project aims are critical to understanding the LXRP program of integrated projects choice.

One underlying assumption was that continuous improvement and innovation diffusion would lead to smarter design and delivery and greater VfM/BV (Chapter 9). Integrating key supply chain participants, such as the design, contractor-delivery and operator with the client representative, helps facilitate improved value-chain (rather than supply-chain) management (Fischer et al., 2017). Learning within and between projects became a LXRP defining feature with the JCC as its prime organisational enabling mechanism. Projects within the program become part of the value-chain.

A second prominent aspect of the LXRP team integration is addressing risk and uncertainty. An advantage of IPD/Alliancing is its availability of multiple insights (technical detail, administrative, and cross-disciplinary) being focussed through the alliance PAA and governance structure to not just manage risks (known-knowns and known-unknowns) but to explore and prepare for uncertainty (unknown-unknowns) in a way that Howell et al. (2010) refer to as framing project decisions through a balanced plan-driven and problem-structuring approach that encourages agility, resilience, and versatility. Walker et al. (2017) also argue that ambiguity, where team participants may mis-interpret a 'known' or mistakenly assume 'facts,' can be avoided through close cross-team and cross-discipline integration facilitating effective questioning assumptions. Mechanisms such as creating a no-blame workplace environment (Lloyd-Walker, Walker and Mills, 2014; Koolwijk, van Oel and Moreno, 2020) and integrating key team participants into the project/program team helps reduce uncertainty-ambiguity through broadening and deepening the project/program expertise pool and marshalling these value-chain resources to respond flexibly to important/critical unforeseen events.

The LXRP and other similar organisations integrate to the extent they deem fits their context would base their team integration consideration on more grounds that the above two, but they fit well with a value-chain capability shaping perspective of project delivery.

5.5   Conclusions

This chapter focussed on how projects achieved project team integration and how the program of projects was integrated. We briefly discussed the general mechanisms for integrating teams then explained the integrated IPD/alliancing project team mindset. We then described the mindset for integrating projects within a program. We probed LXRP constraints and opportunities, focussing on how organisational learning and innovation diffusion was facilitated. This led to a discussion on how project alliance teams were recruited and how those participating in the corporate functions integrated the projects into a coherent program.

We conclude that *project* integration was achieved through:

1. Its recruitment of project alliance teams, initially through an alliance syndicate standard expression of interest approach with united team shortlisting based on participating teams' demonstrated abilities and capabilities to form a united project alliance team;
2. Teams typically spend 15–20 weeks under an Alliance Development Agreement to develop the TOC plan in conjunction with the road and rail service operator team and collaborative interaction with the PO representative, informed by a LXRP reference design to guide VfM. The TOC was also subject to an external viability and probity process;
3. PO approval of the TOC led to formal integration of the successful syndicate with the facility operator and PO representative signing the PAA. This team would have collaborated, jointly working on the TOC plan gaining confidence, trust and team building through that process;
4. The PAA specified several important integration mechanisms;

    a. Project and not individual team organisational performance was monitored with gain/pain sharing arrangement to incentivise an integrated united-team mindset.
    b. The best-for-project mindset was enshrined in the PAA reinforcing an integrated focus clarifying participant fiduciary duty towards project success.
    c. PAA specific participant behavioural requirements clarified inter-participant integration mechanisms such as a no-blame workplace and consensus decision making that further solidified integrated team action;
    d. the JCC played a significant part in integrating projects within the program and different disciplines and interest groups to develop a LXRP community spirit; and

5. Substantial co-location, joint communication facilities and integrated APT transparency and accountability cemented an integrated approach for project alliances.

Achieving integration *across projects* within the program was achieved by several important devices:

1. The matrix organisation that linked corporate cross-project consistent policy and action illustrated in Chapter 3, Figure 3-3.
2. The establishment and maintenance of the JCC with its numerous sub-groups of specialists across the program forming effective communities of practice (Chapter 3).
3. The JCC provided a particularly powerful integrative mechanism linking projects within the program; and
4. The stakeholder engagement program entity that facilitated internal stakeholder engagement.

Principally, important outcomes of cross team/discipline integration were:

1. Improved continuous project delivery performance within and across projects.
2. Improved risk and uncertainty/ambiguity management.
3. Improved resilience and agility in response to unanticipated events.

The main contribution made in this chapter is to frame how APTs are recruited to develop integrated teams of specialist that facilitate project delivery collaboration within a program context. The LXRP provides a rare opportunity to explore how a program of projects integration may apply where social and knowledge capital is generated and exploited across the projects within the program. While the LXRP Figure 3-3 matrix organisation, discussed in this chapter, may appear complex and difficult to manage, evidence suggests that the LXRP integration approach is effective. We also discussed some of the human mechanisms used to integrate APT's both at the project and program level.

Notes

1 www.laingorourke.com/projects/australia/south-eastern-program-alliance/
2 See https://bigbuild.vic.gov.au/jobs/training-programs/graduate-programs

References

Aaltonen, K. and Turkulainen, V. (2023). "The use of collaborative space and socialisation tensions in inter-organisational construction projects." In *Construction Project Organising*, Addyman, S. and H. Smyth (eds), pp. 167–184. Wiley.
Alhava, O., Laine, E. and Kiviniemi, A. (2015). "Intensive big room process for co-creating value in legacy construction projects." *Journal of Information Technology in Construction* (ITcon). 20 (11): 146–158.
Artto, K., Martinsuo, M., Gemünden, H.G. and Murtoaro, J. (2009). "Foundations of program management: A bibliometric view." *International Journal of Project Management*. 27 (1): 1–18.
Auditor-General of the Australian National Audit Office (2000). *Construction of the National Museum of Australia and the Australian Institute of Aboriginal and Torres Strait Islander Studies* Canberra, Australia.
Barton, R., Aibinu, A.A. and Oliveros, J. (2019). "The value for money concept in investment evaluation: Deconstructing its meaning for better decision making." *Project Management Journal*. 50 (2): 210–225.
Clark, G.L. (2011). "Fiduciary duty, statute, and pension fund governance: The search for a shared conception of sustainable investment." https://ssrn.com/abstract=1945257

Clegg, S.R., Loosemore, M., Walker, D.H.T., van Marrewijk, A.H. and Sankaran, S. (2023). "Construction cultures: Sources, Signs and solutions of toxicity." In *Construction Project Organising*, Addyman, S. and H. Smyth (eds), pp. 3–16. Chichester, West Sussex, John Wiley & Sons.

Davies, A., Gann, D. and Douglas, T. (2009). "Innovation in megaprojects: Systems integration at London Heathrow Terminal 5." *California Management Review*. 51 (2): 101–125.

Davies, A. and Mackenzie, I. (2014). "Project complexity and systems integration: Constructing the London 2012 Olympics and Paralympics Games." *International Journal of Project Management*. 32 (5): 773–790.

Davis, P.R. and Walker, D.H.T. (2020). "IPD from a participant trust and commitment perspective." In *The Routledge Handbook of Integrated Project Delivery*, Walker, D.H.T. and S. Rowlinson (eds), pp. 264–287. Abingdon, Oxon, Routledge.

Department of Finance and Treasury. (2006). *Project Alliancing Practitioners' Guide* Department of Treasury and Finance G. o. V. Melbourne, Australia.

Department of Health (2012). *The ProCure21+ Guide – Achieving Excellence in NHS Construction*, Service N.H. Leeds, UK, Department of Health.

Department of Infrastructure and Transport (2011). *National Alliance Contracting Guidelines Guide to Alliance Contracting*, Department of Infrastructure and Transport A.C.G. Canberra, Commonwealth of Australia. www.infrastructure.gov.au/infrastructure/nacg/files/National_Guide_to_Alliance_Contracting04July.pdf

Doherty, S. (2008) *Heathrow's T5 History in the Making*, Chichester, John Wiley & Sons Ltd.

Drescher, M.A., Korsgaard, M.A., Welpe, I.M., Picot, A. and Wigand, R.T. (2014). "The dynamics of shared leadership: Building trust and enhancing performance." *Journal of Applied Psychology*. 99 (5): 771–783.

Drouin, N. and Turner, J.R. (2022) *Advanced Introduction to Megaprojects*, Northampton, UK, Edward Edgar Publishing

Eccles, N.S. (2018). "Remarks on Lydenberg's 'Reason, Rationality and Fiduciary Duty'." *Journal of Business Ethics*. 151 (1): 55–68.

Fischer, M., Khanzode, A., Reed, D. and Ashcraft, H.W. (2017) *Integrating Project Delivery*, Hoboken, NJ, Wiley.

Howell, D., Windahl, C. and Seidel, R. (2010). "A project contingency framework based on uncertainty and its consequences." *International Journal of Project Management*. 28 (3): 256–264.

Institute of Civil Engineers (2021). *Project 13 Framework* On-line UK.

Koolwijk, J.S.J., van Oel, C.J. and Moreno, J.C.G. (2020). "No-blame culture and the effectiveness of project-based design teams in the construction industry: The mediating role of teamwork." *Journal of Management in Engineering*. 36 (4): 04020033.

Kurtz, C.F. and Snowden, D.J. (2003). "The new dynamics of strategy: Sense-making in a complex and complicated world." *IBM Systems Journal*. 42 (3): 462–483.

Lawrence, P.R. and Lorsch, J.W. (1967). "Differentiation and integration in complex organizations." *Administrative Science Quarterly*. 12 (1): 1–47.

Lloyd-Walker, B.M., Walker, D.H.T. and Mills, A. (2014). "Enabling construction innovation: The role of a no-blame culture as a collaboration behavioural driver in project alliances." *Construction Management and Economics*. 32 (3): 229–245.

Martinsuo, M. and Killen, C.P. (2014). "Value management in project portfolios: Identifying and assessing strategic value." *Project Management Journal*. 45 (5): 56–70.

Morris, P.W.G. (2013) *Reconstructing Project Management*, Oxford, Wiley-Blackwell.

Patanakul, P. and Pinto, J.K. (2017). "Program Management." In *Cambridge Handbook of Organizational Project Management*, Sankaran, S., R. Müller and N. Drouin (eds), pp. 106–118. Cambridge, UK, Cambridge University Press.

PMI (1996) *A Guide to the Project Management Body of Knowledge*, Sylva, NC, USA, Project Management Institute.

PMI (2008) *The Standard for Program Management*, Newtown Square, PA, Project Management Institute.

PMI (2016) *Governance of Portfolios, Programs, and Projects : A Practice Guide*, Newtown Square, PA, Project Management Institute.

PMI (2017) *A Guide to the Project Management Body of Knowledge*, 6th Edition, Sylva, NC, USA, Project Management Institute.

Ross, J. (2008). "Price competition in the alliance selection process." PCI Alliance Services, infrastructure delivery Forum. Perth, WA, Main Roads Department of Western Australia. www.alliancingassociation. org/Content/Attachment/FOM%20June%209-reasons-for-single-DCT_C-J.%20Ross%202008.pdf

Saukko, L., Aaltonen, K. and Haapasalo, H. (2022). "Defining integration capability dimensions and creating a corresponding self-assessment model for inter-organizational projects." *International Journal of Managing Projects in Business*. 15 (8): 77–110.

Smalley, M., Lado-Byrnes, L. and Howe, M. (2004). *NHS Wales: Construction Procurement Review – Selection of a preferred option for the NHS in Wales*, Ashford T.C.W.C. a. B. Cardiff, Wales.

Snowden, D. and Rancati, A. (2021). *Managing Complexity (and Chaos) in Times of Crisis. A Field Guide for Decision Makers Inspired by the Cynefin Framework*, Luxembourg, Publications Office of the European Union.

Snowden, D.J. and Boone, M.E. (2007). "A leader's framework for decision making." *Harvard Business Review*. 85 (11): 69–76.

Szulanski, G. (2000). "The process of knowledge transfer: A diachronic analysis of stickiness." *Organizational Behavior and Human Decision Processes*. 82 (1): 9–27.

van Marrewijk, A., Ybema, S., Smits, K., Clegg, S. and Pitsis, T. (2016). "Clash of the titans: temporal organizing and collaborative dynamics in the Panama Canal Megaproject." *Organization Studies*. 37 (12): 1745–1769.

Victorian Auditor-General's Office (2017). *Managing the Level Crossing Removal Program*, Melbourne.

Victorian Auditor-General's Office (2020). *Follow up of Managing the Level Crossing Removal Program*, Printer V.G. Melbourne.

Walker, D.H.T. (2020). *The Integrated Side of Integrated Project Delivery*. Melbourne.

Walker, D.H.T., Davis, P.R. and Stevenson, A. (2017). "Coping with uncertainty and ambiguity through team collaboration in infrastructure projects." *International Journal of Project Management*. 35 (2): 180–190.

Walker, D.H.T. and Hampson, K.D. (2003). "Project Alliance Member Organisation Selection." In *Procurement Strategies: A Relationship Based Approach*, Walker, D.H.T. and K.D. Hampson (eds), pp. 74–102. Oxford, Blackwell Publishing.

Walker, D.H.T., Hampson, K.D. and Peters, R.J. (2002). "Project alliancing vs project partnering: A case study of the Australian National Museum Project." *Supply Chain Management: An International Journal*. 7 (2): 83–91.

Walker, D.H.T. and Lloyd-Walker, B.M. (2015) *Collaborative Project Procurement Arrangements*, Newtown Square, PA, Project Management Institute.

Walker, D.H.T. and McCann, A. (2020). "IPD and TOC development." In *The Routledge Handbook of Integrated Project Delivery*, Walker, D.H.T. and S. Rowlinson (eds), pp. 581–604. Abingdon, Oxon, Routledge.

Walker, D.H.T. and Rowlinson, S. (2020). "The Global State of Play of IPD." In *The Routledge Handbook of Integrated Project Delivery*, Walker, D.H.T. and S. Rowlinson (eds), pp. 41–66. Abingdon, Oxon, Routledge.

Walker, D.H.T., Vaz Serra, P. and Love, P.E.D. (2022). "Improved reliability in planning large-scale infrastructure project delivery through Alliancing." *International Journal of Managing Projects in Business*. 15 (8): 721–741.

Whyte, J., Davies, A. and Sexton, C. (2022). "Systems integration in infrastructure projects: seven lessons from Crossrail." *Proceedings of the Institution of Civil Engineers – Management, Procurement and Law*. 175 (3): 103–109.

Winter, M. and Smith, C. (2006). *EPSRC Network 2004–2006 Rethinking Project Management Final Report*, Manchester.

6   LXRP Project-Program Collaboration

Derek H.T. Walker, Peter E.D. Love, Mark Betts and Juliano Denicol

6.1   Introduction

How did the LXRP achieve effective within-project and cross-project collaboration in its program of projects and its organisational structure for each of its projects within the program?

This chapter explores team collaboration in both project alliances and the program of projects. It explains how and why project teams were able, and committed, to collaborate with a unified sense of purpose that aligned with the projects strategic objectives and explains how collaboration was achieved across projects within the program. It extends understanding how LXRP teams collaborate.

Chapter 2 outlines the LXRP complexity context and Chapter 3 explains the LXRP purpose, strategy, and how that led to developing a suitable governance framework. Chapter 4 clarifies the rational for the IPD/Alliancing delivery decision. This chapter linking to Chapter 5, confirms effective integration being highly dependent on effective team mutual trust and commitment. It is also linked to Chapter 7 as a central part of an effective workplace culture.

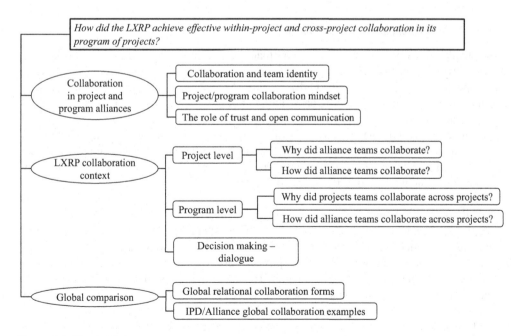

Figure 6-1 Chapter 6 structure

DOI: 10.1201/9781003389170-6

We next discuss general alliance project collaboration principles, framing these against LXRP case study research findings, followed by discussion on the necessary collaboration capability antecedents.

6.2   Collaboration in Project and Program Alliances

Collaboration infers deep levels of team coordination and making group decisions as a unified purposeful unit. Early management theorist visualised managing projects as best organised by the division of work by workflow task-specialists efficiently organising their assemblies to be coherently integrated. Managing projects with simple or standard repetitive processes may function effectively in controlled environments using rudimentary project management (PM) techniques, however, as Snowden's Cynefin Framework identifies (Kurtz and Snowden, 2003; Snowden and Boone, 2007; Snowden and Rancati, 2021), once project activities reach a complex state featuring many unknown-unknowns and unknowable-unknowns, it is vital to access diverse expertise, freely sharing insights contributed in planning and executing projects. The Cynefin framework's premise maintains *co-ordination* by a governing hierarchical authority for complex projects is ineffective and often produces unexpected negative consequences. Snowden et al. suggests that highly complicated, complex and chaotically turbulent situations need expert advisors and decision makers closely *collaborating* as a united committed team.

Ali and Haapasalo (2023) in a study of Finnish hospital alliance projects, trace a development of relationship levels on alliance projects moving from *cooperation* at a base level focussing on alignment of interests, through *control* mechanisms being developed to build and support that cooperation, leading to *coordination* mechanisms to entrench the relationship leading to collaboration. They characterise *collaboration* as being a dynamic engagement process based on high levels of mutual understanding and shared vision and goals. The main message is that for collaboration to exists, there needs to be supporting cooperation, control and coordination mechanisms in place.

6.2.1   *Collaboration and project team identity*

Conceptually, effective collaboration is more than the dictionary definition of just working together. It involves intense dialogue by committed participants who work jointly, each with varying perspectives on an issue, sharing understanding of how norms, processes, rules and identity help integrate teams and team participants into a cohesive unit to solve problems, make decisions and take concerted action as a united team (Bedwell et al., 2012).

Key to effective collaboration is recognising and managing shared identity. Chapter 11 discusses internal (project delivery participants) and external stakeholders (government departments and authorities, local community, end-users etc.) identity in more depth. Identity, according to Cornelissen et al. (2007) classify identity as:

1. *Social,* where individuals feel aligned with a group/organisation's general values and beliefs.
2. *Organisational,* where individuals have shared understanding of an organisation and how it operates and functions, and individuals feel part of that organisation; and
3. *Corporate,* where the organisation (e.g., project) serves a higher-level entity such as a government department or is part of an integrated program of projects. Individuals may/may not share corporate values and aims but they understand that they work within this system.

Individuals collaborating on projects may perceive themselves with multiple identities. Identity, according to intensive qualitative research findings from an educational establishment case study by Gioia et al. (2010), is forged or constructed, but how? Their thematic data analysis reveals an 8-stage process. It starts with a clearly articulated vision that effectively binds people to a core proposition that they can share and relate to, as argued by Christenson (2007) as necessary for enabling project management success.

The expressed vision is subjected to interpretation and contextualisation through individual and group dialogue negotiating if and how they can form meaning about what the vision means and does not mean (to them). This involves testing and experimentation about its validity and legitimacy through exploring vision practicality and contextual fit, defining its boundaries and absorbing feedback. The result is a vision that people who engage with this process, accept as legitimately fitting with their norms and aspirations. It becomes internalised and, within project delivery contexts, institutionalised (Scott, 2014). Identity and legitimisation sets the stage for motivated team collaboration to ultimately successfully realise the accepted vision (Christenson and Walker, 2004).

Collaboration flourishes or withers, depending on the organisational workplace culture and climate – whether it invites diversity of ideas and views or is a hierarchical and micro-controlling workplace climate. Chatman and O'Reilly (2016, p. 205) highlight organisational culture and climate being linked but different. Culture "begins with an emphasis on the norms and values that provide signals to people about how to act and feel ... it is explicitly focused on shared meaning, values and norms as sources of collective identity and commitment." Culture is prescriptive norms about expectations. However, climate is about how strongly those norms are understood and adhered to (Schneider, Ehrhart and Macey, 2013). This idea is linked to commitment. Commitment is perceived at three levels (Meyer and Allen, 1991). A *continuance* level has people wishing to continue working, complying and conforming to expected norms. A *normative* level implies a sense of duty, ethics or reciprocity, while an *affective* level is identity-based – a 'want to' desire. An alliance climate is governed by the project alliance agreement (PAA) behaviour culture. People are committed at one/several of those levels. They may collaborate because they want to (*affective*) but also feel obliged to (*normative*) while there will always be *continuance* commitment based on PAA behavioural requirements. This climate impacts dialogue effectiveness and collaboration quality.

Linked closely to identity and commitment concepts, is *purpose,* extensively discussed in Chapter 11 regarding motivating internal alliance project team (APT) stakeholders to collaborate with a deep understanding of the project (and program) purpose (Ali and Haapasalo, 2023) and how they contribute to that purposeful outcome (Hackman and Oldham, 1976). Figure 6-2 illustrates how collaboration emerges from effective dialogue. A compelling vision liberates free consent and willingness to share ideas, collaboration and engage in dialogue.

Consent is an underexplored concept in a social, business and ethical context (Schaber and Müller, 2018), it links to perceived legitimacy (Schnüriger, 2018). We tend to assume for example when project alliance participants consent to PAA requirements that they truly support espoused dialogue and collaboration behaviours. But how can we be sure? Schnüriger states that consent allows behaviours that may, under many circumstances be unacceptable, to become a norm. Organisational participants's fiduciary duty[1] requires loyalty to their employer's interests. The PAA integrates participant organisations into the project alliance entity, thus transferring participant organisational duty to the alliance organisation. This requires and enables participant performative collaboration consent, it allows *and expects all* alliance participants, to challenge assumptions and offer their expert perspective through dialogue that enables *affective* participant commitment to the project vision and purpose and value co-generateration. This requires careful organisational participant recruitment and selection processes (Chapters 5 and 7). An important pre-requisite for effective collaboration is participant trust, commitment, and inherent

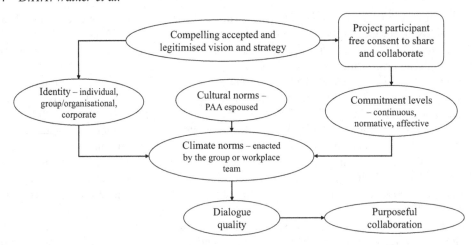

Figure 6-2 Explaining how high-quality dialogue happens

or developed dialogue abilities. Both the workplace organisational governance mechanisms and consenting participants must be competent to give consent (Schnüriger, 2018), in this case to engage in dialogue and knowledge sharing.

People and group norms are influenced by their sense of identity. Chatman and O'Reilly (2016, p. 199) argue that "norms can be parsed into three distinct dimensions: (1) the content or what is deemed important (e.g., teamwork, accountability, innovation), (2) the consensus or how widely shared norms are held across people, and (3) the intensity of feelings about the importance of the norm (e.g., are people willing to sanction others)." Each person's identity carries with it a sense of values and norms, perceived proper process, authority, legitimacy that contribute to their attitude and behaviour towards collaborating with others. People might identify as a construction methods technical expert, an employee of company X, a member of professional organisation Y. They may have had cross-disciplinary experience and live in the area affected by this project or affected by its intended outcome. They may have similar identifications and psychological and expertise-based influences that impact their propensity for collaboration.

Each participant will consider their behaviour and action depending upon their organisational and corporate identity. Their organisational loyalty will depend upon the organisation they *believe* they belong to (their home-based company or the project delivery team?) and their understanding of how the project decision outcome may impact their corporate identity (wider program of projects). In summary, they might express their identity as:

I work on a program of integrated projects of which my project is one of several integrated alliance project work packages (my corporate identity)
I work on Alliance Project P (my organisational and group identity)
I am mainly working on this project paid for by the project alliance (I will return to my home-organisation Contractor K on completion of P)
My expertise is A, B, C (perhaps substantial cross-disciplinary/role experience)
My professional affiliation is X and that has an ethics code, and my past education and experience leads me to believe Y, Z
I will personally benefit from the project's intended benefit.

There are important differences between Project Alliancing and conventional project delivery forms discussed in Chapter 4. One significant PAA difference is the expectation that project participants will feel free to examine and challenge assumptions. Identity brings many 'best/ optimum/ethical/only' assumptions about the way to act. The PAA specifies behaviours require low power/information asymmetry between teams/individuals. Everyone should respect the expertise of others and make consensus-based decisions (Ross, 2003; Department of Infrastructure and Transport, 2011). Therefore, each participant will expect when discussing issues to make decisions, that they should contribute their expertise to make a best-for-project decision. Because all participants are paid by the project owner (PO) through full reimbursement to their home-based organisation for salary etc., each participant's fiduciary duty is to the project and not its home-based organisation. The PAA obliges everyone to make best-for-project decisions and not one that benefits any other interests. This releases waves of creativity through collaboration, especially with elimination of professional standing and perhaps perceived power differentiation based on discipline identity features. Reducing information asymmetry is a significant factor in improved collaboration because it legitimatises placing all relevant information 'on the table.' The PAA substantially supports collaboration through its governance measures. All parties may have different perspectives on issues to be decided upon and are free to offer their insights (guided by their perceived identities) and to question assumptions of other participants. Therefore, a fuller, richer, and more informed discussion can take place.

Figure 6-2 illustrates collaboration quality emerging from dialogue. Walker et al. (2022, p. 4) argue that the value of dialogue across the project cycle comprises:

Information exchange – increasing access to data, information and knowledge; Innovation – encouraging new approaches and ideas; Organisational learning – transfer of learning across boundaries; Psychological – inducing a mentally safe, respectful and rewarding workplace experience; Relationship building – facilitating teams to understand and respect each other; Monitoring and control – facilitating a more collective shared view of responsibility and accountability; and Effectiveness – through minimising waste and rework.

Dialogue is more than debate or discussion it is

not about advocacy where one party wins a debate. Rather it is about exchanging perspectives on an issue to suspend assumptions and explore options that may otherwise have never been considered. It therefore is a value adding tool because often surprising outcomes are achieved that are superior to any one person's or group's starting position.

(Senge, 1990, p. 241)

Therefore, meeting participants can engage in open, unhindered discussion using their multiple perspectives to reach a superior best-for-project outcome that would not otherwise be the case. The group identity focus shifts motivation towards reaching a best-for-project solution. Decisions are not made on perceived individual power, possession of critical unshared knowledge, or biased thinking, because purposeful dialogue ensures APT participants make more fully considered decisions.

6.2.2   *Project/Program collaboration mindset*

Walker and Lloyd-Walker developed the Collaboration Framework-based 50 interviews with IPD-Alliancing expertise (2020b). They found that establishing a workplace system of platform facilities, behaviours that engender a culture in which the participants feel psychologically safe to engage in dialogue (through questioning assumptions and offering alternative perspectives) and supporting processes and routines is central to collaborative inter/cross-team IPD/Alliancing. Taking an institution theory perspective (Scott, 2014), they illustrate the Collaboration Framework as 16 elements linked to Scott's (2014) three institutionalisation pillars: regulatory (that determine the rules, regulations and legitimacy constraints); normative (that set the legitimate behaviours and workplace culture); and cultural-cognitive (that explains how people use their cultural norms to interpret the regulative pillar items). They argue the need for this to be institutionally anchored into how Alliancing and other IPD forms, collaborate on projects.

Figure 6-3 illustrates the first five elements as providing a platform facility to support collaboration across teams within the integrated APT that shares a clear motivation to collaborate, a joint governance structure that each APT participant adheres to, an integrated risk mitigation approach, a joint communication approach and shared platform, and substantial co-location. The next five elements relate to shared requited alliance behaviours that form the cultural norms of the APT including authentic leadership, a balance of trust and control, a shared common best-for-project mindset, and a no-blame workplace culture. The active component of IPD/Alliancing collaboration comes from the last six elements comprising the processes, routines and means for consensus decision making, an incentivisation mechanism to reinforce rewards or penalties being levelled at projects rather than individual team performance, a focus on mechanisms to promote and support learning and continuous improvement, pragmatic learning-in-action processes, transparency, and team mutual dependence and accountability.

Framework element measurements can reveal the extent to which an individual or team identifies, through shared understanding and values, with other APT participants. These element characteristics are legitimised by the PAA for Alliancing projects, the Integrated Form of Agreement (IFoA) in many North American IPD contracts and other contract forms such as the New Engineering Contract (NEC) NEC3/4 in the United Kingdom (UK). The IPD/Alliancing

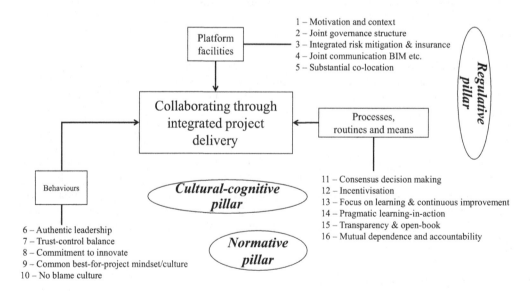

Figure 6-3 The Collaboration Framework

(Source: Walker and Lloyd-Walker, 2020b, p. 27)

literature, relating specifically to collaboration, maintains that collaborating participants share collaboration values and have a common understanding about identifying themselves with their integrated team to effectively work together.

Figure 6-3 illustrates the shaping of a Project Alliance mindset. It provides greater clarity about the three major collaboration components and 16 identified elements. Walker and Lloyd-Walker explain each Collaborative Framework component and element in detail (2015; 2020c; 2020a; 2020d). Platform facilities provide a structure and set of collaboration protocols (the PAA) through facilities and infrastructure providing what Scott (2014) refers to as the regulatory pillar of institutional theory. The behaviours component contain elements that espouse cultural norms such as reducing perceived and real power/information asymmetry and hierarchy, specifying higher order commitment through its best-for-project mindset expectations form the basis for what Scott (2014) refers to as the normative pillar and that component with the third component supports Scott's cultural-cognitive pillar (2014).

Institutional theory explains how the project alliance culture becomes locked-in and institutionalised. This reinforces the common project identity in its participants reinforcing dialogue and collaboration as *the natural* workplace practice. It institutionalises collaboration and a best-for-project mindset. PAA governance and other elements in the Collaboration Framework support and/or guide participants into a different level of working together beyond co-ordination and cooperation.

We now explore the best-for-*program* program mindset. In this mode we refer to a greater need for participant *program* identity. Typically, project participants need to visualise their contribution to a project outcome and how their efforts enhance connected and integrated projects within a program of projects to produce a best-for-program outcome.

Often contract requirements and project participants' views of their fiduciary duty do not extend to a program of projects so this presents a specificity and focus dilemma. Being oversensitive to program outcomes may water down or detract from focussing on more immediate concerns and governance performance performance monitoring measures may solely stress project performance. Naturally, under those circumstances resources and energy is focussed on the project.

Explicitly linking a program outcome through clear strategic program-wide goals is essential if that is the PO's aim. Chapter 3 (see Table 3-1) discusses the project strategy in detail. Table 6-1 illustrates selected examples of how a program's purposeful mindset extends a best-for-project mindset.

The program mindset encompases outcomes rather than outputs. Public infrastructure megaprojects often have a transformational social, environmental and economic core vision e.g., the automated light rail transit *Réseau Express Métropolitain* (REM) in Montréal Québec described in Drouin and Turner (2022, Chapter 5). Potentially, having only a project focused mindset may lead to missed opportunities as explained by Gil and Pinto's (2018) analysis of several high profile projects where projects experienced both scope and value creep, in linked and complementary projects when stakeholders merge their projects with an existing project, e.g., the 2012 London Olympics. These megaprojects are consistently referred to as *projects*, yet they might more realistically be referred to as programs of (loosely or tightly) linked and coherent projects within a program of projects.

The Terminal 5 Heathrow project (T5) was delivered successfully within time and cost (Doherty, 2008), however it drastically failed at its opening when the baggage handling system was first used (Brady and Davies, 2010), being considered as a less important back-end part of the T5 project. Had the project been considered, and managed, as a set of inter-related and closely integrated projects in a coherent program, then this vital (from a user's perspective) project may

Table 6-1 LXRP strategic objectives, purpose, and project/program mindset

Strategic objective	Project mind-set	Program mind-set	Program identification
Seperating road and rail at critical junctures.	Removing a specific project's rail crossings and associated station and precinct works.	Ensuring road and rail safety through grade seperation. Improving raod and rail traffic flow and efficiency.	Achieving transport network-level flow, safety, efficiency, and intermodal cross-project interfaces.
Implementing a Metropolitan Network Modernisation Program.	Project scoped crossing removals, station rebuilds and signalling and intermodal connection.	Linking projects in a sequence scoped to minimise network disruption during construction.	Linked to other MTIA[2] projects, improving road/rail/freight intermodal transport to accommodate projected demand expansion.
Improving urban amenity	High quality urban project design improving each precinct as livable, desirable, and economically/socially vibrant.	Quality of life and health improvement for residents and road/rail passengers. Creating urban regeneration development.	Transformation of station precinct areas as valuable social, economic and environmental urban assets.
Improving integrated land use along rail corridors to create vibrant community hubs.	Exploring property residential and commercial development enhancement opportunities.	Transformational change leaving a positive legacy of linear parklands, recreational facilities, and healthy social assets.	Transformation of barren rail-track land and underdeveoped precints into social and economic assets.

have been managed with greater focus on its many interface systems. At the end of projects there is a tendency for those with vital knowledge of the projects history to have left it, moved to other duties and are reticent to spend time on de-briefing, completing documentation such as as-build information, and other vital data. Such seemingly less interesting work is often neglected. A *program* mindset may go a long way to remedying this problem.

If a program of projects is 'marketed' as the outcome in which each project is a *vital* component, then project participants are more likely to view back-end or front-end project works as important pieces of a complex jigsaw. If those projects have a shared project and program identity, then all these projects may be more coherently managed. Mark Wild (2022), CEO of Crossrail a webinar organised by University College London and IPMA, mentioned that one Crossrail identity mistake was an over-emphasis on the engineering parts of Crossrail at the expense of later commissioning and signally vital project components.

Crossrail could have been more explicitly broken up into legitimately integrated large projects (e.g., stations, connecting tunnels, signalling, rollingstock, training and on-boarding train operational staff). These were broken into several very large project work packages (due to the scale of Crossrail and capacity of individual syndicates of delivery organisations to manage the works as one package) but they could have been managed and marketed as a more integrated program of projects. Crossrail Ltd (CRL) acted as the delivery authority and chief systems integrator in a similar, but slightly different, manner to LXRP. However, greater focus on the program meta-systems integrator nature may have enabling clear and targeted identity messaging and dialogue to internal and external stakeholders. Muruganandan et al. (2022) in their Crossrail study observed that several of the main project delivery participants did not have a strong best-for-Crossrail mindset and that CRL had not effectively managed system interfaces between participants and project phases (particularly the commissioning and transition phase) well enough to fully integrate the various systems into a coherent whole with ensuing challenges to deliver to cost and time. Initial phases were highly successful but, as with T5, the final phase failed expectations.

A best-for-*program* mindset is as vital for successful outcomes, as a best-for *project* mindset is to ensure individual projects within the program receive sufficient attention. Each project is considered as legitimate and vital for the program success. This enables development of both organisational-level (project) identification as well as program-level (corporate) identification of the relevant stakeholders.

6.2.3   Trust, communication and governmentality in alliance collaboration

Bond-Barnard et al. (2018) undertook an extensive study using structural equation modelling (SEM) to analyse data from an international survey of 151 project practitioners, demonstrating significance links between project team trust and collaboration, enhancing potential project success. Their study also explored factors supporting collaboration with: "the physical proximity between its team members; the commitment the team members have towards the project; conflict between the team members where less conflict improves the collaboration; the degree of coordination in the project team; the strength of the relationships between team members and other stakeholders; and a balance of intrinsic and extrinsic incentives" (2018, p. 449). These quantitative findings concur with other qualitative data studies in analysing trust and commitment in a project alliance (Davis and Walker, 2020), strategic alliance context (Das and Teng, 2001) and other relationship-based project delivery forms (Gil, Pinto and Smyth, 2011; Dewulf and Kadefors, 2012).

Trust is developed and sustained through dyadic relationships, constantly tested and validated. Trusting and trusted behaviour feed a sense of work-group identity, confidence, validity, and legitimacy. Seminal work on the nature of trust by Mayer et al. (1995) describe trust in terms of *ability* (trust a person's or team's capacity and willingness to do what was promised), benevolence (good will and reasonable intentions), and *integrity* (consistently doing what was promised). This helps explain why people may trust or distrust each other and why *ability* trust also extends to organisational support and discipline perspective (Rousseau et al., 1998; Müller, 2017b). Person X may trust, or distrust Y because they trust/distrust their organisational or discipline 'home.'

This reasoning suggests that communication may be open as is intended and required in Alliancing, or guarded through choosing to use cautious phrasing, misrepresentation, or avoiding discussing a subject. This is where dialogue, explicit expectations, and legitimacy of challenging assumptions is vital to building and retaining trust. Attempts to use communication to distract, rather than illuminate, become more recognisable in Alliancing situations. APTs require power/information asymmetry reduction that leads to legitimacy in challenging assumptions, exposing deceitful behaviours, and opening discussion. Therefore, trust more readily flourishes when combined with *affective* commitment to best-for-project solutions to challenges.

Clegg has written an impressive body of work on power use/abuse pertinent to construction project work. He recently explored toxic construction culture and workplace climates (Clegg et al., 2023) and also about alliance project power and rules within (Clegg et al., 2002). Power is applied by various parties and prescribed through governance arrangements that set rules, requirements, performance criteria and their measurement. Müller (2017a) notes that teams interpret these rules and how they are legitimately applied. It is their governmentality, mentality about governance, that shapes decision making and action.

Müller posits three ways in which power is used to control employee behaviour and action (Müller, 2017b): *outcome control*, based on governance expectations and how outcome performance is assessed; *behaviour control*, appropriate for competent respected professionals where governance guides legitimate ways to achieve the expected outcomes; and *clan control*, based on people's desire on guidance to perform to their socially identified peer group – building on professional or discipline norms. People's perception of legitimate behaviour to power and authority is also impacted by their governmentality paradigm. They may assume an agency theory interpretation, being a PO agent and therefore, acting in both their own and the PO interest with variable results when facing a conflict of interest. Alternatively, they may assume a stewardship relationship in which their interests are focused on mutual trust in producing the agreed outcome (e.g., best-for-project) by mutually benefiting without resorting to opportunistic behaviour (Davis, Schoorman and Donaldson, 1997). Alliancing mobilises a *stewardship* governmentality with combinations of the three control mechanisms governed by the PAA and the project participant selection and skills development process (see Chapter 7) that selects participants who naturally want to work within an alliance clan/team workplace climate.

We argue that trust, communication, power and governmentality impact identity perceptions, motivation, and intentions about planned collective and individual behaviour. This directly depends on the achieved collaboration extent in an alliance/IPD project. Fishbein and Ajzen (2009) studied behaviours and their antecedents using the Theory of Planned Behaviour (TPB) (Ajzen, 1991). This may be applied to understand collaborative behaviours within an Alliancing context. Following Wynn et al's. (2021) adaptation of the TPB within a PM Alliancing context, Figure 6-4 illustrates how collaboration is influenced by trusting behaviours.

The project context shapes the starting point for collaboration. The PAA is a critical influence as it prescribes the drivers of effective collaboration such as expected behaviours and all elements illustrated in Figure 6-2, together with way in which the Alliance project or program of projects treats internal stakeholder engagement (Chapter 7 and 11). This sets a particular tone influencing and shaping espoused project participant norms and identity as discussed above in this chapter. Project participant attitudes are also shaped by the project context.

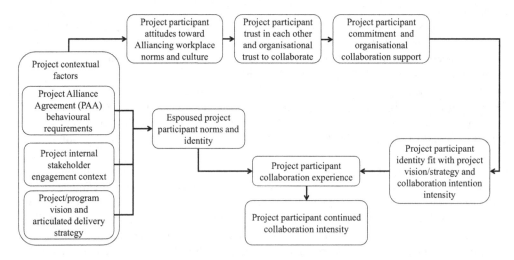

Figure 6-4 Collaboration intensity within an Alliancing context

Project participant attitudes towards the Alliancing workplace norms and culture influences the extent of trust that may be generated between project participants (cross-team and cross-discipline) that in turn influences participant commitment to trust and collaborate with other participants. This commitment is also impacted by the institutional support provided by project and program alliance organisations. These regulate the intensity of participant collaboration intent.

The enacted level and intensity of collaboration is experienced by participants as being positive or negative. Positive experiences prove useful and rewarding and we observe, intellectually stimulating. Negative experiences may not necessarily result in a cessation in collaboration if causes and barriers are quickly identified and remedied but continued negative experience exhausts trust, commitment, and credibility in the value of collaboration to participants and dialogue.

The role of internal stakeholder engagement cannot be overestimated (Chapter 11). The purpose of such engagement is to ensure clarity of communication between the program/project level leadership and participants to reinforce their sense of LXRP identity, to keep them informed about how the overall program is progressing, and to enhance participant stakeholder sense of achievement through their contribution to an important infrastructure program.

IV22 made some interesting observations on having embedded internal stakeholder communication team members. He stated that due to the burnout hazard and more general mental wellbeing considerations, that having these experts was vital in coping with workplace stress points such as the impact of COVID, coping with fatigue when maintaining momentum under severe performance pressure. Expert communication and mental health and morale support was appreciated by the workforce and reduced churn and increased job satisfaction and trust in the project leadership team.

The impact of effective collaboration is to allow hidden or unobvious issues to be exposed and addressed and to enhance the participant collaboration experience leading to strengthened and deeper mutual trust and greater willingness to continue collaborating.

6.3   The LXRP Collaboration Context

Chapter 2 outlines the LXRP general context in detail. This section focusses on *why* and *how* collaboration takes place at both the *project* and *program* level.

The LXRP substantially achieves cross-project, and within project team, collaboration. This is explained by a designed-in range of integration and collaboration mechanisms. There is a constant tension between emphasising a program and project *identity* to avoid distraction from immediate issues. Feldman argues that "Collaboration can create capacity for addressing not only the current problem but also those that follow. New ways of understanding collaboration can help us achieve these potentials" (2010, p. S159). This is a dialogue byproduct where participants explore an issue from multiple perspectives and gaining insights for the present situation and creating mental models of future states.

LXRP's Alliancing *project* delivery choice is predicated on a belief that Alliancing effectively drives innovation and continuous improvement (Love et al., 2016; Love and Walker, 2020), Chapter 9 disusses continuous improvement and innovation in detail. Figure 6-2 illustrates how the project alliance workplace culture-climate supports dialogue, underpining collaboration. One aspect of this is the well-understood LXRP's cultural vision and delivery strategy objective of its commitment to continuous improvement, another is its commitment to better quality decision making and collective action through having a more comprehensive and broader range of expertise applied to those activities. Figure 6-4 shows how this is aided by inter and intra-team participant trust. Figure 6-3 illuminates Collaboration Framework elements that indicates the 16 elements that help explain how Alliancing collaboration occurs.

6.3.1   *Project alliance team collaboration*

Here we illustrate an example of an LXRP Alliance project collaboration based upon interview supported quotes, project documentation and other LXRP supplied information.

Collaboration was vital during construction of LXRP Alliance Project package 1 works, 'the 37-day blitz,' that provided program level collaboration and provided organisational learning that was later adopted program-wide on other projects where a trench-approach to the road-rail grade separation was adopted. The 37-day blitz LXRP example is documented in detail in a research study undertaken by one of the authors[3] (Walker, Matinheikki and Maqsood, 2018). This removal was the first LXRP alliance program package that was completed, announced on 1 August 2016, so that railway services would be fully operational during footy finals and the Caulfield Cup racehorse meeting (Litterick, 2017, p. 9). Shortly after taking a delayed site possession, it was decided to replan the detailed project logistics to accommodate a shorter project duration because the end-date had been fixed and the original plan could not achieve the time-critical result. This required a massive effort in replanning the project logistics and construction method. It also required considerable effective close collaboration between APT participants. The plan was successfully delivered, according to IV02:

> It took the team just 37 days over June/July to remove three dangerous and congested level crossings, excavate the rail corridor, lower the rail line under the roads and start building three new stations the construction challenge was to dig out 250,000 cubic metres of soil in 10 days and pour 35,000 cubic metres of concrete in six days.

Not only was the scale and scope of the work extraordinarily technically challenging, but a totally innovative approach to sheet piling wall trench wall supports was devised that enabled the excavation plan to work and the entire logistics of excavation and soil removal and concrete delivery was planned to accommodate a busy and congested road system. The above cited report provides greater detail on many aspects, but this collaboration illustration focusses upon critical collaboration examples.

The APT comprised multi-discipline and multi-team alliance participants. Thus, multiple perspectives and multiple expertise skill sets were focussed on developing the plan to move from undertaking the works while trains were non-operational during the early hours of the morning to totally closing the line and bussing passengers, transferring them for the affected rail stretches while undertaking trench works for lowering rail line stretches below the road grade and constructing a bridge at road grade level for the road crossing and removal of boom-gates. Why collaborate? The PAA locks in a fixed price TOC[4] and so it was up to the APT to ensure they find a way to achieve the time/cost targets or else suffer painshare losses to their organisation's profit result. This was a planning and delivery task that would not be possible under conventional project delivery approach or without deep, committed, and effective cross-disciplinary collaboration. How did they collaborate?

The team members included a specialist rail operator team who had deep and relevant expertise in rail safety operations. One member, who was pivotal in suggesting an innovative sheet piling approach that saved time and solved various logistical challenges had also worked in a civil contracting organisation and was able to understand the challenges from a civil engineering as well as rail safety perspective. Other team members from a specialised road operator group as part of the alliance had access to decades of road traffic historical statistical data and had traffic modelling skills. They were able to contribute to the logistic plan of material and worker movements to and from the site and were able to model their data to find optimum truck and people traffic flows to facilitate smooth workflow, reduce disruptions to daily road/pedestrian traffic that might otherwise occur and to forward plan and facilitate any required permits for traffic disruption and detours. The design team had motivated and resilient expert staff on the alliance team as did the engineering and signalling mechanical/electrical contractor alliance participants. This represented a highly focussed and expert team that were able to develop a workable plan and execute it successfully. Naturally, some aspects of planned approach needed resilient adjustment as circumstances warranted but their *commitment, resilience,* and *identity* as *the team* to achieve this outcome prevailed.

This example illustrates vital collaboration and alliance team integration features:

1. The team was highly integrated and identified as the APT with a best-for-project mindset.
2. The team was multi-disciplinary and able to question assumptions and explore options from a wide variety of perspectives. They could better consider cause-and-effect and downstream implications of various options under consideration by having that breadth and depth of perspective.
3. The application of skills from the facility operators, rail and road, proved crucial to the plan.
4. This blitz approach subsequently helped model similar approaches across the program to other LXRP project alliance packages using a similar approach to managing rail-road traffic contexts; and
5. The alliance approach required low power-information asymmetry and respect for inter-disciplinary expertise within a safe no-blame workplace environment that welcomed learning-in-action and 'crazy' ideas about what might be achieved.

6.3.2   *Program alliance team collaboration*

Our second example illustrates project team collaboration in developing the Target Outturn Cost (TOC) and time/delivery plan for another Alliance Project package of works.

The second example we chose to illustrate collaboration is at the project front-end TOC development process that is unique to alliance and IPD. Chapters 4 and 5 explain the Alliance delivery choice in detail and the role of a fixed time/cost TOC that binds all APT participants to a common joint performance outcome. This illustrates the why and how of collaboration in Alliancing.

Walker and McCann (2020) explain that the normal TOC development process takes between six to nine months of intense collaborative teamwork of the project design, construction delivery, PO representative and often the facility operator (FO) teams working as a united integrated collaborative team. Walker et al. (2022) explain how this is a more reliable way to develop a reasonable and credible forecasted project cost-time. The LXRP TOC development approach evolved due to its longer term nature as a program of projects where lessons learned from one project was passed onto following projects. IV16 offered a key insight, central to the 'why choose Alliancing' question, is that strongly performing APTs are valuable resource assets. Disbanding them at a project's completion represents potentially discarding a significantly valuable social capital, intelectual capital, and scare resource capital asset. Additionally, when we question the rationale to form new APTs we return to the value-for-money concern, how can we ensure that any APT has developed a delivery solution that is competitive? The LXRP reasoned, because its program spanned over a decade and its scope was well in excess of the initial 50 crossing removals and associated works to currently 110 removals, that is had highly reliable benchmark performance data on project performance to justify a more nuanced approach to the competitive TOC development approach favoured by most government project owners (Department of Infrastructure and Transport, 2011).

Interviewee IV10 noted that

> We will need to differentiate the LXRP PAA where it recognises the program of projects. PAA KRA/KPI's[5] have both a project focus, and separately, a series of program focused measures. The PAA provides for an Initial Works Package (IWP) and subsequent Additional Works Packages (AWP) [linked to the initial project alliance package] – Each AWP has its own TOC[6] and is a unique project but fits into and is managed concurrently by the project's AMT/ALT. Separate AWP Organisations are established – the lead of an AWP is more attuned to the traditional Alliance Manager.

This unique Alliancing arrangement has several cross-project implications. First, it enabled continuity of both project team and participation beyond completion of the original concept of five Project Alliance work packages. The AWPs permitted new project TOC development, leveraging from as-built documentation form previous Project Alliance packages that could be amended and enhanced and validated as best-value through refernce comparisson and benchmarked agains the LXRP's body of project completion data. Thus, continuous improvement cross-project knowledge could be harnessed.

Second, scarce human resources and their knowledge and expertise could be retained and leveraged. The AWP innovation in Project Alliance recruitment and selection considerably enhanced several collaborative effectiveness measures. While AWP team membership may evolve from the original APT participants, project identity was largely retained, and program identity reinforced with wider project exposure within the program.

Third, retaining many participants resulted in retained and enhanced inter-participant trust and familiarity with 'the system.' Typical TOC development period was reduced to 15–20 weeks. Re-use of knowledge from one project to another, more effective trust and collaboration and familiarity with the TOC process led to enhanced program-level collaboration. By later stages of the program, the Joint Coordination Committee (JCC) linking experts from across the program in what may be understood as both a high-level program coordination mechanism and at its subsidiary levels a series of integrated communities of practice (Chapter 3 and 9) to share technical, process and innovation diffusion knowledge, had been established and institutionalised as part of the program culture.

This example highlights one of many illustrations of *program* collaboration that could be cited.

6.4   Global Project/Program Collaboration Context

In the UK, there is substantial material written about the successful practices employed to deliver Heathrow Airport Terminal 5 in London (Davies, Gann and Douglas, 2009). T5 had challenges when transitioning the built asset to operations (Zhang et al., 2023) which created a temporary narrative of failure around the project. In fact, if a deeper analysis is conducted, there were many pioneering initiatives during the execution of the T5 program, which stimulated the next generation of megaprojects in London, namely: London 2012 Olympics, Crossrail, Thames Tideway Tunnel, and High Speed 2 (Denicol, Davies and Pryke, 2021).

T5 was delivered following a new approach for collaboration, influenced by extensive market analysis of previous international airports and megaprojects. Such investment in front-end research highlighted significant risk to the financial stability of the client (British Airports Authority, at the time) if the traditional practices of outsourcing the risk to the supply chain were followed to deliver T5. Therefore, T5 invested in creating a new approach of collaboration, incentivising the supply chain to work collaboratively, and integrated with the client to deliver the program. There were formal and informal mechanisms in place to enable such collaboration, with the T5 Agreement being mentioned as the formal component that established the foundation for such new organisational behaviour.

BAA was praised for acting as an intelligent client, an adjective that was used multiple times to reflect the ambition for and the implementation of new practices to drive different outcomes. T5 understood early on that collaboration and integration were key to enable the successful delivery of the program, often highlighting the need to tackle the fragmentation and structural complexity of construction supply chains. This highlights the need for clients to act as the system of systems integrators, but crucially to design the supply chain architecture in the front end (Denicol, Davies and Pryke, 2021). Such design of the systems architecture is foundational, yet it needs to be met with the formal mechanisms to enable actions, which are often constrained by contractual agreements. In this sense, the T5 agreement was influential and innovative in promoting new behaviours.

The practices from T5 influenced the next generation of megaprojects in London, there was a deliberate effort to disseminate the learnings through written material, engagement with professional bodies, and wider professional networks. Many major and megaproject professionals transferred to the next complex megaproject in the UK, the 2012 London Olympic Games, working as conduits and ambassadors of T5 practices (Grabher and Thiel, 2015). Yet, it is important to highlight the structural differences between T5 and London Olympics. Whilst T5 had British Airports Authority (BAA) as a permanent owner and operating a network of airports in the UK, including the ongoing operation at Heathrow Airport, London 2012 Olympics had to set-up a new client organisation from scratch, therefore with no embedded organisational memory as BAA (Davies and Mackenzie, 2014; Denicol, 2020). This is an important structural feature, which often has implications to structures implemented by clients to engage with the supply chain.

London 2012 Olympics established the Olympic Delivery Authority (ODA) to act as the client, structuring and coordinating the arrangements with the supply chain (Stefano et al., 2023). Considering the lack of knowledge, a new entity, and the timeframe to deliver the Olympic Games, ODA selected a consortium of three firms to act as its delivery partner at client-side, providing the augmentation of capacity and capability. Three companies were organised as a joint venture and formed the CLM delivery partner, namely: CH2M, Laing O'Rourke, and Mace. The composition of the joint venture provided a complementary set of skills and organisational background to ODA, CH2M was recognised by its program management capabilities (Denicol and Davies, 2022), whilst Mace and Laing O'Rourke had a track of record in construction management. The remit of the delivery partner was to work collaboratively with the client (ODA) and manage the program, being directly responsible for the integration of all Tier 1 contracts (e.g., Olympic Stadium, Aquatics Centre). ODA worked integrated with CLM yet had an important role in managing the relationships and interfaces outside the project, engaging external stakeholders and shielding the project from external disturbances.

The establishment of new standalone entities to act as the megaproject client was a common feature of all subsequent projects in London (Denicol, Davies and Pryke, 2021), namely: Crossrail, Thames Tideway Tunnel, and High Speed 2. The purpose of such client organisations was to act as the ultimate program manager for the megaproject, integrating the works delivered by multiple Tier 1 contractors. Considering the lack of capabilities, delivery partners were employed by all clients, working at client side together with employees of the new client organisations (Denicol and Davies, 2022). Regardless the granularity and size of the packaging adopted, the client and delivery partner are the entities with overall program-wide visibility for coordination and integration of interfaces. For instance, Crossrail's procurement strategy created several Tier 1 contracts to be managed, whilst Thames Tideway Tunnel and High Speed 2 organised to create larger Tier 1 work packages, resulting in less interfaces to be managed by the client organisation (Muruganandan et al., 2022). There is a clear evolution of the size of contracts/work packages, which are now consolidated in geographical regions and sections of the project. This signals in part market and economic conditions, as Crossrail was procured close to the financial crisis, but also the challenges for the client organisation of building capabilities to integrate several Tier 1 interfaces. This is even more pronounced in client organisations that are established specifically to deliver one megaproject.

In the case of Thames Tideway Tunnel, the client organisation working with a delivery partner to manage three Tier 1 joint ventures (West, Central and East sections). The client appointed a fourth Tier 1 contract to a technical systems integrator, responsible for working with the three joint ventures across the tunnel route, ensuring technical consistency and prepare the asset for operations. Thames Tideway Tunnel put in place an Alliance structure, aiming to incentivise the parties to work collaborative between them at the horizontal level, beyond their usual vertical contractual links to deliver specific work packages. The principles to promote horizontal collaboration and learning to impact megaproject performance (Denicol, Davies and Krystallis, 2020) are sound and in line with the degrees of complexity observed in megaprojects, which are in addition to the structural fragmentation of the construction industry. Yet, considering the low profit margins of the construction industry, it is important that clients consider the balance between horizontal and vertical incentives when setting up alliance structures. Although the horizontal incentives might be in place to promote collaboration, if there is not enough reward in the horizontal, commercial organisations might be more incentivised to keep their top professionals involved in managing the vertical links of their contract/work package.

6.5   Conclusions

Crossrail and T5 in the UK have been discussed as adopting similar relationship-based approaches to collaboration as IPD/Alliancing. These, however, were managed as a project rather than a program of projects. We suggest that they may have been better delivered if coherently logical packages of works could have been identified as *projects* and their integration into a *program* established through a strategy designed to ensure cross-project collaboration. A *program* approach embeds systems thinking and systems interface integration that requires high level collaboration between delivery participants and phases in the program of works. Several conceptual underpinnings for effective collaborative dialogue have been explored and explained.

Importantly we identified project purpose and show how that affects personal identity, legitimacy, trust. Together they impact effective dialogue influencing collaboration quality. This, when considered with Chapter 5 integration aspects, helps explain how the LXRP achieved effective collaboration. The chapter also draws upon global project collaboration insights focussing on several UK megaprojects. There are many similar features of those projects and the LXRP that help explain project collaboration, however, a main difference is the way that the LXRP was strategically established (Chapter 3) as a program of highly integrated projects with a PAA requiring effective collaboration.

Notes

1 Fiduciary duty is the duty of an employee to be loyal and promote the interests of the employee see Clark, G. L. (2011). Fiduciary Duty, Statute, and Pension Fund Governance: The Search for a Shared Conception of Sustainable Investment https://ssrn.com/abstract=1945257

2 Melbourne Transport Infrastructure Structure (MTIA) see Chapter 2 context for more details.
3 www.researchgate.net/publication/330837569_level_crossing_removal_program_package_1_nurtur-ing_innovation_in_complex_alliance_delivery_projects
4 See Chapter 3 – The Target Outturn Cost (TOC) binds participating teams to a fixed time and cost with their profit margins held at stake with an incentivised gain-pain sharing system based on the final cost/time and other key result area (KRA) measures.
5 Key result areas (KRA) and key performance indicators (KPI), Alliance management Team (AMT) and Alliance Leadership Team (ALT) are governance arrangement see Chapter 3 for more detailed explanation.
6 See Chapter 3 – The Target Outturn Cost (TOC) binds participating teams to a fixed time and cost with their profit margins held at stake with an incentivised gain-pain sharing system based on the final cost/time and other key result area (KRA) measures.

References

Ajzen, I. (1991). "The theory of planned behavior." *Organizational Behavior and Human Decision Processes*. 50 (2): 179–211.

Ali, F. and Haapasalo, H. (2023). "Development levels of stakeholder relationships in collaborative projects: Challenges and preconditions." *International Journal of Managing Projects in Business*. 16 (8): 58–76.

Bedwell, W.L., Wildman, J.L., DiazGranados, D., Salazar, M., Kramer, W.S. and Salas, E. (2012). "Collaboration at work: An integrative multilevel conceptualization." *Human Resource Management Review*. 22 (2): 128–145.

Bond-Barnard, T.J., Fletcher, L. and Steyn, H. (2018). "Linking trust and collaboration in project teams to project management success." *International Journal of Managing Projects in Business*. 11 (2): 432–457.

Brady, T. and Davies, A. (2010). "From hero to hubris: Reconsidering the project management of Heathrow's Terminal 5." *International Journal of Project Management*. 28 (2): 151–157.

Chatman, J.A. and O'Reilly, C.A. (2016). "Paradigm lost: Reinvigorating the study of organizational culture." *Research in Organizational Behavior*. 36: 199–224.

Christenson, D. (2007). Using Vision as a Critical Success Element in Project Management Doctor of Project Management, DPM, *School of Property, Construction and Project Management*. Melbourne, RMIT.

Christenson, D. and Walker, D.H.T. (2004). "Understanding the role of 'vision' in project success." *Project Management Journal*. 35 (3): 39–52.

Clark, G.L. (2011). "Fiduciary duty, statute, and pension fund governance: The search for a shared conception of sustainable investment." https://ssrn.com/abstract=1945257

Clegg, S.R., Loosemore, M., Walker, D.H.T., van Marrewijk, A.H. and Sankaran, S. (2023). "Construction cultures: Sources, Signs and solutions of toxicity." In *Construction Project Organising*, Addyman, S. and H. Smyth (eds), pp. 3–16. Chichester, West Sussex, John Wiley & Sons.

Clegg, S.R., Pitsis, T.S., Rura-Polley, T. and Marosszeky, M. (2002). "Governmentality matters: Designing an alliance culture of inter-organizational collaboration for managing projects." *Organization Studies* 23 (3): 317–337.

Cornelissen, J.P., Haslam, S.A. and Balmer, J.M.T. (2007). "Social identity, organizational identity and corporate identity: Towards an integrated understanding of processes, patternings and products." *British Journal of Management*. 18 (s1): S1–S16.

Das, T.K. and Teng, B.-S. (2001). "Trust, control, and risk in strategic alliances: An integrated framework." *Organization Studies*. 22 (2): 251–283.

Davies, A., Gann, D. and Douglas, T. (2009). "Innovation in megaprojects: Systems integration at London Heathrow Terminal 5." *California Management Review*. 51 (2): 101–125.

Davies, A. and Mackenzie, I. (2014). "Project complexity and systems integration: Constructing the London 2012 Olympics and Paralympics Games." *International Journal of Project Management*. 32 (5): 773–790.

Davis, J.H., Schoorman, D.F. and Donaldson, L. (1997). "Towards a stewardship theory of management." *Academy of Management Review*. 22 (1): 20–48.

Davis, P.R. and Walker, D.H.T. (2020). "IPD from a participant trust and commitment perspective." In *The Routledge Handbook of Integrated Project Delivery*, Walker, D.H.T. and S. Rowlinson (eds), pp. 264–287. Abingdon, Oxon, Routledge.

Denicol, J. (2020). "Managing megaproject supply chains: Life after Heathrow Terminal 5." In *Successful construction Supply Chain Management: Concepts and Case Studies*, Pryke S. (ed.), pp. 213–235. Hoboken, NJ, Wiley-Blackwell.

Denicol, J. and Davies, A. (2022). "The megaproject-based firm: Building programme management capability to deliver megaprojects." *International Journal of Project Management*. 40 (5): 505–516.

Denicol, J., Davies, A. and Krystallis, I. (2020). "What are the causes and cures of poor megaproject performance? A systematic literature review and research agenda." *Project Management Journal*. 51 (3): 328–345.

Denicol, J., Davies, A. and Pryke, S. (2021). "The organisational architecture of megaprojects." *International Journal of Project Management*. 39 (4): 339–350.

Department of Infrastructure and Transport (2011). *National Alliance Contracting Guidelines Guide to Alliance Contracting*, Department of Infrastructure and Transport A.C.G. Canberra, Commonwealth of Australia. www.infrastructure.gov.au/infrastructure/nacg/files/National_Guide_to_Alliance_Contracting04July.pdf

Dewulf, G. and Kadefors, A. (2012). "Collaboration in public construction: Contractual incentives, partnering schemes and trust." *Engineering Project Organization Journal*. 2 (4): 240–250.

Doherty, S. (2008) *Heathrow's T5 History in the Making*, Chichester, John Wiley & Sons Ltd.

Drouin, N. and Turner, J.R. (2022) *Advanced Introduction to Megaprojects*, Northampton, UK, Edward Edgar Publishing.

Feldman, M.S. (2010). "Managing the Organization of the future." *Public Administration Review*. 70 (s1): s159–s163.

Fishbein, M. and Ajzen, I. (2009) *Predicting and Changing Behavior: The Reasoned Action Approach*, London and United States, Taylor & Francis Group.

Gil, N. and Pinto, J.K. (2018). "Polycentric organizing and performance: A contingency model and evidence from megaproject planning in the UK." *Research Policy*. 47 (4): 717–734.

Gil, N., Pinto, J.K. and Smyth, H. (2011). "Trust in relational contracting and as a critical organizational attribute." In *The Oxford Handbook of Project Management*, Morris, P.W.G., J.K. Pinto and J. Söderlund (eds), pp. 438–460. Oxford, Oxford University Press.

Gioia, D.A., Price, K.N., Hamilton, A.L. and Thomas, J.B. (2010). "Forging an identity: An insider-outsider study of processes involved in the formation of organizational identity." *Administrative Science Quarterly*. 55 (1): 1–46.

Grabher, G. and Thiel, J. (2015). "Projects, people, professions: Trajectories of learning through a mega-event (the London 2012 case)." *Geoforum*. 65: 328–337.

Hackman, J.R. and Oldham, G.R. (1976). "Motivation through the design of work: Test of a theory." *Organizational Behavior and Human Performance*. 16 (2): 250–279.

Kurtz, C.F. and Snowden, D.J. (2003). "The new dynamics of strategy: Sense-making in a complex and complicated world." *IBM Systems Journal*. 42 (3): 462–483.

Litterick, S. (2017). *Level Crossing Removal Project Burke, North, McKinnon, Centre*, Melbourne, Project Management Institute Victoria, Australia. www.melbourne.pmi.org.au/Documents/Managing%20Projects%20PMI%2028March%202017.pdf

Love, P.E.D., Teo, P., Davidson, M., Cumming, S. and Morrison, J. (2016). "Building absorptive capacity in an alliance: Process improvement through lessons learned." *International Journal of Project Management*. 34 (7): 1123–1137.

Love, P.E.D. and Walker, D.H.T. (2020). "IPD Facilitating Innovation Diffusion." In *The Routledge Handbook of Integrated Project Delivery*, Walker, D.H.T. and S. Rowlinson (eds), pp. 393–416. Abingdon, Oxon, Routledge.

Mayer, R.C., Davis, J.H. and Schoorman, F.D. (1995). "An integrated model of organizational trust." *Academy of Management Review*. 20 (3): 709–735.

Meyer, J.P. and Allen, N.J. (1991). "A three-component conceptualization of organizational commitment." *Human Resource Management Review*. 1 (1): 61–89.

Müller, R. (2017a). "Organizational project governance." In *Governance and Governmentality for Projects: Enablers, Practices, and Consequences*, Müller, R. (ed.), pp. 24–37. Abingdon, Oxon, Routledge.

Müller, R. (2017b). "Organizational project governance." *Governance Mechanisms in Projects*, Müller, R. (ed.), pp. 173–180. London, Routledge.

Muruganandan, K., Davies, A., Denicol, J. and Whyte, J. (2022). "The dynamics of systems integration: Balancing stability and change on London's Crossrail project." *International Journal of Project Management*. 40 (6): 608–623.

Ross, J. (2003). *Introduction to Project Alliancing*. Alliance Contracting Conference, Sydney, 30 April 2003, Project Control International Pty Ltd.

Rousseau, D.M., Sitkin, S.B., Burt, R.S. and Camerer, C. (1998). "Not so different after all: A cross-discipline view of trust." *Academy of Management Review*. 23 (3): 393–405.

Schaber, P. and Müller, A. (2018). "The Ethics of Consent – An Introduction." In *The Routledge Handbook of the Ethics of Consent*, Schaber, P. and A. Müller (eds), pp. 1–5. Georgetown, Canada, Taylor & Francis Group.

Schneider, B., Ehrhart, M.G. and Macey, W.H. (2013). "Organizational climate and culture." *Annual Review of Psychology*. 64 (1): 361–388.

Schnüriger, H. (2018). "What is consent?" In *The Routledge Handbook of the Ethics of Consent*, Schaber P. and A. Müller (eds), pp. 21–31. Georgetown, Canada, Taylor & Francis Group.

Scott, W.R. (2014) *Institutions and Organizations*, Thousand Oaks, CA and London, Sage.

Senge, P.M. (1990) *The Fifth Discipline – The Art & Practice of the Learning Organization*, Sydney, Australia, Random House.

Snowden, D. and Rancati, A. (2021). *Managing Complexity (and Chaos) in Times of Crisis. A Field Guide for Decision Makers Inspired by the Cynefin Framework*, Luxembourg, Publications Office of the European Union.

Snowden, D.J. and Boone, M.E. (2007). "A Leader's framework for decision making." *Harvard Business Review*. 85 (11): 69–76.

Stefano, G., Denicol, J., Broyd, T. and Davies, A. (2023). "What are the strategies to manage megaproject supply chains? A systematic literature review and research agenda." *International Journal of Project Management*. 41 (3): 102457.

Walker, D.H.T. and Lloyd-Walker, B.M. (2015) *Collaborative Project Procurement Arrangements*, Newtown Square, PA, Project Management Institute.

Walker, D.H.T. and Lloyd-Walker, B.M. (2020a). "Behavioural elements of the IPD collaboration framework." In *The Routledge Handbook of Integrated Project Delivery*, Walker, D.H.T. and S. Rowlinson (eds), pp. 315–344. Abingdon, Oxon, Routledge.

Walker, D.H.T. and Lloyd-Walker, B.M. (2020b). "Characteristics of IPD: A framework overview." In *The Routledge Handbook of Integrated Project Delivery*, Walker, D.H.T. and S. Rowlinson (eds), pp. 20–40. Abingdon, Oxon, Routledge.

Walker, D.H.T. and Lloyd-Walker, B.M. (2020c). "Foundational elements of the IPD collaboration framework." In *The Routledge Handbook of Integrated Project Delivery*, Walker, D.H.T. and S. Rowlinson (eds), pp. 168–193. Abingdon, Oxon, Routledge.

Walker, D.H.T. and Lloyd-Walker, B.M. (2020d). "Processes and means elements of the IPD collaboration framework." In *The Routledge Handbook of Integrated Project Delivery*, Walker, D.H.T. and S. Rowlinson (eds), pp. 454–481. Abingdon, Oxon, Routledge.

Walker, D.H.T., Love, P.E.D. and Matthews, J. (2022). "The value of dialogue in Alliancing Projects." In *The 9th International Conference on Innovative Production and Construction*. Melbourne, Australia, CIB. https://virtual.oxfordabstracts.com/#/event/public/1970/program?session=35425&s=600

Walker, D.H.T., Matinheikki, J. and Maqsood, T. (2018). *Level Crossing Removal Authority Package 1 Case Study*, RMIT University Melbourne, Australia.

Walker, D.H.T. and McCann, A. (2020). "IPD and TOC development." In *The Routledge Handbook of Integrated Project Delivery*, Walker, D.H.T. and S. Rowlinson (eds), pp. 581–604. Abingdon, Oxon, Routledge.

Walker, D.H.T., Vaz Serra, P. and Love, P.E.D. (2022). "Improved reliability in planning large-scale infrastructure project delivery through Alliancing." *International Journal of Managing Projects in Business*. 15 (8): 721–741.

Wild, M. (2022). "Delivering the Elizabeth Line." www.youtube.com/watch?v=p9QFy1HnFwA

Wynn, C., Smith, L. and Killen, C. (2021). "How power influences behavior in projects: A theory of planned behavior perspective." *Project Management Journal*. 52 (6): 607–621.

Zhang, X., Denicol, J., Chan, P.W. and Le, Y. (2023). "Designing the transition to operations in large inter-organizational projects: Strategy, structure, process, and people." *Journal of Operations Management*.

7 Delivering Through High-Performance Workplace Culture

Jane Magree, Christina Scott-Young and Beverley Lloyd Walker

7.1 Introduction

How did LXRP develop high-performing teams through creating a transformational workplace culture?

The chapter builds upon previous chapters to explain how the LXRP workplace culture was initiated, nurtured and maintained. It represents a transformation compared to the business-as-usual conventional project delivery approach (Chapter 4). The LXRP is also unusual in that it crafted a program workplace mindset where project participants share an integrated workplace culture that is consistently common across all projects. Chapter 3 explained how the LXRP developed its program/project strategy to achieve its intended purpose and vision.

Biesenthal et al. (2018) highlight, through using an institutional theory lens, how project strategy is delivered. Institutional theory (Scott, 2014), explained in Chapter 2 from the LXRP perspective and Chapter 6 from a collaboration viewpoint, illustrates how people's norms and guiding cultural stance shape their interpretation of governance mechanisms such as rules, templates and guidelines – shaping their workplace culture. IPD/Alliancing institutionalises behaviours and attitudes towards knowledge sharing, trust and commitment, and collaborative dialogue (Hall and Scott, 2019). The IPD/Alliancing approach, adopted by the LXRP, has established a workplace culture through its Project Alliancing Agreement (PAA) that requires specific collaborative behaviours, forming project participants' norms. One important implication is that collaboration and dialogue can increase productivity (Chapter 6), leading to learning and continuous improvement (Chapter 9). Other chapters in this book provide a theoretical explanation of concepts effectively adopted and adapted by the LXRP in contributing to its workplace culture.

Understanding the LXRP workplace culture and how it was designed and has evolved is explored in detail next. The LXRP workplace culture is explained using illustrative interviewee quotes and drawing upon internal LXRP documents. The chapter's conclusions link to Chapter 8, which describes how LXRP leadership and its human capital management approach contributed to a high-performance workplace culture and teamwork. This chapter's contribution explains how the LXRP's cross-team/cross-project integrated collaborative workplace culture was developed and sustained.

7.2 Salient Concepts and Theory

This chapter focuses on delivering high performance culture, enabled by high performance work systems combined with high-performance teams. What exactly is a high-performance team?

DOI: 10.1201/9781003389170-7

O'Neill and Salas (2018) guide high-performance teams, stating that:

> First, teams must deliver on stakeholder objectives at the highest level of quality. Second, teams must mature into increasingly capable work units over time. Third, teams must enable their members to continuously develop and grow their capabilities over time. This definition of high performance teamwork is intentionally broad and global, as teams can be productive but, if they do not maintain viability, their members will eventually burn out.
>
> (p. 326)

'Highest level of quality' refers to meeting or exceeding identified stakeholder value propositions. Chapter 11 discusses how the LXRP engaged with its internal and external stakeholders and articulated the program purpose through its strategy, clearly identifying their expected outcomes, including the legacy to be left post-program. The level of work team capability development and how it is grown and nurtured is discussed in Chapter 8. It is now established that high-performance teams are transparent about their impact on delivering the program strategic objectives, and they are supported to achieve outcomes that meet and exceed strategy expectations. This implies that the strategy-outcomes fit must be carefully crafted to ensure the program's purpose is unambiguous. In LXRP's case, the purpose was clear from its business case (Victorian State Government, 2017) and was communicated through its Blueprint (LXRP, 2023a) vision statement "To deliver Great Change, transforming the way Victorians live, work and travel." The strategy defines what it refers to as the 5-Greats, which will be discussed in more detail later. High performance is, therefore, explicitly linked to connection of systems and people.

Organisational culture is a key component of everyday organisational functioning and is a concept that people intrinsically understand and experience within an organisation. It is a concept that has been considered to have deep impact on organisational performance and productivity (Akpa, Asikhia and Nneji, 2021).

It is the sense making and control gaining mechanism that guides and shares mindsets and behaviours. Although there is no consensus on a definition of organisation culture, a literature review concludes that there are a number of common characteristics. This includes the concept of 'sharing' within groups – shared understanding of purpose, mindsets and behaviours. Culture is considered a social construct, and it is multi-dimensional and multileveled including artefacts and symbols (Akpa, Asikhia and Nneji, 2021).

Many different definitions of culture have been proposed. Here we define organisational culture as "a set of core values, behavioural norms, artefacts and behavioural patterns which govern the way people in the organisation interact with each other and invest energy in their jobs and the organisation at large" (van Muijen et al., 1999, p. 250). This definition points to two layers of culture the visible behaviour of people in the organisation and the deeper layer formed by invisible norms and values. These contribute to the underlying causes of behaviour. Senior leaders are foundational to culture creation and embed and transmit culture in the thinking, feeling and behaviour of people in the organisation (Den Hartog and Verburg, 2004). This chapter explores how leadership established the powerful connection between high performance work systems and the people of LXRP through structure, mindsets and processes.

Mindsets are powerful beliefs that profoundly impact how people choose to behave. Dweck's (2012) study of early childhood learning and behaviour development identified two types of mindset (*fixed* and *growth*) that explain people's default assumptions about their capability to learn and change. Mindset (1) *fixed-minded* exemplifies the status quo, involving a rigid adherence to a particular path, unquestioningly following that path regardless of the context. Mindset (2) *growth-minded* is governed by 'truth/facts' being assessed contextually and assumptions about 'truth/facts' being assumed to be contestable, and the belief that change and innovation are possible. Growth mindset has a focus on development, challenge and effort with results following. Fixed mindset focuses on permanent traits which cannot change. When contextualised in terms of project management at LXRP a growth mindset of thinking and acting differently is consistently espoused by leaders and reinforced through structures and processed explained here. This collective mindset has helped LXRP to evolve and grow as it has continuously improved over the course of eight years.

For example, a fixed or closed mindset could see the LXRP purpose as just fixing a problem with performance, as seen in terms of the iron triangle – cost/time/ quality (Atkinson, 1999). The growth-mindset approach taken by LXRP early on when developing the LXRP program Blueprint cast the program as transformational, one that pragmatically fixed an immediate problem but also added significant social value. Similarly, a fixed mindset would design hierarchical fashion strategy-supporting governance arrangements based mainly on command-and-control mechanisms. A growth mindset supporting governance design would feature low power-information asymmetry and empowerment and trust people's sense of accountability.

Transactional outcomes such as productivity improvement are essential, as are transformational outcomes such as social value and long-term legacy (see Chapter 3, Figure 3-2). A growth-mindset is important when open to considering transactional aspects. For example, the LXRP not only promotes a manufacturing mindset (off-site manufacture, as-built design sharing across projects within the program (Chapter 10), and modularisation) that implies that potential modularisation in removing 110 rail crossings presents replication logic opportunities and unique challenges. The context should also be considered, and critical thinking should be applied to adapting rather than adopting innovation and standard construction infrastructure practices. One LXRP published strategic aim was to benefit from adopting a production mindset. This would take advantage of the repetitive nature of the crossing removal work and prompt a workplace value of sharing knowledge and innovation *across* the program, with significant gains to be made.

Construction is critical for the world economy but has a long record of poor productivity, with productivity only increasing by 1% from 1997 to 2017, according to McKinsey Global Institute (2017), by moving to a manufacturing-style production system with repeatable delivery methodology, the industry could achieve a five to 10 times productivity boost. Adapting a manufacturing mindset (LXRP, 2023b, p. 26) is one important mindset feature of its workplace culture.

Using the term *workplace culture*, we draw upon early literature exploring the nature of culture. Schein's theory of organisational culture focuses on the basic underlying organisational assumptions, espoused values and key artefacts within an organisation, and how their linkages can influence people's engagement and performance. Schein (2010) suggested that organisational culture serves the dual roles of adaptation to changes in the environment external to the organisation and enabling internal integration of open mindsets and processes that contribute to high performance.

Schein (2010) identified culture as being the "basic assumptions and beliefs that are shared by organisational members" (p. 9). At the surface level, culture is observed through artefacts that define that culture. The PAA can be seen as an artefact that shapes values, as the LXRP's Blueprint explained shortly. Beneath that surface are the groups' espoused values that guide behaviours. These manifest as artefacts that 'badge' them. At the deepest level, underlying assumptions (the mindset) justify and legitimise the espoused values as representative of the cultural group. New (as in a project) cultures can be designed and crafted by influencing a group to: adopt certain assumptions; espouse particular values as valid and consistent with the assumptions, and create artefacts that badge those values.

Culture is described as the result of embedding what a founder or leader has imposed on a group. Culture is thus created, embedded, evolved, and ultimately manipulated by leaders (Schein, 2010, p. 3). The dynamic processes of culture creation and management form the essence of leadership, and "leadership and culture are two sides of the same coin" (p. 3). The LXRP leadership team valued adaptability and crafted it to shape its workplace culture to fit its think and act differently mindset. Interestingly, the mindset at one level had a productivity increase focus through having a manufacturing mindset to modularise and take advantage of scale and continuous improvement that may be interpreted as being focused on VfM, but this was balanced with a broader social value, and legacy outcome perspective focuses on providing an output-outcome purpose duality. LXRP's leadership will be examined in more depth in Chapter 8.

7.3   LXRP High-Performance Workplace Culture

This chapter extensively cites quotes from interviewees to support the LXRP workplace culture discussion of how key artefacts, the espoused values and underlying assumptions generated high performance through the connection of structure, mindsets and practices together. Interviewee data from LXRP staff is utilised by using their storytelling of how the culture of LXRP was created, evolved and sustained.

In 2015, the then-CEO created the newly formed LXRP executive leadership team. He targeted high-performing people considered high-calibre talent with a high aptitude for creative responses to challenges (Chapter 8). This leadership team recognised the opportunity to disrupt traditional transport infrastructure mega-project success barriers. Barriers include:

- Teams that constantly formed and disbanded when projects started and stopped, resulting in knowledge and learning being lost – the eternal beginner syndrome (Flyvbjerg, Budzier and Lunn, 2021).
- The industry faced skill shortages and low diversity through being male-dominated (Clegg et al., 2023) and not conducive to work-life integration due to long hours and high pressure with its intense short-term demands on people, which in some cases led to associated mental health challenges.
- Procurement and surety of funding on individual projects did not encourage long-term investment in capability and innovation (Wolstenholme, 2009).
- The life of a single project or contract was too short to realise benefits or recoup investment costs (LXRP, 2023b).

The LXRP's executive leadership team also recognised that existing systems and processes for government delivery authorities and transport infrastructure delivery models were more suited to stand-alone projects rather than LXRP's enduring 15-year program of works. In addition, they understood that traditional project management procurement approaches would not deliver the outcomes required, as explained in detail in Chapter 4.

Therefore, from the outset, LXRP management set up a distinct delivery structure through a performance-based *modus operandi* with clear lines of accountability and decision-making criteria. This was strategically planned and executed at the LXRP's establishment. LXRP designed and created an ecosystem of consistent high performance that flows from the organisation's leadership, through the teams and individuals within and across projects in the program.

As outlined earlier, at the heart of every successful and dynamic organisation lies a well-defined culture that shapes its members' values, behaviours, and interactions. LXRP built a high-performing ecosystem through its relentless and enduring performance expectation at the organisational, team and individual levels. The high-performance culture of LXRP has emanated from the top through supportive leadership and the creation of a responsive organisational structure (see Chapter 3). Support included empowering team participants and the freedom to make mistakes, acknowledge and reflect on them through dialogue with colleagues, and learn from experience in a blame-free constructive way. For example, interviewee IV17 said:

> The underpinning of the success of any team is the senior leadership support. With level crossings, the senior leaders are supportive of being able to challenge the status quo. That's probably key in allowing us to deliver some of the initiatives that we've delivered. Not only did they allow you the ability to explore things, but if things didn't work and the outcome wasn't what you were looking for, there was no criticism. There is still support. We could say 'we tried that, and it didn't go as we thought and we need to pivot' and look down another avenue. Support is not just from a financial sense but also from a resource sense.

The above illustrates employee empowerment and creativity with growth mindset leadership, and critical thinking that permeated the organisation.

To understand how this high-performance culture was created, it is essential to know how key building blocks of this ecosystem were put together and held in place by a keystone generally not utilised in short-term stand-alone infrastructure projects. This keystone is the centre of the business operating approach of the LXRP, otherwise known as the LXRP Blueprint (LXRP, 2023a). It was developed and has grown from business administration theory and best practices gained from experience in the public and private sectors by LXRP's leadership. From this bespoke LXRP Business Operating Approach, encompassing the Blueprint and five key building blocks illustrated in Figure 7-1, the repeatable LXRP ecosystem delivering performance-based

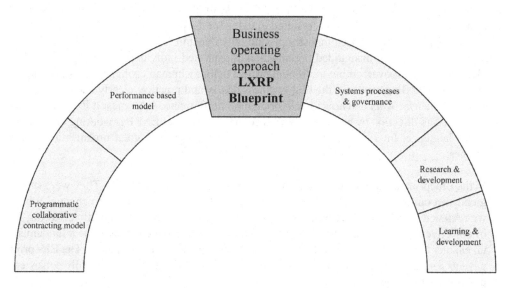

Figure 7-1 LXRP Business Operating Approach

(Source: Magree 2024[1])

outcomes has grown. The outcome is an organisation that represents a dynamic and results-oriented government organisation that is already creating a legacy of reshaping industry productivity dynamics.

The ecosystem can be visualised as an arch with its supportive building blocks held together by the Blueprint keystone. The programmatic collaborative contracting model focused on the program of alliance additional work packages (AWPs), which were tightly integrated, as described in Chapter 5, into the program through governance mechanisms described in Chapter 3, and the leadership focus on cross-project collaboration and dialogue described in Chapter 6. The performance-based model locked in expectations of program participant responsibility and accountability through the systems processes and governance such as the PAA and other governance mechanisms such as key performance indicators (KPIs), the Joint Coordination Committee (JCC) and incentivisation through gain-pain sharing (see Chapter 3). The open growth-mindset workplace culture was also supported by special entities created by the LXRP (discussed shortly) for research and learning development.

The LXRP Business Operating Approach keystone block was created through the LXRP Blueprint artefact. This Blueprint is strategic and combines the best of contemporary business practice theory with a programmatic (program of integrated projects) approach to collaborative contracting. The Blueprint is a key artefact brought to life through a highly capable leadership team, including their delivery partners and employees. The Blueprint approach is designed to successfully disrupt traditional models of infrastructure project delivery and build a legacy of lasting change for infrastructure delivery. The remaining sections of this chapter spotlight the *why* aspect of this approach, the *what* part of the keystone and its building blocks, and how the approach has been embedded and sustained.

7.3.1   *LXRP workplace culture Blueprint influence*

A clear and consistent strategic direction with vertical and horizontal alignment of performance expectations was perceived as foundational to success. Many LXRP interviewees point to this manifestation through the LXRP Blueprint. This sets the strategic foundation for a high-performing ecosystem of projects with the shared purpose of transforming the way Victorians live, work and travel. This ecosystem has resulted in value for the people of Victoria and is leaving a legacy of reshaping construction industry productivity dynamics.

The LXRP Blueprint has guided organisational, team and individual performance and has resulted in the LXRP overcoming traditional project delivery human capital barriers to enhance productivity. The Blueprint uses the best business theory and practice, clearly and consistently articulating the *why*, *what* and *how* of delivering high performance outcomes. It forms the guiding purpose and direction by building people's connection to what LXRP is working on through a shared language. Interviewee IV16 illustrates what distinguishes great organisations from good ones:

> In the Blueprint, the strategic objectives are connected all the way through. They were strategically set out at the start. The Blueprint is pretty enduring and hasn't had to change. Whilst we review it on a 12-monthly basis as an executive leadership team it's a pretty enduring set of objectives that underpin our mission and vision. I have seen the Blueprint be a fundamental enabler to work more efficiently and is equally understood by the Alliances. The Blueprint gave us a structure to get the processes in place. What the Alliances see is a really consistent position to drive consistent outcomes.

Key elements of the LXRP Blueprint answer seven key business operating fundamentals:

1. Why do we exist? The LXRP Mission.
2. What is important to us? The Values.
3. What do we want to create? The Vision.
4. How do we want to get there? The Strategy.
5. How do we monitor and measure success? The Performance Outcomes.
6. What do we need to do? The Plans and Corporate Programs.
7. What do I need to do? Personal Development Plans – People and Performance.

The mission, values, vision, and strategy create a system of ideas that forms the shared purpose, values, direction, and meaning through a common language that everyone in the organisation can connect with. This Blueprint is designed to be the foundation for high performance and the tool that allows LXRP and its people to shape their future and achieve their mutual aspirations.

Each Figure 7-2 Blueprint element is explained in more detail.

Mission – LXRP's why

LXRP's Mission is "the Level Crossing Removal Project exists to expertly deliver transport infrastructure projects for the Victorian Government" (LXRP, 2023a, p. 1). This is LXRP's enduring purpose and the fundamental reason for its existence. It is the perpetual guiding direction for the organisation.

Values – What is important to LXRP?

LXRP's values are commonly known and understood through the acronym CARES. CARES articulates what is important to LXRP as an organisation and guides all project members' behaviours, actions and decisions. They are: Creativity – innovate, learn, share; Accountability – passion, promise deliver; Relationships – connect, listen and engage; Empowerment – lead, decide, act; and Safety – everyone, everywhere, every day.

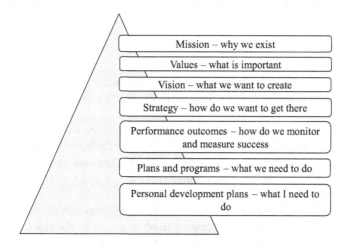

Figure 7-2 LXRP Blueprint for success

Values are espoused in all large organisations and can be seen on their websites. What is unique about the values of LXRP is that they are consistently espoused, lived by leadership, and are commonly understood. They are articulated by the LXRP's people and are reinforced consistently with recognition and reward mechanisms. The fact that they are actively and visibly espoused by leadership is a key element in high-performance culture (Schein, 2010). The LXRP values underpin what people do. They have become embedded as part of everyday practice. They are part of the culture and are expected of everyone, no matter their role. The values are articulated through leadership communication, people and performance levers, and reward and recognition programs. Everyone is expected to adopt these values and demonstrate them daily in their work and relationships. Strong cultural values promote innovation, internal flexibility, better utilisation of human capital and enduring strategic goals in the organisation (Akpa, Asikhia and Nneji, 2021). LXRP achieves all this by being embedded as espoused values by leadership, reinforced and enacted in guiding decision-making through reward and recognition, and within the divisional and individual annual plans summarised on a single page. As one interviewee (IV21) observed:

> Culture is what happens when you turn your back, it's when no one's looking at what happens. Culture is largely driven by what's what gets recognised and rewarded. So people will very quickly conform to what gets recognised and rewarded. I think the fact that from a very early space, there was a very consistent recognition on a quarterly basis of the CARES values.

Strong cultural values have been cited as promoting innovation, internal flexibility, better utilisation of human capital and enduring strategic goals in an organisation (Akpa, Asikhia and Nneji, 2021).

The LXRP values are the foundation for its culture. The values are encouraged and reinforced through the quarterly CARES recognition and reward program that recognises people who put the values into action, as revealed by another interviewee:

> The values are very clearly articulated and recognised every quarter. Through our recognition program people are publicly held up for what they have done behaviourally against those values. Everybody can say CARES, it's creativity, accountability, relationships, empowerment and safety. I've never been able to just rattle them off so readily [before this program].
>
> (IV23)

Several interviews noted that the values form a decision-making framework that helps people and guides their decisions from a new starter to the chief executive level, and one stated that:

> A couple of the values that really resonate with our people are around empowerment and creativity. Empowerment is in the sense that the values help guide your decision-making. We encourage decisions to be made at the lowest level of the organisation possible. People have a framework in which to do that through these values and a safety net if they're getting nervous about the decision. So that philosophically means if you can do something today, then don't put it off till tomorrow. Therefore, this prevents the organisation from grinding to a halt because all the decisions need to go up to the chief executive. The creativity value gives people permission to do things differently to what they might have done before – think how can we do it better, and ultimately lead to localised change or industry change. Creativity is a fundamental core value and so what we do day in and day

out hopefully is leading to some form of industry change. Creativity is beyond just what we're doing today on an individual project. It's about how we're trying to progress and that is a core element that is really, really important for the industry.

(IV18)

Vision – What LXRP wants to create

LXRP's vision is: "Delivering great change – transforming the way Victorians live, work and travel" (LXRP, 2023a). All program participants clearly understand this, as noted by at least one interviewee.

What the CEO did extraordinarily well that companies all over the world often failed to articulate well, is the 'why.' Why we [LXRP] exist is actually about changing the way Victorians live, work and travel. I think there are very few organisations in the world that can articulate that straight off the tongue from the executive team right through to the graduate that comes in.

(IV23)

What stands out about this particular vision is how people have connected to and internalised it. They can tell you what LXRP is working on achieving. While they may not recognise it as the vision, they can tell you that this is LXRP's purpose and what everyone is collectively working on.

Strategic objectives – How will LXRP get there?

A clearly understood strategy by everyone in the organisation that guides their mindsets and practices is a powerful enabler for a high-performing ecosystem that ultimately delivers upon the strategic objectives.

The LXRP Blueprint articulates the essence of the business operating approach as five strategic objectives known as the 5-Greats: Great network, Great places, Great partnerships, Great engagement, and Great people.

The 5-Greats are fundamental in driving organisational, team, and individual performance. These five strategic objectives are commonly understood by the internal corporate people within LXRP and the five Alliances. Many interviewees referred to the value of having the 5-Greats as a guide to their behaviour and decision-making.

When a critical mass of people in an organisation speaks a shared language, connects with a shared meaning and set of values and has absolute clarity of its direction and desired outcome it can truly achieve great things.

(IV24)

We can articulate what's important to us through the 5-Greats and it resonates through all of our decision making through the whole organisation and into our alliance teams and as a consequence, we have much greater consistency in decision making and how people respond to, managing issues.

(IV16)

We could have a strategic Blueprint, and some posters around and that's it. But this is not what it is. When we are in meetings working with a delivery partner on a solution for a level crossing, we are thinking in the back of our minds – how does this contribute to Great Places or our value of Great Relationships, for example. And so, we have these prompters that as leaders we are always throwing out there as a reminder to people that this is the path we are on. This is how we work together and behave. If there is not something we are aligned to, then everybody in the organisation is empowered to say oh, hang on – I thought we were in the business of delivering great places, or being creative and empowering our people or whatever it is. It is the fact that it is spoken about. We have framed our templates, tools and our systems and proformas – you name it – around those elements.

(IV19)

For example, when we had our presentations in the development phase and our delivery partners actually played the 5-Greats back to us. 'This is how we are going to deliver great places.' So, the Blueprint is something that infuses out into the alliances. The 5-Greats are not just written once. And you forget about it. The Alliances are clear about what the client is looking for. They may or may not have the Blueprint on their wall but they are clear.

(IV17).

It is the clear articulation of the 5-Greats that underpins decision-making and that drives the actions of the alliances forward through shared understanding. The WHAT of the strategic objectives is clear through the 5-Greats, and it is up to individual directors (see Chapter 3, Figure 3-3) and their teams to determine HOW they will execute these.

Chapter 8 will explore key levers that underpinned the GREAT people strategic objective in more detail.

Performance Outcomes – How LXRP monitors and measures success against the strategic objectives

Performance indicators designed to assess LXRP's achievements against intended benefits have been clearly articulated. These are anchored in the Blueprint and underpinned by a hierarchy of business planning tools that creates vertical and horizontal alignment across LXRP – both internally within and across each corporate division, as well as across the five Alliances. Implementation plans and programs detail the initiatives and actions the people of LXRP need to implement to achieve the strategic objectives articulated in the 5-Greats and, through the vision and values facilitating teams to plan their work in a way that actively and systematically strives to achieve the strategic objectives. Several plan types were developed to achieve this aim.

Annual divisional plans – called divisional "plans on a page" – are 12-month implementation plans that articulate how each division will progress toward the strategic objectives of the Blueprint. These plans answer the question – what do *we* need to do in our team? The plans-on-a-page incorporate key areas of focus including a commitment to well-being and building LXRP's body of knowledge to be shared to enable repeatable standard delivery mechanisms. Every key division within LXRP has a page plan aligned with the organisation's strategic objectives, providing consistency in approach. The plans on a page are powerful artefacts that practically bring the 5-Greats to life.

Accountability for the plans is generated by directors sharing these plans with the CEO and with each other. They are reviewed annually, and what to keep, drop, or pivot on is discussed. Research has shown that consistency in the organisation reflects the effectiveness of embedding a high-performance culture by improving the alignment of individual and team behaviour with organisational strategic objectives (Akpa, Asikhia and Nneji, 2021).

The annual divisional plans are refreshed each year and the goals within them are reviewed with a 'pivot, park or pursue' approach.

Annual individual plans called 'my activity plans' – help individuals answer questions about their personal actions needed to be done to help achieve an individual's team plan, which in turn helps LXRP achieve its organisational strategic goals. The aim of the individual plans is to connect what they need to do with what the team needs to do through its plan which in turn helps the organisation achieve its 5-Greats strategic objectives through vertical alignment.

The embedding of the template of the plans and the annual revisiting and refreshment of the plans is a clear example of the embedded structure and accountability mechanisms. Culture is created by what leaders pay attention to, and these plans on a page are a key focus.

At the corporate level, five program-wide plans were designed and developed. Five corporate plans and programs have been progressively rolled out to focus, enable and accelerate organisation-wide action. They are key enablers for helping LXRP successfully deliver on its program and meet its requirements as a public sector agency. These are: HIVE – improvement and innovation – innovation focus; KIT – knowledge insights and tools – knowledge focus; LEADSAFE – Everyone safe and well – health and safety focus; LXU – Learning and Development – learning focus; and MOSAIC- diversity and inclusion – inclusion – social outcomes focus.

These have evolved and serve as an important governance mechanism. They all link to the JCC, and program participants actively maintain their sustainability and program-wide reach.

7.3.2   *LXRP five building blocks*

Figure 7-1 illustrates the five building blocks supporting the key keystone element of the business operating approach.

The five key building blocks are:

1. Programmatic collaborative contracting model
2. Performance-based model
3. Learning and development
4. Research and development
5. Systems, processes and governance

These building blocks were embedded to create a transformative organisational culture that formed part of LXRP's business operating approach. When asked, "What underpins successful delivery at LXRP?" interviewees consistently referred to the essence of these five building blocks. Interviewees considered that the programmatic approach enabled a solid and consistent pipeline of work for the Alliances, eliminating the traditional start–stop-start of major projects, while ongoing work was 'theirs to lose.' The programmatic approach contributed to building trust and collaboration to design solutions that broke down traditional probity walls, which resulted in utilising valuable and scarce resources most effectively and efficiently.

The mindsets and practices within these five building blocks are fundamentally connected. They are the yin and the yang. They both underpin and, at the same time, enable the LXRP Blueprint to develop the Business-based Operating Approach. These pillars are examined here to share actionable insights to adopt or adapt the success of LXRP in other delivery agencies and project contexts.

Schein (2010) points to leadership and culture being two sides of the same coin. At LXRP, it is the mindsets and the basic underlying assumptions of leaders that set the tone for how people behave, together with the enabling practices that manifest these building blocks.

It is not enough to simply identify the building blocks, it is the powerful *combination* of the mindsets and practices within them that gives them their power through the artefact of the LXRP Blueprint, the espoused and lived CARES values that drive decision-making, and the underlying assumptions of being accountable that drive the high-performance culture of LXRP.

Building Block 1: Programmatic collaborative contracting model[2]

> All contracting is relationship contracting, and the alliancing model is the model we use here. Really it is about the mindset you bring. Having the right people[3] from day one is what makes it successful.
>
> (IV18)

The programmatic collaborative contracting model was developed by the LXRP executive leaders, who intrinsically understood that working together and collaborating drives better outcomes. A model mindset of collaboration was created (Chapter 6). However, the main difference between LXRP and project alliancing is that LXRP has integrated its projects tightly into a project program that aligns project alliance aims with the overall program. Moreover, the program strategy was to split the 15-year program into a sequence and series of integrated AWPs. Releasing the program through AWPs varying from several hundred million dollars to a billion-dollar value allowed each project to be scaled and scoped to be manageable and not overburden any AWP syndicate. The packaging of projects in this way as a pipeline of works and the performance reporting governance mechanisms also allowed significant gathering of benchmark data so that the project owner could be reassured that VfM was being attained. AWP syndicates were incentivised to be competitive in their delivery solutions, and the program procurement process maintained integrity. Once in place, the programmatic model fostered the growth of a collaborative mindset through the practices created from it. The result was a significant perceived benefit to project performance. Many interviewees pointed to the nature of the programmatic work pipeline as instrumental in enabling this collaborative mindset. This is supported by interviewee comments.

> A couple of things that have enabled us is definitely the framework of the alliance, and having five alliances, and giving them certainty of pipeline so they're committed. They've got a culture of looking at the horizon and investing in those discretionary things and not having that project lifecycle of a start, a middle and a finish, and people tailing off and disappearing into the sector again and looking for the next thing. We've got people that have hung around and gone from one to the next. And so, this I think the biggest legacy.
>
> (IV15)

> The pipeline model in and of itself drives very different behaviours than when it's a bespoke individual project rather than a program.
>
> (IV18)

> The Programmatic Collaborative Contracting model drove change at scale and speed: "it creates so much more change."
>
> (IV21)

> It's the change of mindset from being a competitor and having to withhold information because this is my edge against you. There is plenty of work here for all of us with the

pipeline of work from the program model. There has been a concerted effort to create that environment so you have the building of trust.

(IV17)

Hence, the Programmatic Collaborative Contracting model facilitated a mindset of collaboration that was manifested through diffusing and leveraging learnings, as noted by several interviewees

LXRP has got a really consistent approach from the top down. All the ranks of people within LXRP as far as they're concerned, they just have one way to work, and it was all about success through collaboration.

(IV22)

An alliance model with a programmatic outlook is a lot more strategic. From day one it's all about how do we leverage off that breadth of work and what can we achieve as an alliance and as part of a program? The program allows leverage [to create innovation and value].

(IV21)

In order to [remain competitive for government-funded projects] we need to be learning from and sharing with each other. At LXRP we asked our Alliances, if you do something exceptionally well, teach everybody else how to do that. Because as a group. As a collective we will get more work coming our way – and it is self-perpetuating. This program will continue on and on. [It is] also demonstrating that if you win, I win. We all win collectively.

(IV19)

The ability to iterate and learn and enable a learning culture and system thinking around learning culture is only because we've had this advantage (from the model), This has given us the ability to actually take that advantage and apply it.

(IV15)

LXRP's success through this programmatic collaborative contracting model is consistent with research that shows alliance-type contracts are more effective for achieving transport infrastructure project success than reimbursement or fixed-price contracts (Rana and Shuja, 2022).

Building Block 2: Performance-based model

The performance-based model is related to organisation, team and individual performance. The mindset that underpins this model is strategic, with sophisticated mechanisms driving performance.

Key drivers include: program alliances compete with LXRP to earn their right to deliver their allocated packages; Robust Project Owner mechanism test value for money, including extensive benchmarking data; The locked-in commercial model for the duration of the alliance; Competitive pricing (measured against benchmark) is based on long term opportunity; and alliance performance is measured against a track record test (LXRP, 2023).

According to interviewee IV15:

The commercial model is clever. It's probably a fairly traditional alliance model incentive on stuff that's important to us. But I think a key difference here with this alliance model and how we're using it. And that has enabled the extraordinary, unprecedented collaboration between the five alliances and the five design houses, the five constructors, and also the five owner teams that are embedded in there.

There is one big key, about this whole thing, that would fall apart without it. We have a very, very capable active client.

The central functions, are really there to help those five alliances, led by the Program Director. So there's a clear accountability line, and that Program Director has embedded active client people.

How LXRP monitors and measures success through its performance-based model is anchored in the Blueprint and underpinned by a hierarchy of business planning tools that creates vertical and horizontal alignment across LXRP – internally within and across each corporate division and the five alliances.

The plans on the page described earlier are a key lever that created accountability for what each team will do to contribute to achieving the 5-Greats.

LXRP also uses incentivisation to reinforce the importance of innovation and its collaborative program-wide diffusion by adopting the accepted (but often ignored) management principles of motivation, i.e., rewarding expected desirable behaviour (Kerr, 1995). According to IV19, "if we genuinely want people to come up with brilliant ideas and share them … we also [need to reward] them for doing that."

Consistent with a multi-level conception of large government projects as described by Daniel (2023), LXRP differentiates the multiple levels of its organisational structure when offering its incentives. Therefore, the incentive scheme encouraging collaborative sharing and adoption of innovations is multi-layered (Chapters 3 and 9). At the higher alliance partner and sub-contractor level, the incentive to collaborate through sharing knowledge and innovation is linked to the continued ability of participants' companies to justify future work on the program. This clearly articulated link reinforces the imperative for the alliance teams to work together as a collective for the mutual common good of all involved in the program of work.

> We need to be learning from and sharing with each other. At LXRP we asked our Alliances, if you do something exceptionally well, teach everybody else how to do that. Because as a group, as a collective we will get more work coming our way – and it is self-perpetuating.
>
> (IV19)

To further encourage a collaborative approach to knowledge-sharing, LXRP has initiated a financial incentive scheme that rewards individual alliance companies for devising new and improved processes. LXRP has socialised this mindset with an 'adopt or adapt' practice reinforced by incentivising alliance partners through its innovation KPIs to collaborate and share their innovative and continuous improvement initiatives (Chapters 3 and 9). Although the financial incentive is small, it appeals to the companies involved. Importantly, the valuing and publicising of improvements to process or implementation are widely acknowledged among program alliance members. Alliance members experience this program-wide focus on innovation and the cross-sharing of new improved initiatives as intrinsically motivational: "When you've got all five [alliances] doing it … it just means so much more and gives you so much more drive" (IV21).

The knowledge-sharing and innovation KPI incentive scheme tapped into individual team members' valuing of intrinsic (intangible) rewards, such as professional pride in achieving improvements that are recognised as the "gold standard" by their project alliance peers. Such intangible intrinsic rewards are known to be particularly motivating for professional knowledge workers (Bailey et al., 2017) like those working on the LXRP (e.g., planners, engineers, project managers, and occupational health and safety officers)

Incentives for desired behaviours implemented at the individual manager and employee level are discussed in detail in Chapter 8.

Building Block 3: Learning and development

LXRP's transformational strategy to build a learning organisation (Senge, 1990) was crafted by the executive team from the outset: "Always challenge! Don't accept business as usual.... This philosophy and behaviour is encouraged from the CEO all the way down to the alliances" (IV19).

A learning organisation mindset was foundational and infused the knowledge management system that was created so that people could easily find, learn and share the body of knowledge created. A culture of continuous learning was fostered through targeted approaches to learning and development for high-potential talent. This leadership capability development approach is described in Chapter 8.

A mindset of being a learning organisation encouraged growth, new learning and continuous improvement at multiple levels of the LXRP ecosystem – at the program level, the project level, the individual level, and the community level (see Chapters 3 and 9 on the JCC's role). Evolutionary improvements have been made in their project management processes, design and planning, and task execution as well as in the social design aspects of the program of projects, including in employee occupational health and safety, community safety and well-being, social inclusion, and in stimulating transformational change within the wider construction industry. The Great People strategic objective is interlinked with this pillar. See Chapter 8 for more detail.

LXRP encouraged a mindset of 'pinch [copy] with pride' where alliance team members were encouraged to utilise the best practices of their alliance partners to escalate the improvement of their individual company's practices and procedures.

> Everyone's quite willing to share anything. It's a nice way to be able to advance industry as well as being able to share that knowledge and actually try to improve and be more efficient and productive and just do things better too.
>
> (IV21).

According to one interviewee, an example of this opportunity to learn from and influence alliance partners and, therefore, the industry is the "safety programs and the videos and the educational awareness programs that we've developed here at the alliance, we've actually rolled out in the other alliances. It's awesome to see other companies just grab the videos you've made and utilise them on their projects" (IV21). This collaborative knowledge sharing with companies that would normally be competitors is a form of *co-opetition* that occurs in strategic alliances formed between competitors in the same industry to achieve mutually beneficial innovation and value-added outcomes (Bagherzadeh, Ghaderi and Fernandez, 2022).

It is unusual for projects to have the reach and impact required to impactfully contribute to the professional development of its people due to the start, stop nature of projects. The programmatic of a strong and sustained pipeline of work and has enabled this systematic approach to professional development, learning, and employee experience.

The Level Crossing's University Program was intentionally created to reflect LXRP's commitment to developing great people and putting them at the centre of how LXRP operates.

Valuing the People and Culture function and empowering this function to design and deliver a suite of professional learning and development tools and resources for all LXRP employees and encompasses programs and experiences to grow capabilities and skills. This is a key lever in supporting the GREAT PEOPLE strategic objective by unleashing people's potential.

The building of human capital and workforce competencies has been embedded in a suite of learning and development programs offered to employees, discussed in Chapter 8.

Building Block 4: Research and development

A mindset of thinking and acting differently has been firmly established through the value of creativity and has also been articulated in the (physical and mental) safety culture behaviour of 'think and act differently.' This mindset and behaviour are recognised and rewarded through intertwined mechanisms.

From the outset, LXRP has purposefully brought together technically capable subject matter experts from different companies, different industries, and different sectors to extend the breadth and depth of knowledge available to all members of the LXRP.

> The Alliances are clear about what the client is looking for. So, it is a blend of internalising the key elements of the Blueprint and Alliances doing their own thing. This is a good thing because you will want some diversity of thinking and you still want people to challenge [and] to explore how we can improve on things.
>
> You can think we have done this before and repeat it time and time again: but it is actually the opposite. It is constantly challenging and saying what went well on the last cycle? What did not work well? What can we improve on?… Don't accept that something has always been this way, because if it's not right, collectively let's try and fix it.
>
> (IV19)

The executive team has strategically created the innovation mindset in LXRP. It is driven by the key results areas (KRAs) and KPIs that reward above and beyond performance that has value to the State and ensure that the government does not pay for poor performance. This focus on exceeding performance expectations has been reinforced through the establishment of continuous improvement KRAs. KRA 1.1 focuses on "adding value through the implementation of continuous improvement and innovation initiatives," and KRA 1.2 encourages the adoption of continuous improvement initiatives (Chapter 9). The KRAs have fostered collaboration and knowledge sharing, as well as the adoption and adaptation of ideas and innovation to improve productivity. These KRAs are designed to drive the achievement of repeatable 'gold standard' delivery.

By establishing an Innovation Committee, the LXRP has reinforced the specific focus on exceptional innovation. Through its cross-company and cross-functional innovation committee membership, LXRP has adopted a focused, high-involvement hands-on approach to effectively foster innovation similar to that adopted on the UK Crossrail megaproject (Crossrail, 2012). Collaboration between alliance members not only promotes individual team member's creativity but also enables project teams to utilise these innovative ideas effectively. A successful innovation process resulting from this alliancing type of contract has enabled the successful achievement of project goals in terms of cost, schedule, scope and quality, as well as producing greater client and team satisfaction, a result similar to Rana and Shuja's (2022) findings.

In addition, the Innovation Committee has empowered operating teams to take up the challenge of continuous innovation. Rather than retaining innovational leadership as solely the role of the specialist Innovation Committee, LXRP has cascaded responsibility for innovation down to the alliance teams and project levels to foster the shared leadership of innovation. Shared leadership involves distributing leadership among team members or networks (Scott-Young, Georgy and Grisinger, 2019), empowering employees working at the project coalface to develop better processes, practices and outcomes.

LXRP's approach also reflects the latest thinking in *agile innovation* by driving innovation down to the individual project level in the program of works, LXRP enabled the harnessing of "the innovative behaviour of employees [which] is the micro-foundation of enterprise innovation" (Yuan and Ma, 2022, p. 145).

LXRP innovation has also been supported by a purpose-built intranet for knowledge sharing and the provision of a formal, searchable knowledge repository. However, some team members valued the immediacy of face-to-face knowledge sharing through the co-location of the five alliance teams.

It's great to have a database and have the extranet and portals for all this information to be stored, but you still get a lot more from people actually talking and collaborating and understanding and knowing each other.

(IV21).

Some LXRP interviewees attributed the successful knowledge diffusion to their co-located office space, which fostered regular face-to-face, informal communication and relationship-building. Co-location of team members promotes accessibility to colleagues, transparency, openness, and trust (Aaltonen and Turkulainen, 2023). As interviewee IV21 noted, "there's no secret discussions happening in separate rooms." Team co-location has long been known to facilitate effective communication, collaboration and team functioning that leads to successful project outcomes (Scott-Young and Samson, 2008; 2009).

Building Block 5: Systems, processes and governance

Rather than a project mindset, due to the programmatic nature of the level-crossing removals, a manufacturing-style mindset of repeatable exceptional standard performance was adopted. This manufacturing mindset is the key to the consistency of performance that is repeatable across other delivery agencies and projects.

Within this program, an internal body of knowledge was purposefully created on the LXRP intranet that enables efficient business and project management, together with a platform called the *Extranet* (Chapter 10) that houses the knowledge and is easily accessible to all LXRP participants, a portal for performance data and insights and a smart search tool that ensures easy information searching.

The development of these two intranet and extranet platforms and the effort to develop a workable taxonomy and extensive body of knowledge reflect LXRP's commitment to building a learning culture where everyone across LXRP can easily find, learn from and share the knowledge developed across the program over eight years and beyond. This is a legacy activity to ensure the knowledge is sustained into the future rather than being lost, as lessons learned have often been in the past. The result has been better consistency, standardisation, and efficiency of project delivery.

The LXRP transformative culture was created and has been reinforced continuously through structure, supported by a clear chain of command, clear values and an operating approach that relentlessly drives high performance. Clear systems, processes and governance are the lynchpin of LXRP's high-performance culture. According to many interviewees, consistent ways of operating, with repeatable processes that are clearly understood and accessible by all together with gated decision-making processes and accountability mechanisms, are critical to LXRP's high-performing culture; and can make or break it. Accountability is one of the CARES values, and it drives the bias for action and the way decisions are made.

7.4 Conclusions

This chapter describes the LXRP high-performance workplace culture, explaining *how* that occurred, and describes its features and characteristics. This may be useful to readers who wish to adapt it to suit their context, perceived project purpose and strategy. Much of what LXRP has done in developing its workplace culture has been done piecemeal elsewhere to varying extents of success.

Important points can be summarised from this chapter's discussion.

1. The enduring artefact of the Blueprint keystone created a common purpose and meaning with its vision and 5-Greats strategic objectives. The Blueprint through its shared language binds teams and projects together with shared understanding and language to help achieve their strategic objectives and at the same time make a positive difference to the world. The teams are working on the same strategic goals and empowered to leave a legacy for Victoria.
2. LXRP's espoused values are unique – not because of what they are, but because of the way they are used to underpin decision making every day through reinforcement mechanisms. They are lived and role modelled by the leaders – genuinely espoused (Schein, 2010).
3. LXRP's high performance culture has rested on the shared assumption established at its inception of thinking and acting differently – a growth mindset. This collective mindset has enabled LXRP to challenge industry norms, evolve and change over the course of its eight years of operation and deliver exceptional performance.
4. The powerful combination of the structure, mindsets and systematised practices work together with the outcome being high performance.
5. Building connection to purpose and meaning through the consistent shared language of the Blueprint creates the foundation for leaders, teams and individuals working together to create a great organisation.

The Business Operating Approach through its combination of the Blueprint artefact, the espoused values and underlying assumptions and mindsets reinforced by leadership contributes to building an ecosystem of high performance utilising a repeatable model of operating that delivers value and successfully challenges the traditional barriers to transport infrastructure megaproject success. IV22 observed that a program is close to running a business rather than just a project. LXRP operates as a business, and decisions are made with this ethos, which is not a short-term ethos typically associated with one-of-a-kind projects. The business operating approach is unique in a project environment. It enables a high-performing ecosystem to be brought to life through the Blueprint the structure, mindsets, and practices of its five supporting building blocks.

This has supported LXRP to drive cultural change in the industry, enhance productivity and leave a legacy in developing future leaders, improving health and safety, social inclusion and innovative solutions, and building a body of knowledge that enables repeatable exceptional standard delivery. As interviewee IV19 noted:

The legacy is a demonstration of what can be achieved when you challenge the norm and when you challenge culture. We can say we have this great solution that someone else can use. We can say we have a great governance model, but I think the overarching legacy is when you get motivated people working together as one team you can show great outcomes.

Notes

1 Magree, J.L. (2024). LXRP Business Operating Approach, unpublished internal working paper.
2 An integrated program of collaborative AWPs.
3 'Right people' refers to a fit between person/task/organisation, i.e., having a compatible mindset.
4 'Gold standard' is a term used by LXRP to denote exception levels beyond expectations and requirements.

References

Aaltonen, K. and Turkulainen, V. (2023). "The use of collaborative space and socialisation tensions in inter-organisational construction projects." In *Construction project organising*, Addyman, S. and H. Smyth (eds), pp. 167–184. Wiley.

Akpa, V.O., Asikhia, O.U. and Nneji, N.E. (2021). "Organizational culture and organizational performance: A review of literature." *International Journal of Advances in Engineering and Management* 3 (1): 361–372.

Atkinson, R. (1999). "Project management: cost, time and quality, two best guesses and a phenomenon, its time to accept other success criteria." *International Journal of Project Management.* 17 (6): 337–342.

Bagherzadeh, M., Ghaderi, M. and Fernandez, A.-S. (2022). "Coopetition for innovation – the more, the better? An empirical study based on preference disaggregation analysis." *European Journal of Operational Research.* 297 (2): 695–708.

Bailey, C., Madden, A., Alfes, K. and Fletcher, L. (2017). "The Meaning, Antecedents and Outcomes of Employee Engagement: A Narrative Synthesis." *International Journal of Management Reviews.* 19 (1): 31–53.

Biesenthal, C., Clegg, S., Mahalingam, A. and Sankaran, S. (2018). "Applying institutional theories to managing megaprojects." *International Journal of Project Management.* 36 (1): 43–54.

Clegg, S.R., Loosemore, M., Walker, D.H.T., van Marrewijk, A.H. and Sankaran, S. (2023). "Construction cultures: Sources, Signs and solutions of toxicity." In *Construction Project Organising*, Addyman, S. and H. Smyth (eds), pp. 3–16. Chichester, West Sussex, John Wiley & Sons.

Crossrail (2012). *Crossrail Innovation Strategy: Moving London forward*, Crossrail London, UK.

Daniel, P.A. (2023). "Modeling relationships of projects and operations: Toward a dynamic framework of performance." In *Research Handbook on Project Performance*, Anantatmula, V.S. and C. Iyyunni (eds), pp. 39–53. Northampton, Edward Elgar Publishing.

Den Hartog, D.N. and Verburg, R.M. (2004). "High performance work systems, organisational culture and firm effectiveness." *Human Resource Management Journal.* 14 (1): 55–78.

Dweck, C. (2012) *Mindset : How You Can Fulfil Your Potential*, London, Constable & Robinson.

Flyvbjerg, B., Budzier, A. and Lunn, D. (2021). "Regression to the tail: Why the Olympics blow up." *Environment and Planning A: Economy and Space.* 53 (2): 233–260.

Hall, D.M. and Scott, W.R. (2019). "Early Stages in the institutionalization of integrated project delivery." *Project Management Journal.* 50 (2): 128–143.

Kerr, S. (1995). "On the folly of rewarding A, while hoping for B." *Academy of Management Executive.* 9 (1): 7–14.

LXRP (2023a). *The LXRP Blueprint*, Melbourne, Level Crossing Removal Project.

LXRP (2023b). *LXRP Model – Enabling transformational infrastructure Version 3 – July 2023 ELT Review*, Melbourne, Level Crossing Removal Project.

McKinsey Global Institute (2017). *Reinventing Construction: A Route to Higher Productivity.* McKinsey & Company Institute M.G.

O'Neill, T.A. and Salas, E. (2018). "Creating high performance teamwork in organizations." *Human Resource Management Review.* 28 (4): 325–331.

Rana, A.G. and Shuja, A. (2022). "Influence of leadership competencies on transport infrastructure projects' success: A mediated moderation through innovative-work-behavior and the project type." *Pakistan Journal of Commerce and Social Sciences.* 16 (1): 1–33.

Schein, E.H. (2010) *Organisational Culture and Leadership*, San Francisco, Jossey Bass.

Scott-Young, C. and Samson, D. (2008). "Project success and project team management: Evidence from capital projects in the process industries." *Journal of Operations Management.* 26 (6): 749–766.

Scott-Young, C.M., Georgy, M. and Grisinger, A. (2019). "Shared leadership in project teams: An integrative multi-level conceptual model and research agenda." *International Journal of Project Management.* 37 (4): 565–581.

Scott-Young, C. and Samson, D. (2009). "Team management for fast projects: an empirical study of process industries." *International Journal of Operations & Production Management.* 29 (6): 612–635.

Scott, W.R. (2014) *Institutions and Organizations*, Thousand Oaks, CA and London, Sage.

Senge, P.M. (1990) *The Fifth Discipline – The Art & Practice of the Learning Organization*, Sydney, Australia, Random House.

van Muijen, J.J., Koopman, P., et al. (1999). "Organizational culture: The focus questionnaire." *European Journal of Work and Organizational Psychology.* 8 (4): 551–568.

Victorian State Government (2017). *Level Crossing Removal Project – Program Business Case – Procurement Strategy Supporting Information*, Government V. Melbourne.

Wolstenholme, A. (2009). *Never Waste a Good Crisis – A Review of Progress since Rethinking Construction and Thoughts for Our Future*, London. 32pp

Yuan, H. and Ma, D. (2022). "Gender differences in the relationship between interpersonal trust and innovative behavior: The mediating effects of affective organizational commitment and knowledge-sharing." *Behavioral Sciences.* 12 (5): 145.

8   LXRP Human Resource Development

*Jane Magree, Beverley Lloyd-Walker and
Christina Scott-Young*

8.1   Introduction

*How does LXRP's leadership ensure they have the Great People and systems to transform the
way Victorians live, work and travel, and leave a positive legacy?*

LXRP's Blueprint (LXRP, 2023) states its mission as "to expertly deliver transport infrastructure projects for the Victorian Government." LXRP's vision is to, via these projects, "deliver
Great Change, transforming the way Victorians live, work and travel." For this goal to be
achieved, the 5-Greats – Great Network, People, Places, Engagement and Partnerships – which
are their strategic objectives, will need to be addressed by having Great People in high-
performing project teams. This cannot happen spontaneously. It has to be crafted and shaped
through a strategy developed through committed, focused, supportive leadership (Cooke-Davies,
Crawford and Lechler, 2009). To do this, LXRP connected purpose, ignited energy and elevated
organisation-wide performance through its leadership. Great people are at the heart of how
LXRP operates, and this chapter provides insights into how great change has been delivered
through the combination of great people who have contributed to building a high-performing
ecosystem enabled by strong structure and processes.

8.2   Creating a High-Performance Culture

Developing the workplace culture required to achieve LXRP's mission and vision needs systems
that attract, motivate and retain high-performing project team members. The LXRP deliver this
mission through coordinated actions linked directly to the LXRP (2023) Blueprint that was conceptualised at the program's beginning and further developed. Chapter 7 explains the Blueprint
elements and the LXRP (2024) model

The staffing (human capital) process – attracting, recruiting, selecting, motivating and
retaining – is vital. Staffing processes that support high performance have been extensively
researched over many years. Well-designed employment practices attract, retain, and motivate
project managers (Ling et al., 2018) by ensuring they experience job satisfaction. A range of
factors contributes to employee engagement, motivation and satisfaction, including job-person
fit, person-organisation fit and person-environment fit and, for project-based organisations,
person-team fit (Kulik, Oldham and Hackman, 1987; Ellis et al., 2022; Hajarolasvadi and Shahhosseini, 2022a; 2022b; Sun et al., 2022).

LXRP's aim of leaving a positive legacy indicates the values of the organisation and will attract,
motivate and retain high performers with the same values. Additionally, leadership style, professional development linked to career opportunities, employment security and rewards and recognition assist to attract, motivate and retain high performing employees (Zheng et al., 2019; Agyekum
et al., 2020; Sergeeva and Kortantamer, 2021; Borg and Scott-Young, 2022; Rana and Shuja, 2022).

DOI: 10.1201/9781003389170-8

8.3   Developing Great People to Fit the LXRP Workplace Culture

8.3.1   Attracting and recruitment

From the very beginning, LXRP targeted those they knew were exceptional talent, having demonstrated both technical and soft skills. LXRP aimed to be a high-performing, informed client, so this calibre of talent was required to establish an engineering organisation that ensured this (LXRP, 2024).

The aim was to have the *right people*[1] in the right jobs at every organisational level. People working in the transport infrastructure industry in Melbourne are often aware of each other's performance and workplace attitude after having studied or worked together. This was the case during the early days of the LXRP. The executive leadership team was formed by targeting well-known, high-performing, technically excellent people in the public and private sectors. The leadership team then created a structure that delivered upon the performance expectation of accountability for decision-making and action. This was driven by an organisational structure of core disciplines (or divisions) focused on highly integrated performance at the program level, as illustrated in Chapter 3, Figure 3–4.

The scope of the program alliance required greater effort in attracting the 'right' people to perform the job and have values that fit with LXRP's values. LXRP's leadership team continued to ensure they had the right people fit from the top down, and where there was not a fit, or performance was poor and was not improved, employment arrangements changed. From the beginning, it was recognised that fair and suitable remuneration was required to attract exceptional people with the skills and capabilities needed. This was critical as a mix of public and private sector people was needed if LXRP was to achieve a crucial part of its strategy: being an informed client. As an engineering organisation, LXRP specifically recruited people from various industry backgrounds with deep technical expertise. They sought "a mix of people" that "had exposures to" (IV18) a range of industries to be that informed client. As IV15 explained, this enabled LXRP to be "one above an active client; we're actually specialised experts in some … key risk areas technically."

Current labour shortages can attract the right people for the job and the organisation, encouraging them to apply for a position more demanding than in times of high labour supply, especially given the program scope. Clear statements about the organisation's values (CARES) support the attraction of the right people by appealing to those whose values match those espoused by the organisation and can lead to increased job satisfaction, performance, commitment and retention (Yusliza et al., 2021).

Employer branding is about creating an image of the organisation that will appeal to potential employees and reinforce current employees' beliefs that they are working for the right organisation. It has been defined as "internally and externally promoting a clear view of what makes a firm different and desirable as an employer" (Lievens, 2007, p. 51). Corporate social responsibility (CSR) can create an image that makes an organisation attractive to potential recruits. It can form part of the organisation's brand by attracting those whose values align with its CSR initiatives. LXRP's mission and overall aim are to reduce level crossing accidents and, thereby, injuries and deaths, reduce greenhouse emissions, provide new and environmentally appropriate train stations and create new social spaces that contribute to community interaction and cohesion. It aims to leave a *legacy* that benefits the infrastructure delivery industry and society. Socially responsible organisations with an attractive employer brand are perceived to have a good reputation and can attract potential applicants (Tkalac Verčič and Sinčić Ćorić, 2018).

Creating an image that will attract the right people to LXRP links to its Blueprint and the actions it drives. The right people have the knowledge, skills and enthusiasm to work in a challenging but rewarding environment. This requires a willingness to grow, develop and form high-performing delivery teams, moving between projects as the pipeline of projects that form the program alliance unfold.

Providing a sense of achievement, of having contributed a legacy of something bigger that will benefit the construction industry for years to come, was expressed by IV17 as:

> the vision is not just to support our program; it's what can we do in the industry ... what are the things that we can ... introduce that the industry will adopt that will continue and maintain within the industry.

For many potential recruits, being involved in a program of projects delivering community benefits can attract them to an employer such as the LXRP. IV18 said about competing for high-quality staff in the current skills shortage environment:

> I find that highly motivating, and I think our people find that highly motivating because it's beyond just what we're doing today on an individual project. [When LXRP was first established] we ... attracted industry leaders and highly capable people to the program.

LXRP believes that employing people who can work in a collaborative, relationship-based procurement environment requires attracting the right people, ensuring the recruitment process gathers applications from suitably qualified people, then selecting those who will fit the LXRP culture and values and linking their efforts to the 5-Greats. Having attracted suitable people, it was important for LXRP to encourage them to apply and assess their LXRP culture fit. Employing the right person,

> with the attitude, with the openness to ... do what we do and operate how we operate ... isn't for everyone ... so having the type of conversations that will help us to know if they will fit in is important.
>
> (IV17)

The likelihood of receiving applications from well-suited applicants is high when qualifications and experience for a role are clear and CSR initiatives support the employer brand. The probability of the person being the right fit for the job, the organisation, and the environment is significantly increased.

Person-job (P-J fit) fit ensures applicants have the knowledge, qualifications, attitudes and skills for the job. Interview questions or tests used in selection can check for P-J fit. Person-job fit and person-team fit (P-T fit) is essential for project team effectiveness (Kristof-Brown, Zimmerman and Johnson, 2005). In a project-based setting, P-T fit can be vital for creating high-performing teams. Incorporating supporting activities to develop desired team behaviours during formation can be combined with testing for team skills to achieve high-performing project teams (Ahiaga-Dagbui Dominic et al., 2020). P-T fit has been found to impact project performance with low levels of fit leading to cost overruns as a result of increases in schedule delays and errors (Hajarolasvadi and Shahhosseini, 2022b). Person-organisation fit (P-O fit) can be essential in achieving performance goals. It occurs when an employee's values, personality, and goals align with their organisation's. Compatibility exists between the employee and the employing organisation when well-matched, resulting in "higher levels of organizational commitment, job satisfaction, job retention, organizational citizenship behaviours, and job performance" (Subramanian, Billsberry and Barrett, 2023). Soltis et al. (2023, p. 451) define P-O fit as "the perceived congruence between individual and organizational values." A fit provides positive outcomes for both the organisation and the individual, and this holds across industries and time (Kulik, Oldham and Hackman, 1987; Astakhova, 2016; Chowdhury, Yun and Kang, 2021). Person-environment (P-E) fit relates to "understanding of employees' emotions, attitudes, and behaviour in the workplace" (Vleugels et al., 2023). P-E fit is concerned with how well employees' values fit with those of the environment in which they work.

People seek out roles in organisations that espouse values that they also hold and stay if the leadership team's actions demonstrate those values. CSR can impact attraction, motivation, behaviours, commitment and retention (Bouraoui, Bensemmane and Ohana, 2020). LXRP's stated aim indicates their Intention to bring about changes in the local community that will improve everyday life. This creates a brand image that communicates their intentions and, at the same time, appeals to those who wish to be part of that improvement to the society in which they live. Improved rail transportation due to level crossing removals is not only about getting people from point A to point B; LXRP aims to reduce travel time and stress levels for drivers and reduce pollution from vehicle fumes. Additional improvements in train station facilities and surrounding community open spaces can form part of the attraction to being involved in delivering positive outcomes for all. They contribute to ensuring a good P-E fit between applicants and the organisation.

Fit is vital for LXRP's success. The Blueprint provides clarity of purpose. There is a clear link between the Mission, Vision, Values and Behaviours, revealing the type of person that will fit their high-performing project teams and support the achievement of LXRP's strategic goals. What does the Blueprint do to help with this process? It goes beyond saying how many people with a particular set of skills, knowledge, and qualifications are required within the teams and at what stage; it talks about ensuring that the espoused values are held constant through the behaviours of all team members. This means that the people attracted to LXRP, those who apply for roles and are selected to become members of the organisation, have exhibited beyond the traditional knowledge skills and attributes (KSAs) their ability to relate to others, work in teams, share knowledge and experiences, and pull together to achieve the overall aims of the LXRP by living the values.

The original Blueprint developed by LXRP leadership, adjusted in response to learning from projects, has created the foundation upon which high performance can be built. IV20 stated that "the core of our success as an organisation is … having a high performing workforce … we not only… hire individuals, but also" develop "those people to their full potential as part of our project life cycle." Table 8-1 illustrates important human capital staffing elements for high-performance teams and how LXRP responded to these.

8.3.2   *Developing LXRP's accountability culture*

Weekly reporting conversations through a structured process were established that flowed through the disciplines at all levels of the organisation. This accountability culture dealt with poor performance and high-calibre, high-aptitude people were identified and targeted for development opportunities. LXRP considered that attracting and retaining this executive leadership team fit was foundational to success. Separation mechanisms enforced a strong fit between individuals and the organisation. LXRP believed that great organisations develop great people by focusing on leadership development, ensuring a fit between people and the organisation, and, where not, being unafraid to manage poor performance.

> we absolutely did bring in a very robust way to manage people out. I'm very proud of the fact that we've got zero escalation to Fair Work[3] and we do it with good. grace.
>
> (IV23)

Table 8-1 LXRP's actions to achieve high-performance teams

Staffing element	Actions	LXRP's actions
Attraction	Employer brand	An employer brand built on CSR/Greater good – community benefit; diversity goals Industry change/improvement CARES

Recruitment	Networking,
	  word-of-mouth and
	  employer brand
		Beginning – Network
		Later – Network graduates, those with matching values,
		  those who see an approach to diversity as attractive
		SEEK,[2] low reliance on, e.g., advertising or
		  consultancies (head hunters)

Selection	Ensuring P-J & P-T fit	Networking approach = knowing the individual's job and
		  team fit
		Grads who have completed holiday placements?
		  Probation
		Supportive professional development

Retention/
  motivation
	Satisfaction
	Security
	Professional development
	Career progression
	Alliancing mindset
	Ensuring P-O & P-E fit
		Autonomy – job design, decision making
		Program of projects offer security
		Professional development linked to career
		Provide a sense of achievement/advancement
		No-blame culture supports satisfaction, achievement
		  of goals, experimentation and learning, even from
		  failures, through well-developed relationships and high
		  levels of trust
		Creates an environment of inclusion

Diversity	Culturally intelligent leadership Government requirements Team performance Greater good	Government requirements guide only; benefits drive Professional development for: career advancement recognition of overseas qualifications indigenous employment, training

Leadership/
  Supervision
	Team culture
	Recognition of
	  contribution Feedback
	Trust
		Leaders that:
		walk around and listen, not talk to
		recognise the contribution to the team
		live the organisation's espoused values
		do what they say they will do

High performing
  teams
	Have high levels of trust
	Commitment to shared
	  vision and team
	Efficient and effective
	  work practices aligned
	  with the organisation's
	  strategy
	Clear strong
	  communication
		Transformational leaders
		Strong commitment to LXRP's vision, strategy, CARES
		  and 5-Greats
		Link all to the Blueprint
		Blueprint is continually reinforced by leadership

Mechanisms involving structure and processes to foster high-performance accountability and manage poor performance were designed, including clearly articulated performance plans with accountability created by expecting the plans to be executed. Probation was extended in the organisation from three to six months, with the option to extend it for a further six months if required.

The accountability structure for account executives was based on assumptions you can't manage everything and ensuring accountability by direct reports. Two questions were asked of direct reports each week. What are your key issues for the week, and what are your key deliverables? These meetings focused on accountability and were consistent with LXRP's values.

Divisional plans comprise a one-page document, refreshed each year. Articulating these plans creates peer accountability. Each year, the CEO asks, "what did you achieve, what you we need to park, and what do we need to pivot?" driving a bias for action: just do, learn, don't make the same mistake again. Act, take the initiative, and have accountability and responsibility. Taking responsible risks is part of the LXRP leadership empowerment ethos.

Leaders counsel their teams to answer three questions as they encourage creativity, thinking and acting differently. Three things to consider before acting:

1. Will we end up on the front page of "The Age" (a newspaper in Victoria)?
2. Could we seriously injure someone?
3. Will it cost money unnecessarily?

Committee-based consensus decision-making is not encouraged at LXRP because it is believed this may lead to decision-making avoidance or lack of decision accountability. However, decisions are made based on available data, responses to assumption questions, and input from multiple perspectives. People are empowered to think and question when decision-making.

> Strategic objectives are connected all the way through so we can articulate what's important to us. We can do it clearly and that resonates through all of our decision making through the whole organisation and into our alliance teams. As a consequence we have much greater consistency and you have (clearer) decision making in how people respond to managing issues.
>
> (IV16).

> I live in all of the sort of values … about empowering people and empowering people around us. What we can only do that if we've got a level of confidence around the decision making criteria that they're going to be using in any decisions or recommendations that they come back with. I've seen it is a fundamental enabler to us to be able to work more efficiently.
>
> (IV16)

8.3.3   Developing LXRP collaboration and trust

Trust is vital for collaboration that engenders genuine dialogue (Chapter 6). Trust building requires team member selection that goes beyond knowledge and skills to personality and the propensity to trust and ensure team members are trained in group collaboration (Ford, Piccolo and Ford, 2017). Project leaders will be pivotal in facilitating team trust development and need strong collaboration skills. Trust building can be time-consuming (Buvik and Rolfsen, 2015). The pipeline of projects within the program alliance leads to organisations working together across various projects over time. This makes spending time on developing deep trust worthwhile. Prior ties can influence the ability of a team to create trust (Buvik and Rolfsen, 2015). The nature of the transport infrastructure industry is such that those with the required specialised skills and an interest in working in the industry have often worked together in collaborative teams; thus, prior ties can aid trust development. Knowledge sharing, commitment, satisfaction, team performance, and strong social networks are outcomes of trust within teams (Buvik and Rolfsen, 2015) and are vital components in achieving LXRP's high-performing teams.

An alliancing mindset established within the PAA requires participants to know how to build relationships to build trust and be open to sharing their challenges. A collaborative working alliance or Integrated Project Delivery environment is established within the agreement. Trust does not just happen. LXRP team members worked together to establish agreed behaviours within a transparent, sharing environment, leading to trust building (Chapters 4 and 6). The culture of LXRP is one of "collaborative thinking, being open and transparent and having mature conversations" (IV19). Further fine-tuning and improving the soft skills of recruits can be incorporated into the onboarding activities, where individuals are introduced to the organisation, its culture and its people.

Adopting an alliance-like culture with high levels of collaboration, trust and knowledge-sharing benefits project performance (Nevstad et al., 2022). A person-job fit applied during the recruitment and selection process can ensure that staff members possess or have the potential to develop high levels of collaborative project team skills. However, staffing activities must also incorporate person-organisation fit to achieve high-performing teams where creativity and innovation occur (Tang et al., 2021).

Collaborative project delivery methodology leads to "the creation of a new working environment," one with "trust-based collaboration and cooperation," which has been found to bring about positive change in performance results of construction projects (Moradi et al., 2021, p. 11). Indeed, they reported several studies across North America, Colombia and Ireland that found that "construction projects undertaken with collaborative delivery models... outperformed those" delivered via traditional methods.

The longer-term relationships that the LXRP's program alliance requires led to the recognition that collaboration and trust would be important for success.

> We took a step back early on and said, hang on, we are going to be working with them for the next two, or three or four years or five years and realised we would need to build trust. So, it meant moving from transaction to relationship because whereas with in a transactional relationship you might 'go to war' if things don't work out, when you are collaborating in an alliance-type relationship it is a case of Your success is my success. Your failure is my failure. Let's do this together.
>
> (IV19)

The sink-or-swim-together culture versus the combative relationships of the past requires people who will communicate with one another, collaborate, and share information because they trust one another (see Chapter 6). The staffing process must consider these qualities to assemble the high-performing teams that LXRP requires.

8.3.4 *LXRP leadership's role*

Great people are at the heart of how LXRP operates.

Well-known leaders, e.g., Mahatma Gandhi or Abraham Lincoln, belong to the Great Man theory based on the belief that leaders are born, not made (Benmira and Agboola, 2021). Leadership has evolved, but attributes that makeup leadership have been studied, identified and described. These include: communication style, integrity, emotional intelligence (EI), trustworthiness, mindfulness, supporting change, and adaptability (Hall, Meyer and Clapham, 2023). These attributes can be linked to performance, but no singularly agreed leadership style suits every situation. Over time, studies of leadership led to Blake and Mouton's managerial grid as it became accepted that leaders may be made rather than born into leadership and that the required behaviours to be a good leader could be learned (Benmira and Agboola, 2021). Situational leadership style developed at this time (Hersey and Blanchard, 1982) where leaders adapted their style to address the needs of the current environment and their team.

Based on research, it was accepted during the mid-20th century that the context or environment within which leadership was being practised needed to be assessed and the appropriate style adapted to fit the situation (Benmira and Agboola, 2021). More recently, leadership discussions have centred around behaviours that can be described as transactional (focusing on cost-benefit exchange), transformational (inspirational, encouraging higher levels of achievement from followers), collaborative (where followers lead one another) and other styles with varying emphasis on the commonly agreed capabilities of a leader (Benmira and Agboola, 2021). In complex projects or programs, especially those continuing over several years and delivering publicly funded outcomes, there are 11 main groups of skills that leaders require: influencing, communicating, team working skills, emotional, contextual, managerial and cognitive skills, and the need for "professionalism, knowledge and experience, project management knowledge and personal skills and attributes" (Bolzan de Rezende et al., 2021, p. 3).

Influencing skills are described as being able to "produce effects on the actions, behaviours, and opinions of others" (Bolzan de Rezende et al., 2021, p. 8). LXRP works with various organisations to deliver major infrastructure to the client and the state government in a demanding environment spanning the lockdowns of COVID-19 and labour and materials shortages after re-opening. This complexity was further increased by the brown-field project nature, with all its inherent uncertainty presenting LXRP leadership with additional issues to cope with. Influencing skills are important in complex environments such as this. Our understanding of leadership now means that we realise that, for instance, trauma teams may confront a time-critical and familiar emergency which requires concentration on process efficiency. In contrast, in other situations, teams may face "unexpected, ambiguous situations that require them to engage in collective sensemaking" (Thommes et al., 2024, p. 2). Adaptive leadership is required here, which Thommes et al. (2024) refer to switching gears to ensure that the appropriate style for the situation is applied. So we see the learning from over 50 years ago combined with current experiences to define leadership slightly differently over time, building on but rarely discarding the theories developed in the past. To achieve the 5-Greats LXRP aspires to requires adopting the most appropriate leadership style for the situation. Adapting the way they lead to suit the situation and using influencing skills to work within a multi-megaproject environment that a major infrastructure program involves is vital.

What capabilities, beyond technical competence and relevant qualifications, are required of project managers in organisations wishing to reap the benefits of delivering projects through collaborative project teams? Successful managers in any industry require both technical expertise and soft skills. EI positively impacts project performance by supporting trust and job satisfaction development (Castro et al., 2022). Its importance in effectively managing the construction industry has been recognised for some time (Love, Edwards and Wood, 2011). There is a link between project managers' communication, problem-solving and leadership skills and project success (Avença, Domingues and Carvalho, 2023, p. 1705). High levels of trust were found to increase collaboration and communication, resulting in improved on-site productivity in the Australian Construction Industry (Chalker and Loosemore, 2016). Transformational and authentic leadership include these elements, and more (Avolio and Gardner, 2005).

Transformational leadership has been found to contribute to project success. A combination of EI and a transformational leadership style has been found to strongly predict public project success (Fareed, Su and Naqvi, 2022). Transformational leaders will live the espoused values of the organisation, which include empowering their employees at LXRP and developing strong relationships with them. As interviewee IV20 said, leaders don't walk around to be seen but to see and to listen.

Links have been found between transformational leadership and elements of the job characteristics model in relation to intrinsic motivation and commitment to goals (Piccolo and Colquitt, 2006). Leaders who exhibit components of transformational leadership will make organisational goals clear for all and live the organisation's espoused values through their actions. Transformational leaders who stimulate the development of trust and job satisfaction in their teams positively impact project success. Employees who experience job satisfaction are less likely to leave their jobs, and trust is a vital element for knowledge sharing within collaborative project team environments. LXRP's Blueprint provides the required level of goal clarity, reinforced by the leadership team, and adapted to fit the various levels and areas of operation. No one leadership style will address the needs of every situation over the long-term operation of a program of projects. Leaders must read the situation they are currently addressing and adapt their leadership style by incorporating those elements of the required leadership styles (Hersey and Blanchard, 1982).

Using what has been identified as the required knowledge, skills, and attributes to work in a collaborative working environment and applying the long-established principles of attraction, motivation, satisfaction, commitment, and retention requires a detailed understanding of role requirements and the long-standing research on recruitment and retention of high performing employees. Managing the ongoing performance of employees then ensures that the desired level of performance continues to be achieved.

8.4   Designing Jobs that Motivate and Provide Satisfaction

Understanding worker motivation has been researched over several years. Though developed fifty years ago, the Job Characteristics Model (JCM) provides a framework for designing jobs which will motivate employees and provide them with job satisfaction (Hackman and Oldham, 1976). Core job characteristics include:

- Skill variety involves the employee performing various tasks using different skills and knowledge.
- Task Identity is when employees see how their work contributes to the project.
- Task significance relates to the extent to which the individual's role affects the lives and work of others.

Other factors come into play and influence the level of motivation and commitment experienced.

- *Autonomy* relates to the level of independence, decision-making freedom and discretion the employee has in scheduling work and deciding how it will be performed.
- *Feedback* refers to how well and appropriately supervisors inform staff of their individual and team performance satisfaction. It also refers to how much information employees receive about how effectively they have completed their tasks.

Some factors moderate the strength of the various influences. The extent to which an individual desires growth, how well they fit with the environment and the leader-member exchange all impact. For instance, how well does the individual react to their supervisor's leadership style? The influence of ability demonstrates the importance of the staffing process in identifying those with the knowledge, skills, and qualifications for the role. The ability to work in an alliancing environment further influences performance potential. The comprehensive program of professional development offerings available for LXRP employees to choose from can assist with further strengthening alliance project team skills. The onboarding process may be a good time for recruits to enhance their alliancing skills further.

Figure 8-1 incorporates findings of Hackman and Oldham (1976) which have been tested by many across a range of industries and cultures since the JCM was first developed (e.g., Piccolo and Colquitt, 2006; DeVaro, Li and Brookshire, 2007; Sever and Malbašić, 2019; Siruri and Cheche, 2021). Examples provided from interviews demonstrate where the JCM is being applied at LXRP.

The willingness to participate and the level of participation of employees in the comprehensive range of professional development offerings has demonstrated a generally high growth need strength across the workforce. When linked to performance monitoring, those with the greatest potential to succeed within LXRP's culture are identified. Those identified can then be connected to appropriate professional development to ensure a talent pipeline is available to support the pipeline of projects. Retention levels have been high, indicating a high level of satisfaction with supervision, including the type and frequency of feedback and leadership style adopted across the various stages of the projects. Security of employment, high levels of trust leading to open communication, knowledge sharing, and a no-blame innovation-supportive environment have also contributed to context satisfaction demonstrated by the commitment to continuous improvement and learning to ensure improved performance from one project to the next within the program alliance.

Professional development of identified talent will support continuous performance improvement. At LXRP, formal learning opportunities are combined with other activities to deepen

Core Job Characteristics	Critical psychological states	LXRP's actions
Skill Variety *Task Identity* *Task significance*	When these three elements are present, employees experience meaningfulness of work.	*Seeing the change that is occurring in their community (IV22) and being involved in decision-making (e.g., the JCC)* *Being involved in something that is leading to positive industry change (IV18)*
Autonomy	Autonomy provides the employee with a feeling of responsibility for outcomes.	*Employees set own performance goals, select supporting professional development (Iv23) Gives people permission to do things differently (IV18)*
Feedback	Feedback provides knowledge of results.	*... we're having a wide impact ... creating a positive public infrastructure ... they're benefitting from new stations and open space we're creating (IV15)*
Moderators Growth need strength Context satisfaction • Satisfaction with supervisor • Satisfaction with the work environment Ability		

Figure 8-1 Application of the Job Characteristics Model at LXRP

(Adapted from: Hackman and Oldham, 1976)

employee involvement in planning and decision-making. This includes involvement in subcommittees of the Joint Coordination Committee (JCC) (see Chapters 3 and 9), where employees can contribute to decision-making within smaller groups where psychological safety combines with commitment to goal setting through involvement. Together with professional development programs, JCC subcommittee participation and taking on higher-level roles at times of need allow employees to complete a variety of tasks and activities of significance to the organisation. Observing the impact of their contribution, employees can reflect upon and recognise the impact of their role in leading to their experiencing growth and autonomy. All contribute to the high motivation and performance levels that LXRP seeks whilst retaining the talent required to achieve their goals.

LXRP's desire to make positive changes to the industry and the community would provide an attractor, motivator and retention tool. This is because "values motivate human action," and people will "direct their actions where it makes sense to them and has some value to them" (Modranský and Lajčin, 2021, p. 351). People who perform work that matches their values and addresses areas they care about will be intrinsically motivated (Marnewick, Silvius and Schipper, 2019). Modranský and Lajčin (2021) also found 'confidence in success' important in motivating project managers. Viewing the program alliance as a valuable contribution to society and performing at a high level would give potential applicants and current employees confidence that they would achieve the success they value in such a collaborative environment.

8.5   Professional Development and Career Progression

The LXRP's Great People strategy aims to unleash people's potential by investing in their skills, capabilities, and well-being. LXRP leadership is committed to developing their people leaders to create great teams. Professional development programs are designed to build people's knowledge, skills, and attitudes, concentrating on the characteristics required to work in a collaborative environment. One way this happens is when people fill each others' roles when they are away. A comprehensive suite of professional development programs also supports building Great People and provides career progression, job satisfaction, commitment and retention. The pipeline model contributes to security for LXRP and their employees. It "drives very different behaviours than when it's a bespoke, individual project rather than a program" (IV17). Hwang et al. (2020) found that job security, or removing the risk of losing their job, was a context factor supporting project manager job satisfaction. Satisfied employees are motivated and committed. The program alliance, therefore, requires people who can demonstrate those behaviours. Employees know they will have continuing employment and potential career progression if they succeed in the environment. A comprehensive suite of professional development programs is available to employees who can move through the four levels of development to support their career progression. As several interviewees noted:

growth is provided via the various development programs.

(IV20)

In a supportive culture which contrasts with past experiences as

when working on shorter term projects many employees would need to change jobs to receive promotion [with project linked contracts encouraging an attitude of] I'm only here for a year, the project will be done and then I'll move on to something else and I'll get … my promotion.

(IV 22)

There is clarity about the … grades and what they mean.

(IV16)

Employees can undertake learning opportunities to support their development and career progression. IV22 provided an example of a graduate engineer who started in 2017 and was now a senior project engineer demonstrating "career progression within that program of works … that's really powerful in terms of being able to attract and retain a really high performing team." Employment retention and the work pipeline motivation was an important consideration at the LXRP's program start with intense competition for skilled human resources from several other Melbourne concurrent megaprojects.

LXRP's *Graduate Program*, while not unique in the industry, has been a key lever in growing and retaining talent due to the programmatic nature of LXRP. With over 4,000 applicants, LXRP takes between 35 and 75 graduates on average from year to year. The graduate program runs for two years, during which graduates rotate into three business areas and four program areas. LXRP developed a COVID-19 safety management strategy that enabled it work throughout the COVID-19 period. The outcome is better connections and relationships across the program and stronger cohesion as a cohort.

Accelerated Development Programs are offered yearly to increase connection and stronger relationships across LXRP, which assists with cross-collaboration and engagement. The outcome is shared understanding and learning with increased trust and better outcomes. This is strategically important to LXRP's programmatic work nature and the growing talent that will remain within the industry.

Performance management ensures that only those who continue to exhibit the LXRP required attributes to perform at the level needed are retained and developed to move to new and more senior roles over time. Through weekly reporting conversations as discussed earlier. As IV23 stated, it was better if leaders "were spending more time talking to their high performers and taking them further than the poor performer." This one change led to the empowerment of employees, increased positive interactions with leadership, the enthusiastic undertaking of development opportunities linked to career progression, and the creation of an overall positive environment.

Multiple megaprojects provide an ideal environment for building program management capabilities over time, with longer-term working relationships providing opportunities for learning from one project within the program to feed into improvements in the next. This includes improving leadership and management capabilities. A study in the UK found that "Megaprojects are sophisticated inter-organisational spaces and offer a platform for capability development based on rich co-creation with multiple partners." They "stimulate the transformation and capability building through the involvement in multiple megaprojects" and "should be delivered through a collaborative relationship *with* clients, not *for* clients" (Denicol and Davies, 2022, p. 505). This has been the approach at LXRP. The relationship between organisations involved in the program alliance and the client has included sharing learning, providing development opportunities for people across the various organisations, and contributing to the objective of bringing about positive change within the industry. Several key roles traditionally filled by non-owner personal have been delivered by facility operator or LXRP personal because they fitted the best person for the job.

The robustness of the Blueprint is unusual in projects because they are usually based on short-term thinking. One of the values is to 'empower people' by supporting the growth and development opportunities available. Decision-making is devolved; hence, it is important that

> people are empowered to get on with making their own decisions at the lowest level possible rather than grinding an organisation to a halt. And I think that's what the Blueprint ... enables.

> (IV18)

Involvement in JCC subcommittees supports decision-making and strategic planning and combines with accelerated development programs to support collaborative working skills.

LXU Level Crossing University was intentionally created to reflect LXRP's commitment to developing great people and putting them at the centre of how LXRP operates. Valuing the People and Culture function and empowering LXU to design and deliver a suite of professional learning and development tools and resources for all LXRP employees and encompasses programs and experiences to grow capabilities and skills supporting the GREAT PEOPLE strategic objective through unleashing the potential of people.

It is unusual for projects to have the reach and impact required to impactfully contribute to the professional development of its people due to the start/stop nature of projects. The programmatic approach (Chapter 7) has enabled the strong and sustained pipeline of work and enabled this systematic approach to professional development, learning and employee experience to support high performance.

8.6   The Role of Employee Diversity

MOSAIC[4] LXRP was developed to address continuous improvement in diversity and inclusion initiatives at the project operational level. LXRP's aim to role model an employment change in the typically white, male construction industry (Clegg et al., 2023) is consistent with the contention that social responsibility should be incorporated into megaprojects (Alotaibi, Edum-Fotwe and Price, 2019) and bring about lasting socioeconomic transformation at the societal level (Daniel, 2023). LXRP recognised the power of inclusion and introduced the MOSAIC corporate program as a way of growing their Diversity, Equity and Inclusion strategy. A dedicated Industry Capability and Inclusion (ICI) team and Capability and Diversity teams mechanism were designed to lead enhanced social inclusion in the construction industry workforce. Its mission is to enable the industry to develop a sustainable, inclusive and diverse workforce whilst delivering some of the largest infrastructure projects in Victoria. By bringing together programs, pathways and initiatives to unlock social inclusion opportunities, LXRP recognised that they could improve the way they work and deliver projects by enhancing their workforce through fostering greater gender diversity, the inclusion of people from Victoria's First Nation, people with different abilities, and those from different socioeconomic and cultural backgrounds, including refugees.

The MOSAIC 'leading through diversity' program was designed to contribute to engaging and retaining women – with a particular focus on women in engineering. This unique program began by helping senior leaders gain insight into unconscious and structural biases and how they could become more inclusive leaders. Senior male sponsors were matched with female spon-sees, and the key outcomes included expanding the women's networks through the advocacy of their sponsors.

Engineering Pathway Industry Cadetship (EPIC) is another program implemented by LXRP. EPIC provides employment pathways to disadvantaged Victorians, mitigating skill shortages in entry-level engineering roles. This Industry-first program aims to bridge the gap faced by new Australians in matching their international qualifications to Australian workforce require-ments. EPIC is an 18-month cadetship for refugee and asylum seeker engineers that provides on-the-job training, support and mentoring while cadets complete a postgraduate Graduate Cer-tificate in Infrastructure Engineering Management qualification, through Swinburne University. The Graduate Certificate, combined with their international qualifications and the workplace experience gained through the program, aims to kick-start their careers in engineering in Vic-toria. The paid cadetship will provide valuable local work experience and access to industry networks – addressing the barriers often faced by new Australians in accessing professional employment.

Diversity has long been linked to superior organisational and team performance and has social value for the community. Working on project teams requires coordinating work distribution and reliance on task performance or interdependence. This occurs at the individual project and another level in an alliance setting. Everyone – across teams – depends on the performance of other teams and each team member. This is part of the 'all in it together' or 'sink or swim' mindset that comes with working in a collaborative environment. Rosenauer et al. (2016) found that higher levels of cultural intelligence led to team processes that enhanced the diversity climate of teams, leading to higher levels of interdependency. For successful nationally diverse teams, interaction between three elements was required: nationality diversity, task interdependence and culturally intelligent leaders. When these are all present, teams' diverse climates and performances improve. COVID related labour shortages exacerbated shortages and for some time limited the ability of organisa-tions to deliver projects. At the same time, the number of international immigrants, many with relevant experience and qualifications, has increased; however, issues relating to recognition of overseas qualifications can limit their employment in a range of roles. Research has explored a range of issues related to diversity and the construction industry, including the low representation of women in the construction workforce, low participation by indigenous people, and a lack of pathways for recent arrivals to gain recognition of overseas qualifications and experience (Choi, Shane and Chih, 2022). The LXRP has provided learning opportunities for those with overseas qualifications to accelerate their progress towards recognition of their qualifications.

LXRP has implemented a range of activities to increase the diversity of its workforce. This was confirmed by a 2017 report that the LXRP (then LXRA) had "developed and implemented innovative programmes to provide opportunities for disadvantaged people or long-term unem-ployed, such as Veterans in Construction and GEN44 (44 internships for university students from under-represented backgrounds)."[5] There continues to be a concentration on supporting these and other related programs within the LXRP.

IV17 stated that with organisational and project team diversity, "you get better outcomes if you've got a broader lens." Diversity for LXRP encompasses many areas, including people from diverse backgrounds and industry experiences. LXRP's environment is one where highly experienced and skilled people from a range of backgrounds are valued. This means not seeking out people only from traditional constructor, designer, or operator roles but "getting a mix of people" (IV17) from many of those different fields, resulting in people with exposure to each of those areas and more.

In line with objectives set out at national and state levels across Australia, employment and development of our indigenous population is an important area that could have been addressed by setting quotas. IV17 stated that Originally, LXRP did have: "Aboriginal engagement targets of $2^{1/}_{2}$%" (IV17). With a largely engineering-based staff, thoughts went to "if we have 100 employees that means … but they changed their thinking. Discussions went to what is our aim here?"

This evolved into a new purpose statement that recognised the aim of their aboriginal engagement initiatives was a social, not numerical, outcome. They realised that by focusing on the desired social outcome, they would achieve and exceed the numerically stated targets for the right reasons.

LXRP has been committed to providing employment and capability development for indigenous Australians. They have worked with the Managing Director of Melbourne construction firm Wamarra,

> who is a proud Wiradjuri man, to implement their social procurement policy. Wamarra has offered sustainable, permanent employment to more than 45 Indigenous Australians through the LXRP's social procurement policies. Indigenous employees are working across several of the projects with the LXRP.[6]

This demonstrates a strong commitment to diversity across a range of areas and has, in turn, provided learning for LXRP and their leadership. The Blueprint was modified to include Storytelling when LXRP leaders learned from their involvement in indigenous employment that demonstrated the value of stories in supporting the development of the culture required for collaborative working, stating that "Storytelling is at the core of the refreshed Strategic Blueprint" (LXRP Blueprint 2023).

8.7   Capability Frameworks

8.7.1   *Global examples*

LXRP's professional development program allows employees to develop new knowledge and skills and then strengthen or deepen them against clearly stated levels of development that link to career progression. Perhaps two comprehensive models, based on research, identify the competencies, experience and soft skills project managers require when working on collaborative project teams. Walker and Lloyd-Walker (2011) developed a taxonomy of skills and experience (hard skills) and characteristics (soft skills) required for collaborative project delivery. The taxonomy was developed based on research in Australia, New Zealand, the United States and Europe with experienced alliance and/or IPD project managers. First, three areas of hard skills are described: Technical skills and experience, Project management skills and experience, and Business skills and experience. These can be further broken down according to level of proficiency, as can the soft skills, which are shown in Table 8-2 below:

The soft skills and leadership characteristics are then broken down into seven levels of leadership attributes and described. Each leadership characteristic shown in Table 8-2 above is then described against the level of proficiency, from Novice through to Proficient Performer. Using the comprehensive descriptors, an individual alone or with their supervisor can identify their current level of proficiency and the attributes or characteristics they need to develop to move to the next level on the path to becoming a Proficient Performer. The levels of proficiency that can be applied to both the hard and soft skills are described next in Table 8-3.

Table 8-2 Soft-skill characteristics

Seven characteristics	Descriptors
Reflectiveness	Being a systems thinker, strategic think-aim-act vs. act-think-aim. Reflectiveness level is high, and knowing the context is crucial.
Pragmatism	Gets on with the job, is politically astute, and works within constraints. Interpreting and re-framing rules to context; how an action is justified as crucial.
Appreciativeness	Understanding the motivations and value proposition of all involved (EI). Judging the most effective response to teams and individuals about their value is the key to influencing others and being influenced by them.
Resilience	Adaptability, versatility, flexibility and persistence. Able to effectively learn from experience. The repertoire of skills and attributes that can be drawn upon is crucial. This is related to the absorptive capacity to learn and adapt. Attitude on how to deal with a crisis "next time" is critical.
Wisdom	Having opinions and advice that is valued, consistent and reliable that others instinctively refer to. To be effective, the key is to be influential based on providing sound advice and being respected for that advice or being an effective broker of wise advice. Judgment of the person brokering advice is crucial.
Spirit	Having the courage to challenge assumptions effectively. Being confident in the value of refining knowledge of context through questioning the status quo or assumed realities is vital to better understanding contexts.
Authenticity	Approachable, trustworthy, and open to ideas, collaboration, discussion, and new ways of thinking. To be an effective broker and "go-to" person, this person must be open-minded and be available when needed. They must be collaborative, have integrity, and be perceived as trustworthy.

(Based on Walker and Lloyd-Walker, 2015)

Individual capability development can be mapped by conducting an initial proficiency assessment and identifying areas requiring development. This may sometimes involve identifying new or different areas that need to be developed to perform a new role. For instance, the model can enable graduates to map their progress, undertaking formal development programs or work assignments to achieve higher levels, thus supporting career advancement.

For teams, capability levels and areas of expertise are further developed into 16 elements enabling, for instance, project partnering capability for an alliance project team to be measured and graphically depicted on a spider diagram (see Figure 8-2). Areas of strength and weakness are visible so that elements requiring attention to achieve the highest level of project partnering can be addressed. When moving to collaborative project delivery from traditional delivery forms, a team's strengths and weaknesses can be mapped and areas of development identified.

The example provided in Figure 8-2 concentrates on areas identified as vital for successful project partnering, specifically in collaborative working environments. Establishing a Joint Governance Structure and developing strengths in Integrated Risk Mitigation and Insurance could address the current deficiencies. This may also be addressed by bringing on to the project team a person with strengths in the areas, enabling all team members to learn from their expertise over time. Research has confirmed the benefits of co-location of collaborative project teams (Antillon et al., 2018). The spider diagram suggests that actions should be taken to increase the co-location of team members. Notably, the graphical representation of the current situation enables leadership to develop a plan to ensure all areas of importance to project partnering are raised to the required level for a high-performing project team.

Table 8-3 Proficiency levels

Level	Experience	Real-time action in context is driven by
Novice	Faces a given problem and situation for the first time	Instructions (training courses, PMBOK® Guide) Learning to recognise objective facts about and characteristics of the situation (models and definitions of project) Learned generalised rules for all similar situations on the basis of identified facts, thus context-independent (project management methodology, procedures) Evaluation of the performance of the skills on the basis of how well the learned rules are followed
Advanced beginner	Has gained some real-life experience	Learning to recognise relevant elements in relevant situations on the basis of their similarities with previous examples (e.g., awareness of a typology of projects) The awareness of the importance of the context of experience; thus making a choice about what are the key elements of the given situation, in addition to context-independent rules (learning from experience, limited reflection, PMBOK® Guide recommendations) Trial-and-error
Competent performer	Amount of experience increases and number of recognisable learned elements and facts becomes overwhelming	Learning from own experience and from others to prioritise elements of the situation Organising information by choosing a goal and a plan Dealing only with a set of key factors relevant to the goal and plan, thus simplifying the task and obtaining improved results Deliberation about the consequences of using own judgement in relation to the given goal and plan (simultaneous subjectivity and objectivity), the relationship of involvement between performer and environment The model of analytical, proficient performer: Elements-rules-goals-plans-decision Ability to think on one's feet (confidence, reflection, choice of action and risk taking)
Proficient performer	Away from cognitivist, analytical rationality (rules, principles and universal solutions) towards perceiving situations rapidly, intuitively, holistically, visually, bodily, relationally	The awareness of interpretation and judgement involved in such decision making, rather than logical information processing and analytical problem solving only Understanding of the situation on the basis of prior actions and experience, acts as deeply 'involved-in-the-world' manager/performer who already knows Reflective understanding and participation in power relations

(Based on Walker and Lloyd-Walker, 2015)

Moradi et al. (2021) identified competencies required for collaborative construction projects based on research conducted in Finland and Norway. Their profile of capabilities of project managers of collaborative construction projects was ranked according to their contribution to successful performance. Twenty-four competencies were grouped into nine competency areas and then ranked according to their contribution to success. They identified difficulties in improving once on the job. Five competency areas are shown below to demonstrate how this matrix may support selecting people for high-performing collaborative teams. What we see in

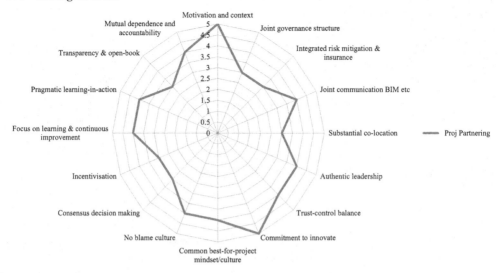

Figure 8-2 Visualisation of project partnering based on the model's 16 Elements

an extract from their model is the identification of the desirable competencies for collaborative project delivery success when selecting new staff. Those who demonstrate trustworthiness, stress tolerance initiative and optimism should be chosen, as attempting to develop these attributes after appointment is difficult.

Moradi et al. (2021) describe four levels of competency:

Key competencies: Central to personality for successful performance
Supportive competencies: Can be improved by training to contribute to successful performance
Hybrid competencies: Contribute to successful performance. Somewhat difficult to improve
Threshold competencies: Minimal requirement for the role

Table 8-4 Project manager competencies for collaborative construction projects

Competency area	Contribution to success	If capability difficult to improve identify in selection process.
Group capabilities	High	Low
Conflict management	High	Medium
Self-assessment		
Trustworthiness		
Stress tolerance	High	High
Initiative		
Optimism		
Management		
Leadership	Medium	Low
Production efficiency		
Relationship building		
Language proficiency	Low	Low
Communication		
Collaboration		
Leveraging diversity		

(Based on Moradi et al. 2021)

Moradi et al. (2021) further developed their competency model for project managers of collaborative construction projects to identify the knowledge, skills, attitudes and potential organisations should seek in recruits. Like Walker and Lloyd-Walker's taxonomy, Moradi et al.'s model can also measure professional development needs and achievement of higher-level capabilities related to potential promotion. Data for this expanded model also came from Norway (Moradi et al., 2021).

Based on the research from which their Matrix Model of Competencies was developed, elements of creative tension were also identified and described by Moradi et al. (2021), e.g., relationship building, collaboration, emotional awareness, etc. These can also be ranked and displayed in a spider diagram format against target levels, similar to that shown in Figure 8-2, where project partnering is the element being measured. The result can be used for both capability frameworks to develop personal development plans, design capability development programs, or inform selection interviews and testing.

Findings from opposite ends of the world reveal the capabilities required for superior performance, confirming the importance of interpersonal skills accompanying the necessary technical competencies (Walker and Lloyd-Walker, 2011; Moradi et al., 2021).

8.7.2   An LXRP example

The bespoke LXRP capability framework is a key foundational tool to articulate the capabilities that fall into three clusters of Thinking, Relating and Interacting and Leadership together with the CARES values. The framework complements technical competency – as engineering excellence underpins it all. Technical excellence provides the challenge to change ability, control, and drive, essential if LXRP is the informed client it strategically sets out to be.

LXRP has developed a sophisticated development framework with the four levels of proficiency expected by LXRP people as they evolve through their career journey.

These capabilities are developed through a learning and development framework that targets the individual's development journey. Individuals connect to their team's broader strategic elements of the 5-Greats through their achievement plan associated with their team's plan. Displayed as a one-page plan, it is easy for individuals to see how their development links through their team to connect and align with the organisational Blueprint plan.

8.7.3   Comparing capability frameworks

The three capability frameworks reveal common areas. All contain elements supporting high levels of performance in collaborative working environments. They also lay out a clear career progression path. Organisations must develop a framework that promotes achieving *their* mission and vision. LXRP's framework ensures a high-performing organisation by creating

Table 8-5 LXRP values

Thinking	Relating and interacting	Leadership	Values (CARES)
Continuous improvement	Influencing and	Visionary and inspiring	Creativity
Strategic and commercial mindset	negotiating	Build high performing	Accountability
Planning and prioritising	Working collaboratively	teams	Relationships
Problem solving and decision	Emotional intelligence	Develops others	Empowerment
making	Effective communication	Adaptability	Safety
Digital and data fluency	Building inclusivity		

(LXRP, 2024)

Table 8-6 Competency maturity level

Foundational	Applied	Accomplished	Leading
Demonstrates a broad understanding of the capability and its importance in the workplace Uses curiosity to drive learning and awareness Is open to new ideas and seeks guidance	Demonstrates the capability consistently in relevant settings Independent but seeks guidance and challenges own thinking when required Encourages and aids others that to improve, driving a learning culture	Uses capability actively to drive outcomes Performs capability to solve problems and non routine issues Challenges norms Identified and implements improvements Develops opportunities to improve capabilities within workforce and coaches others	Demonstrates the capability at a highly comprehensive level Provides strategic oversight and direction on organisational goal to drive long term planning and solutions Provides and enables workforce upskilling opportunities

(LXRP, 2024)

Table 8-7 CARES

Thinking	Relating and interacting	Leadership	CARES values and culture
Continuous improvement	Fluencing and negotiating	Visionary and inspiring	Creativity
Strategic and commercial mindset	Working collaboratively	Builds high-performance teams	Accountability
Planning and prioritising	EI	Develops others	Relationships
Problem solving and decision making	Effective communication	Adaptability	Empowerment
Digital and data fluency	Building inclusivity		Safety

(LXRP, 2024)

high-performing teams. This is done by supporting individual employees to develop high personal performance. LXRP's framework has been implemented program-wide for consistency across all projects within the program. It contains elements that research has identified are vital for developing high-performing teams.

8.8   Conclusions

Information gathered through interviews with LXRP staff revealed a comprehensive and all-encompassing set of coordinated activities linked to the Blueprint. These activities are driven by a leadership team that lives the organisation's espoused values, which has led to developing of a workplace culture that has driven the achievement of the 5-Greats. They have gone beyond delivering projects against traditional success criteria to ensuring they leave a positive legacy. The way that Victorians live, work, and travel has already been transformed and will be further improved as the pipeline of projects proceeds.

Has LXRP delivered the level-crossing removals, and is it providing a positive legacy?

Chapter 12 discusses performance and legacy. In 2021, the then Minister for Transport Infrastructure, Jacinta Allan, said, "Our level crossing removal project isn't just getting rid of those dangerous and congested boom gates – we're delivering new train stations, more open space for communities and new pedestrian infrastructure."[7]

The program continues to deliver projects, with the 75th crossing removal occurring on 25 March 2024 and the rail network reporting a reduction "from an average of 21 incidents a month in 2017 to just five incidents a month in 2023" as a result of the level crossing removals.[8]

Additionally, the Infrastructure Sustainability Council awarded John Holland, a company involved in the program of projects that achieved "the highest sustainability rating and green star rating for works completed at Moreland and Coburg stations in Melbourne."[9]

It has also been reported that the benefits the Melbourne public has received from the removal of level crossings include the creation "of linear parks and connected quiet streets for safer walking and cycling," and the overall program of crossing removals has led to increased rail services and had a positive impact on the city's transport system in general.[10]

These outcomes have been achieved through having a strong, effectively focused leadership team that created the Blueprint, the program alliance's strategic plan and the values that create and sustain their culture were established. This is evidenced by several key staff's career progression, for example the CEO being progressed to lead the Victorian Infrastructure Delivery Authority (VIDA). LXRP's experience provides an example from which others can learn and adapt the learning to suit their organisation. There is also a legacy for the transport infrastructure industry. It has been transformed and will continue due to the initiatives LXRP implemented. These will benefit the industry, organisations that operate within it, commuters and the Victorian public in general.

Notes

1 Right people fit the profile required of high performing teams.
2 The talent recruitment organisation see https://talent.seek.com.au/
3 Fair Work is an Australian Government Act see www.fairwork.gov.au/
4 www.mosaicco.com/Diversity-and-Inclusion
5 www.globalrailwayreview. com/article/63582/melbournes-huge-level-crossing-removal-project-saving-lives-improving-journeys/
6 https://bigbuild.vic.gov.au/news/level-crossing-removal-project/supporting-indigenous-employment-at-lxrp
7 www.globalrailwayreview.com/article/123367/melbourne-level-crossing-removal-continues/
8 www.premier.vic.gov.au/75-dangerous-and-congested-level-crossings-gone-good
9 https://infrastructuremagazine.com.au/2022/06/09/level-crossing-removal-project-receives-highest-sustainability-rating/
10 https://msd.unimelb.edu.au/research/projects/completed/level-crossing-removals-learning-from-melbournes-experience

References

Agyekum, K., Kissi, E., Danku, J.C., Ampratwum, G. and Amegatsey, G.S. (2020). "Factors driving the career progression of construction project managers." *Journal of Engineering, Design and Technology.* 18 (6): 1773–1791.

Ahiaga-Dagbui Dominic, D., Tokede, O., Morrison, J. and Chirnside, A. (2020). "Building high-performing and integrated project teams." *Engineering Construction & Architectural Management.* 27 (10): 3341–3361.

Alotaibi, A., Edum-Fotwe, F. and Price, A.D.F. (2019). "Identification of social responsibility factors within mega construction projects." *World Academy of Science, Engineering and Technology International Journal of Economics and Management Engineering.* 3 (1): 14–27.

Antillon, E.I., Garvin, M.J., Molenaar, K.R. and Javernick-Will, A. (2018). "Influence of interorganizational coordination on lifecycle design decision making: Comparative case study of public–private partnership highway projects." *Journal of Management in Engineering.* 34 (5): 05018007.

Astakhova, M.N. (2016). "Explaining the effects of perceived person-supervisor fit and person-organization fit on organizational commitment in the U.S. and Japan." *Journal of Business Research.* 69 (2): 956–963.

Avença, I., Domingues, L. and Carvalho, H. (2023). "Project Managers soft skills influence in knowledge sharing." *Procedia Computer Science.* 219: 1705–1712.

Avolio, B.J. and Gardner, W.L. (2005). "Authentic leadership development: Getting to the root of positive forms of leadership." *The Leadership Quarterly.* 16 (3): 315–338.

Benmira, S. and Agboola, M. (2021). "Evolution of leadership theory." *BMJ leader.* 5 (1): 3–5.

Bolzan de Rezende, L., Blackwell, P., Denicol, J. and Guillaumon, S. (2021). "Main competencies to manage complex defence projects." *Project Leadership and Society.* 2: 100014.

Borg, J. and Scott-Young, C.M. (2022). "Contributing factors to turnover intentions of early career project management professionals in construction." *Construction Management and Economics.* 40 (10): 835–853.

Bouraoui, K., Bensemmane, S. and Ohana, M. (2020). "Corporate social responsibility and employees' affective commitment: A moderated mediation study." *Sustainability.* 12 (14): 5833.

Buvik, M.P. and Rolfsen, M. (2015). "Prior ties and trust development in project teams – A case study from the construction industry." *International Journal of Project Management.* 33 (7): 1484–1494.

Castro, M., Barcaui, A., Bahli, B. and Figueiredo, R. (2022). "Do the project manager's soft skills matter? Impacts of the project manager's emotional intelligence, trustworthiness, and job satisfaction on project success." *Administrative Sciences.* 12 (4): 1–16.

Chalker, M. and Loosemore, M. (2016). "Trust and productivity in Australian construction projects: A subcontractor perspective." *Engineering, Construction and Architectural Management.* 23 (2): 192–210.

Choi, J.O., Shane, J.S. and Chih, Y.-Y. (2022). "Diversity and inclusion in the engineering-construction industry." *American Society of Civil Engineers.* 38 (2): 02021002.

Chowdhury, M.S., Yun, J. and Kang, D. (2021). "Towards sustainable corporate attraction: the mediating and moderating mechanism of person-organization fit." *Sustainability.* 13 (21): 11998.

Clegg, S.R., Loosemore, M., Walker, D.H.T., van Marrewijk, A.H. and Sankaran, S. (2023). "Construction cultures: Sources, Signs and solutions of toxicity." In *Construction Project Organising*, Addyman, S. and H. Smyth (eds), pp. 3–16. Chichester, West Sussex, John Wiley & Sons.

Cooke-Davies, T.J., Crawford, L., H. and Lechler, T., G. (2009). "Project management systems: Moving project management from an operational to a strategic discipline." *Project Management Journal.* 40 (1): 110–123.

Daniel, P.A. (2023). "Modeling relationships of projects and operations: Toward a dynamic framework of performance." In *Research Handbook on Project Performance*, Anantatmula, V.S. and C. Iyyunni (eds), pp. 39–53. Northampton, Edward Elgar Publishing.

Denicol, J. and Davies, A. (2022). "The megaproject-based firm: Building programme management capability to deliver megaprojects." *International Journal of Project Management.* 40 (5): 505–516.

DeVaro, J., Li, R. and Brookshire, D. (2007). "Analysing the job characteristics model: New support from a cross-section of establishments." *International Journal of Human Resource Management.* 18 (6): 986–1003.

Ellis, F.Y.A., Amos-Abanyie, S., Kwofie, T.E., Amponsah-Kwatiah, K., Afranie, I. and Aigbavboa, C.O. (2022). "Contribution of person-team fit parameters to teamwork effectiveness in construction project teams." *International Journal of Managing Projects in Business.* 15 (6): 983–1002.

Fareed, M.Z., Su, Q. and Naqvi, N.A. (2022). "The impact of emotional intelligence, managerial intelligence, and transformational leadership on multidimensional public project success." *IEEE Engineering Management Review.* 50 (4): 111–126.

Ford, R.C., Piccolo, R.F. and Ford, L.R. (2017). "Strategies for building effective virtual teams: Trust is key." *Business Horizons.* 60 (1): 25–34.

Hackman, J.R. and Oldham, G.R. (1976). "Motivation through the design of work: test of a theory." *Organizational Behavior and Human Performance.* 16 (2): 250–279.

Hajarolasvadi, H. and Shahhosseini, V. (2022a). "Assignment of engineers to constructions project teams based on Person-Team Fit." *International Journal of Construction Management*. 22 (15): 2895–2904.

Hajarolasvadi, H. and Shahhosseini, V. (2022b). "A system-dynamic model for evaluating the effect of person-team fit on project performance." *Journal of Construction Engineering and Management*. 148 (11): 04022126.

Hall, M.L., Meyer, C.K. and Clapham, M.M. (2023). "Leadership theories and styles understood and synthesized." *Journal of Business and Behavioral Sciences*. 35 (3): 93–102.

Hersey, P. and Blanchard, K.H. (1982). "Grid® principles and situationalism: Both! A response to Blake and Mouton." *Group & Organization Management*. 7 (2): 207–210.

Hwang, B.-G., Zhao, X. and Lim, J. (2020). "Job satisfaction of project managers in green construction projects: Influencing factors and improvement strategies." *Engineering, Construction, and Architectural Management*. 27 (1): 205–226.

Kristof-Brown, A., Zimmerman, R.D. and Johnson, E.R. (2005). "Consequences of individual's fit at work: A meta analysis of person-job, person-organization, person-group, and person-supervisor fit " *Personnel Psychology*. 58: 281–342.

Kulik, C.T., Oldham, G.R. and Hackman, J.R. (1987). "Work design as an approach to person-environment fit." *Journal of Vocational Behavior*. 31 (3): 278–296.

Lievens, F. (2007). "Employer branding in the Belgian Army: The importance of instrumental and symbolic beliefs for potential applicants, actual applicants, and military employees." *Human Resource Management*. 46 (1): 51–69.

Ling, F.Y.Y., Ning, Y., Chang, Y.H. and Zhang, Z. (2018). "Human resource management practices to improve project managers' job satisfaction." *Engineering, Construction, and Architectural Management*. 25 (5): 654–669.

Love, P., Edwards, D. and Wood, E. (2011). "Loosening the Gordian knot: The role of emotional intelligence in construction." *Engineering, Construction and Architectural Management*. 18 (1): 50–65.

LXRP (2023). *The LXRP Blueprint*, Melbourne, Level Crossing Removal Project.

LXRP (2024). *LXRP Model – Delivering Great Outcomes* (Draft), Melbourne, Level Crossing Removal Project.

Marnewick, C., Silvius, G. and Schipper, R. (2019). "Exploring patterns of sustainability stimuli of project managers." *Sustainability*. 11 (18): 5016.

Modranský, R. and Lajčin, D. (2021). "Correlation of motivation and value orientation of project managers." *Emerging Science Journal*. 5 (3): 350–366.

Moradi, S., Kähkönen, K., Klakegg, O.J. and Aaltonen, K. (2021). "A competency model for the selection and performance improvement of project managers in collaborative construction projects: Behavioral studies in Norway and Finland." *Buildings*. 11 (4): 11010004.

Nevstad, K., Karlsen, A.S.T., Aarseth, W.K. and Andersen, B. (2022). "How a project alliance influences project performance compared to traditional project practice: Findings from a case study in the Norwegian oil and gas industry.." *The Journal of Modern Project Management*. 9 (3): 139–153.

Piccolo, R.F. and Colquitt, J.A. (2006). "Transformational leadership and job behaviors: The mediating role of core job characteristics." *Academy of Management Journal*. 49 (2): 327–340.

Rana, A.G. and Shuja, A. (2022). "Influence of leadership competencies on transport infrastructure projects' success: A mediated moderation through innovative-work-behavior and the project type." *Pakistan Journal of Commerce and Social Sciences*. 16 (1): 1–33.

Rosenauer, D., Homan, A.C., Horstmeier, C.A.L. and Voelpel, S.C. (2016). "Managing nationality diversity: The interactive effect of leaders' cultural intelligence and task interdependence." *British Journal of Management*. 27 (3): 628–645.

Sergeeva, N. and Kortantamer, D. (2021). "Enriching the concept of authentic leadership in project-based organisations through the lens of life-stories and self-identities." *International Journal of Project Management*. 39 (7): 815–825.

Sever, S. and Malbašić, I. (2019). Managing employee motivation with the job characteristics model. *DIEM: Dubrovnik International Economic Meeting*, Sveučilište u Dubrovniku. 4: 55–63

Siruri, M.M. and Cheche, S. (2021). "Revisiting the Hackman and Oldham job characteristics model and Herzberg's two factor theory: Propositions on how to make job enrichment effective in today's organizations." *European Journal of Business and Management Research*. 6 (2): 162–167.

Soltis, S.M., Dineen, B.R. and Wolfson, M.A. (2023). "Contextualizing social networks: The role of person-organization fit in the network-job performance relationship." *Human Resource Management*. 62 (4): 445–460.

Subramanian, S., Billsberry, J. and Barrett, M. (2023). "A bibliometric analysis of person-organization fit research: significant features and contemporary trends." *Management Review Quarterly*. 73 (4): 1971–1999.

Sun, H., Dai, Y.-Y., Zhang, C., Lee, R., Jeon, S.-S. and Chu, J.-H. (2022). "The impacts of conditions and person-organization fit on alliances performance: And the moderating role of intermediary." *PloS One*. 17 (12): e0275863–e0275863.

Tang, Y., Shao, Y.-F., Chen, Y.-J. and Ma, Y. (2021). "How to keep sustainable development between enterprises and employees? Evaluating the impact of person-organization fit and person-job fit on innovative behavior." *Frontiers in Psychology*. 12: 653534–653534.

Thommes, M.S., Uitdewilligen, S., Rico, R. and Waller, M.J. (2024). "Switching gears: How teams co-construct adaptive leadership style transitions in dynamic contexts." *Organizational Psychology Review*. 0: 1–29.

Tkalac Verčič, A. and Sinčić Ćorić, D. (2018). "The relationship between reputation, employer branding and corporate social responsibility." *Public Relations Review*. 44 (4): 444–452.

Victorian State Government. (2017). *Level Crossing Removal Project – Program Business Case*, Government V. Melbourne.

Vleugels, W., Verbruggen, M., De Cooman, R. and Billsberry, J. (2023). "A systematic review of temporal person-environment fit research: Trends, developments, obstacles, and opportunities for future research." *Journal of Organizational Behavior*. 44 (2): 376–398.

Walker, D.H.T. and Lloyd-Walker, B.M. (2011). *Profiling Professional Excellence in Alliance Management Summary Study Report*, AAA Sydney.

Walker, D.H.T. and Lloyd-Walker, B.M. (2015) *Collaborative Project Procurement Arrangements*, Newtown Square, PA, Project Management Institute.

Yusliza, M.Y., Noor Faezah, J., Ali, N., Mohamad Noor, N.M., Ramayah, T., Tanveer, M.I. and Fawehinmi, O. (2021). "Effects of supportive work environment on employee retention: the mediating role of person–organisation fit." *Industrial and Commercial Training*. 53 (3): 201–216.

Zheng, J., Wu, G., Xie, H. and Li, H. (2019). "Leadership, organizational culture, and innovative behavior in construction projects: The perspective of behavior-value congruence." *International Journal of Managing Projects in Business*. 12 (4): 888–918.

9 LXRP Continuous Improvement and Innovation Diffusion

Peter E.D. Love, Andrew Davies, Derek H.T. Walker and Mark Betts

9.1 Introduction

How does the LXRP develop its continuous improvement and innovation diffusion strategy across projects in the program?

This chapter provides an appreciation of appropriate strategic continuous improvement practices and the adoption rationale. We take a specific program/project alliancing perspective of an alliance project team (APT) that is integrated (Chapter 5) and highly collaborative (Chapter 6). We also explain how non-alliance supply chain project participants fitted into the LXRP continuous improvement strategy. This chapter links to Chapter 10, which explores current digital innovation and compares it with best practices in the infrastructure delivery sector. Figure 9-1 illustrates the chapter's structure.

We begin with a brief theoretical introduction to continuous improvement and innovation diffusion, followed by the LXRP case study discussion to introduce how the workplace culture (Chapter 7 and 8) supported and encouraged continuous improvement across the program of projects. At this juncture, it is worth pointing out the meaning of culture and climate here. Culture

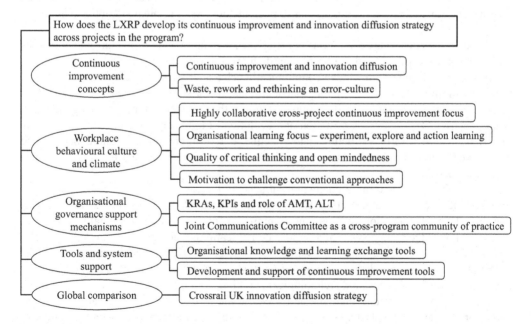

Figure 9-1 Chapter structure

DOI: 10.1201/9781003389170-9

involves the pattern of thought, emotion, and action, which can shape how an organisation responds to problems (Westrum, 2014) and climate is "the shared meaning organisational members attach to the events, policies, practices, and procedures they experience and the behaviours they see being rewarded, supported, and expected" (Ehrhart, Schneider and Macey, 2014, p. 69). Support mechanisms ensuring continuous improvement team/participant norms being reinforced by organisational governance are then discussed. Detail description of how continuous improvement was supported by LXRP-specific tools and systems to ensure the strategy follows. Finally, we offer a brief comparison with the innovation and continuous improvement approach adopted by a similar megaproject undertaken under comparable relational project delivery forms.

9.2   Continuous Improvement Concepts

The early 2000s witnessed growing interest in construction management and engineering system innovation (Slaughter, 2000); innovation diffusion (Peansupap, 2004; Peansupap and Walker, 2005b); complex project innovation contexts (Davies and Brady, 2000; Brady and Hobday, 2011); and how these might contribute to continuous improvement, responding to organisational amnesia (Othman and Hashim, 2004) through ineffectively enabling knowledge diffusion. Seminal innovation diffusion theory by Rogers (1983) was applied to investigate its application to the construction industry (Peansupap and Walker, 2005a) and, more specifically, in alliancing projects (Love et al., 2016).

Szulanski's (1996; 2003) knowledge management research examined how knowledge and learning moves in organisations, how it is transferred, and how the return on knowledge capital can be maximised to create a competitive advantage. Organisations develop processes and routines to transfer knowledge, which can be difficult due to its 'stickiness'. Knowledge stickiness is the inability or unwillingness to transfer knowledge. Tacit knowledge, which is unexplainable and built into our bodily automatic response systems, is difficult to transfer and codify due to knowledge stickiness. However, the success of knowledge transfer is contingent not only upon various kinds of contextual variables but also upon the process of implementation and internalisation (Li and Hsieh, 2009).

'Sticky knowledge' shares common characteristics with innovation diffusion and continuous improvement concepts. Szulanski (1996) identified several contributing 'stickiness' factors arising from sender-receiver knowledge communication characteristics and perceived validity and reliability of knowledge being offered. One additional factor stands out – a *fertile* versus *barren* organisational context. A fertile context *facilitates* the development of knowledge transfer, while a barren one *hinders* the gestation and evolution of knowledge transfer. Differentiating factors include their formal structures and systems, sources of coordination and expertise, and how these frame behaviours. Thus, reducing 'sticky knowledge' is crucial for continuous improvement activities.

Figure 9-2 illustrates three fundamental influencers for continuous improvement activities through 'sticky knowledge' reduction. We explain how this works using theoretical mechanisms that describe how innovation and continuous improvement may be organisationally driven.

Early knowledge management and organisational learning theorists (Nonaka and Takeuchi, 1995; Wiig, 1997; Crossan, Lane and White, 1999) perceived organisation learning as a '4-Is' process that explains knowledge and innovation. It starts as individual-level *intuition* about potential new ideas, processes, and technological adoption. At a group, team or workplace level, ideas are *interpreted* through dialogue and socialisation of observations to create hypotheses and theories. These concepts may then be *integrated* into organisational routines and *institutionalised* (Scott, 2014); people use their cultural-cognitive capabilities to interpret and reconcile

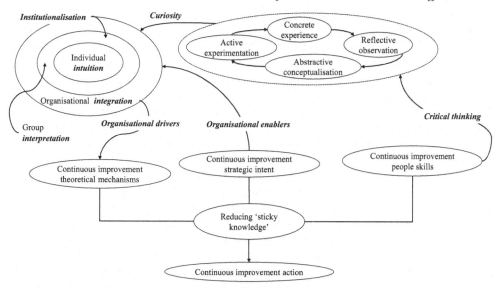

Figure 9-2 Continuous improvement through reducing sticky knowledge

the workplace culture with organisationally-prescribed regulatory framework mechanisms (routines, rules, heuristics, and workplace agreement-contract requirements). Knowledge flows (fluidly/stickily) across organisations moving from individuals to groups and then to organisations, depending upon flow enablement.

Kolb's (1984) learning theories help explain how a cycle of curiosity may be created to create knowledge. People experiment with a new (improved) approach and experience a consequence. They reflect upon that result to make an abstraction leading to hypotheses that may be tested in a cyclical loop or socialised through group level dialogue. Accepted ideas are integrated into the organisational management and governance mechanisms and become institutionalised at the organisational level. Järvinen and Poikela (2001) argue that the 4-Is and individual-group-organisation knowledge diffusion directional processes concepts explain how the application of power and influence impacts the extent of socialising organisational learning (Lawrence et al., 2005).

From a learning and knowledge diffusion perspective, we know that experience, reflection, sharing and exploring perceptions and debating through dialogue are all important continuous improvement strategy ingredients (Senge, 1990). Strategy be then rafted with organisational support, technology platforms and collaborative people behaviours. Critical thinking, and motivation to critically analyse, are pivotal for continuous improvement to envisage improvement ideas (McCuen, 2023). People must possess the requisite knowledge, skills, attributes, and experience and be adequately supported with education and training to abstraction reflection upon experience and conceptualise plausible theories and explanations to promote continuous improvements (Chapter 7). Figure 9-2 illustrates the link between the people component of continuous improvement to organisational learning theory.

Figure 9-2 also presents the vital organisational learning component supported by the organisation and its governance mechanisms. Chapter 3 explains the LXRP's governance, including innovation and improvement key result areas and the organisational enablers that institutionalise continuous improvement strategy (Scott, 2014). Alliancing supports a fertile knowledge and innovation transfer environment through its Project Alliance Agreement (PAA) behavioural and mindset requirements. The LXRP has a conscious strategy for cross-inter-project knowledge transfer having built technology platforms, routines, and administrative structures to create, share and diffuse organisational learning across and between program alliance projects. A barrier to performance is waste (*Muda* 無駄 in Japanese), which can adversely influence the alliances' performance. The most significant waste impacting a project's costs, schedule and safety is rework.

Much has been written about continuous improvement since Imai's (1986) seminal book *Kaizen*. Early theory evolved from quality control and quality management literature strongly focused on continuous improvement through waste reduction/elimination. Non-value activities in construction projects have been widely examined under the lenses of: lean construction (Ballard and Howell, 1997; Koskela, 2000; Love et al., 2021; Lehtovaara, Seppänen and Peltokorpi, 2022; Fireman et al., 2023); material and other physical resources; managerial and administrative effort; system inefficiency; rework; productivity loss through poor safety; over-design 'gold-plating'; knowledge retention; and inefficient transport/storage. Lean construction theory and principles have underpinned development of IPD in North America

Continuous improvement can be driven by adopting and adapting new ideas, diffusing them throughout the organisation, and reducing waste and rework. While waste reduction represents an important part of the construction industry's continuous improvement focus, it is also vital that systems be developed to facilitate lessons learned to be entrenched in corporate memory and diffused across organisations and projects. This recognition led to related literature strands focusing on knowledge management and organisational learning from a continuous improvement perspective (e.g., see Love, Fong and Irani, 2005).

The ability of an organisation to learn is reliant on continuous improvement. Setting learning goals, engaging in dialogue, possessing a willingness and openness to corrections and feedback, and reflecting on errors are needed to enact an effective continuous improvement strategy. However, in construction, reflecting on errors and how they result in rework has not been a practice traditionally given required credence by organisations. The type of error culture within an organisation and a project sets the tone for how people respond, share information, and deal with errors and their consequences. So, the rework context has been typically deemed as 'uncomfortable knowledge' (i.e., denied, dismissed, deflected, or diverted) or a 'zemblanity' (i.e., being unpleasant yet unsurprised) by construction organisations – the presence of error and rework is associated with managerial incompetence or wrongdoing (Love et al., 2019). An error-prevention culture is the norm in construction. Here, construction organisation endeavour to prevent errors in their projects due to their negative consequences; it is natural to take this position. When an organisation's culture focuses on error prevention, people work hard to ensure they do not occur and even worry about committing them. People may become stressed as soon as problems arise, fearing they will be blamed and reprimanded. Consequently, people may hide their errors or unreluctantly report them. For example, senior management often considers non-conformances a sign of a poorly performing project. Thus, non-conformances are often not recorded or dealt with through informal means, hindering continuous improvement. Knowing and sharing error experiences is critical for learning and stimulating innovation.

Errors in the workplace cannot be avoided, regardless of the routines and activities to be performed. Rather than trying to prevent errors, we need to accept that errors happen. But the inability to accept errors happen has contributed to construction organisational ineptness in learning and innovating to improve their productivity and performance. Low-profit margins and clients' reliance on traditional procurement practices hinders their progression, acting like a 'windshield' deflecting risk onto construction organisations. This situation has led the Australian Constructors Association (2017), and many others worldwide (Latham, 1994; Walker and Lloyd-Walker, 2015; Walker, Love and Matthews, 2023) to call for greater use of collaborative delivery strategies to trigger the cultural and behavioural change needed to bring out improved productivity and performance outcomes in projects.

Alliancing facilitates collaboration and cooperation between contracted parties of the owner participant (i.e., non-owner participants, NOPs) in projects. Risk (i.e., positive opportunities and negative threats) are shared through an incentive scheme between NOPs (i.e., risk and reward through a gain-share/pain-share regime). Key features of an alliance include:

- an integrated team;
- a joint governance framework; and
- decisions are made on a 'best-for-project' basis with a no-blame philosophy where there is an implicit assumption that 'errors happen'.

These key features have been found to provide the building blocks for establishing an error management culture in one of the LXRP's program alliances (Love and Matthews, 2022b; Love and Matthews, 2022c; Love and Matthews, 2022a). The organisational error management practices enacted in the alliance examined were: (1) communication about errors; (2) sharing error knowledge; (3) helping in error situations; (4) quick error detection and damage control; (5) analysing errors; (6) coordinating error handling; and (7) effective error handling. The enactment of these practices enriched the flow of information within the alliance and its projects to the point where people actively sought it out to help them with their problem-solving, enabling them to discover and understand why and how errors materialise, facilitating dialogue to manifest and learning to be engendered, which contributed to its efforts to mitigate rework. Communication about errors is the most important error management practice as it facilitates information flow and acts as the impetus from which others emerge (van Dyck et al., 2005). According to Westrum (2014, p. 58), "by examining the culture of information flow, we can get an idea of how well people in the organisation are cooperating and also, how effective their work is likely to be" in ensuring quality and safe operations.

In summary, we now better understand continuous improvement from a systems perspective. To improve productivity, we need to address waste and rework, introduce new materials, work and administrative processes and be able to ensure that knowledge gained about these aspects should be effectively shared and diffused throughout organisations and across industry sectors. Achieving continuous improvement requires the following:

1. Strategic intent, specifically articulated through key result areas (KRAs) managed through governance mechanisms.
2. A supportive continuous improvement focussed workplace culture; and
3. Tools, processes, and technology support systems.

The LXRP specifically stressed continuous improvement in its program of project delivery strategy and addressed it through its governance mechanism, as discussed in Chapter 3.

9.3   Workplace Behavioural Culture and Climate

We discussed workplace culture and climate in Chapters 6 and 7. Schneider et al. (2013) argue that workplace culture is fuelled by prescriptive norms about expectations but that climate is about how strongly they are understood and adhered to. This section explores how the LXRP, through its strategy and ensuring that strategy was applied in practice, shaped, and guided the evolution of a supportive culture and climate to support and enshrine continuous improvement.

Key behaviours that support effective continuous improvement relate to attitudes and identity linked to the Theory of Planned Behaviour (TPB) (Ajzen, 1991) (Chapter 6). Attitudes prevalent in alliancing are anchored in pragmatism and practicality with the benefit of multiple team participant perspectives enabling greater knowledge about potential and likely consequences to planned action due to the multi-disciplinary character of alliance teams. This can be argued to lower the risk associated with experimentation. APT participants are protected by the PAA for no-blame workplace environment requirements resulting in increased critical thinking that questions assumptions and explores unconventional options (Lloyd-Walker, Walker and Mills, 2014). Thus, evidence of continual experiments based on continuous improvement and learning conducted by teams that identify as *a single team accountable and responsible for the whole project's success*, indicates a workplace culture and climate that supports an open-minded organisational learning mindset. LXRP's Joint Coordination Committee (JCC) (Chapters 3 and 9) also demonstrates an innovative program-wide attitude towards these attitudes and planned behaviours. The PAA's incentive arrangements promote rewarding innovation sharing.

Table 9-1 Workplace culture and climate illustrations

Focus on:	Interviewee quote	Comments
Highly collaborative cross-project continuous improvement.	"We are incentivised to share any innovations, particularly technology, we identify with the JCC and other alliances. We've got a continuous improvement strategy that is looking at how we can minimise waste – it's called the War on Waste" (Engineering Manager)	This is a good example of 'desorptive capacity', an additional dynamic capability that enables the safe transfer of knowledge from an alliance to others within the LXRP.
Organisational learning focus – experiment, explore, and action learning.	"We've begun to informally share experiences with our rework across projects and with sub/c. We've started rolling out a program, a kind of community of practice" (Continuous Improvement Lead)	Within an alliance, a community of practice was established as part of its continuous improvement strategy to initiate dialogue between each of its projects and subcontractors. The alliance had recognised the important role its subcontractors could play in stimulating continuous improvement.
Quality of critical thinking and open-mindedness	"We introduced the 5-whys into our non-conformance reporting to better understand why they occurred. There is also a suggestion for learnings" (Quality Manager)	The '5-whys' process was introduced and added to non-conformance reports to understand better why an event arose to help learning occur.
Motivation to challenge conventional approaches	"The language we use in this project is collaborative. The culture is nurturing. You know, there's a support network, and there is a sort of selflessness that comes through as a result of the culture. It is enjoyable to be in this culture where you know it's very supportive" (Design Manager)	The alliance strove to be inclusive, foster a psychologically safe workplace, and establish a culture where people could openly discuss problems, particularly errors and rework. When problems occurred, people helped each other to arrive at a solution.

We focus on four indicators of workplace behavioural culture and climate to illustrate how the LXRP implements its continuous improvement strategy from the human behaviour aspect. Table 9-1 identifies the four foci with illustrative quotes from interviewees after an alliance workshop and our comments to explain these foci.

What stands out from Table 9-1 is that, inherently, alliances' workplace culture is markedly different from traditional project delivery methods discussed in Chapter 4, such as Design and Construct (D&C) and design-bid-build (DBB). IV14 explained how alliance innovation learning is diffused, reinforced, and fine-tuned throughout the LXRP and NOP's home-based companies resulting from LXRP collaboration. The organisation's staff development policy is to cyclical rotation of half their staff from the LXRP to other alliance non-LXRP projects they were involved with delivering. This led to the construction organisation learning from its LXRP experience and from other projects, enabling them to further experiment with their knowledge gained across various contexts. Reinforcing this point IV14 stated:

we take that knowledge, and we share that knowledge onto our projects and the great part of it is we're able to evolve.... So hopefully our future projects will also evolve, and the other half are on are going to go to other projects that I've got on.... we are implementing things in Queensland that we've learned here in Victoria.

In addition, IV14 discussed how his home organisation is diffusing learning from LXRP governance mechanisms, such as the JCC KRA management process. This involves systematically creating and registering innovation and continuous improvement initiatives that incentivise innovation sharing. The JCC innovations register provides not only an innovation description but also video clips demonstrating the 'how to apply it' information. A case in point is digital engineering (DE), where the alliances are encouraged to learn from one another and share their experiences with technology adoption. Such inter-organisational learning is a Key Performance Indicator (KPI), offering opportunities for NOPs to also diffuse their knowledge throughout their home-based organisation:

(LXRP) KPI initiative process there where you come up with, you know, initiatives that you create and that they are willing to pay for and that you share with across the entire LX alliance.... I have the opportunity to see what [NOP B] is doing and go – Let's implement that on our project and get rewarded for that.

Chapter 6 discusses alliancing collaboration. Critically, the LXRP had a cross-project strategy implemented through impacted workplace norms and identities that encouraged behaviours that would enact motivated knowledge-sharing across project teams within the program.

The LXRP have embraced their 5-Cares that express the workplace value approach in their LXRP Blueprint 2023. These *creativity, accountability, relationships, empowerment,* and *safety* values express an important workplace culture guidance system supporting innovation and continuous improvement.

9.4   Organisational Governance Innovation Diffusion Support Mechanisms

The PAA specifies numerous governance provisions targeted to enable and support cross-project knowledge sharing. Chapter 3, Section 3.4 details the PAA governance arrangements. We highlight four relevant innovations.

9.4.1   KRAs-KPIs and PAA innovation incentivisation

The KRAs are specified in the PAA to focus the APT mind on how its performance will be assessed and treated through PAA incentivisation. Time and cost KRAs are relatively simple to operationalise into KPIs measuring KRA performance. The KRAs are specified in the PAA, and the Target Outturn Cost (TOC) process identifies and refines KPIs that are agreed upon by all APT participants.

Cost and time targets are set through the TOC process explained in Chapters 4 and 6, are straightforward, fixed, and variance between the project end cost and handover time results in APT gain or pain sharing. Innovation and continuous improvement KRAs and KPIs are more complicated. One KRA and associated KPI is for innovation, with IV02 commenting:

> there are two components to it: we're measured on an idea generation, so are we generating new ideas and ways of doing things? We're also measured on how well we implement those ideas and how well we share those ideas across the program, and others get the benefit of them. As an example, if we come up with a new way to pour concrete or something, and we reckon it saves us 2% per job, if we share that with the other alliances and they also save 2% per job, we're rewarded for that.

This quote illustrates how KRAs and KPIs are used as continuous improvement support mechanisms. The tool used to record innovations in a program-wide innovation database, explain and measure their benefits is discussed further in this chapter.

9.4.2   Alliance Leadership Team (ALT) and Alliance Management Team (AMT) role

These roles are governance support mechanisms enabling continuous improvement. The ALT role is vital. As explained in Chapter 3 and illustrated in Figure 3–4, The ALT comprise senior management individuals from each APT participating organisation who provide a kind of 'board of directors' role on alliance projects. Individual ALT members usually take an interest in a specific KRA and its associated KPI, and at the regular ALT meetings, the project alliance manager (AM) is held accountable for reporting performance on each KRA and KPI.

The AMT is responsible for gathering and reporting performance for each AMT and ALT meeting. The AM leads the AMT and comprises senior on-site representatives from each team participant. This role is like that of traditional project delivery site management teams. The AMT conducts its accountability and responsibility PAA requirements through its KRA and KPI reporting. Continuous improvement and innovation diffusion These KRAs and KPIs play a critical role in the AMT, and APT focus on formalised and incentivised innovation and continuous improvement.

9.4.3   The JCC's role

The JCC, described in Chapter 3 and illustrated in LXRP's Figure 3–6,[1] is an innovative governance body that effectively integrates projects and project participants across the program to facilitate innovation diffusion and continuous improvement. The JCC's focus is continuous improvement and innovation to deliver improved outcomes defined by their 5-Greats. The JCC has a critical facilitating role at three levels.

At the strategic level the Project Directors, their Alliance General Managers, and Alliance Management Teams develop strategic direction for 5-Greats' delivery. At the tactical level, a range of JCC sub-committees exist to oversee and manage the innovation KRA/KPI incentivisation system to encourage, monitor, review and approve innovation-improvement initiatives and to ensure that each initiative is documented and recorded in the LXRP custom internally developed extranet, available for all LXRP participants. Initiative documentation includes descriptions and 'how-to' short videos etc. This formalises the system and makes it accessible, transparent, and fair. At the operational level three, cross-project operational-level groups (working groups, discipline champions and numerous communities of practice) are established to meet at varying periods and are composed of different numbers of people. For example, the quality management sub-JCC meets annually because although quality management is of prime concern, it is not a specific KRA. However, it has a community of practice (COP) that holds spasmodic events across the LXRP program. Other sub-JCCs include DE, rework reduction, sustainability, safety, community engagement and many other specialities. They often meet using MS Teams software, have innovation submission evaluation sub-committees and interact across the program with alliance work packages (AWP) members and LXRP corporate and discipline specialists who review and approve innovation and continuous improvement performance KPIs for innovation initiation and adoption (Chapter 12, Figure 12–2). The scope and scale of the JCC concept suggest that this is a significant innovation.

It performs two main roles. First, it links the five AWPs to the LXRP program as a form of program steering committee at the project alliance director level. This group discusses and commits to innovation and continuous improvement initiatives. It may, for example, decide to implement a DE process or fund the development or implementation of technologies such as augmented reality (AR), building information modelling (BIM), uncrewed aerial vehicles/drones – UAVs) and robotics. At this level, it enables and fulfils the corporate program-wide innovation diffusion and continuous improvement agenda. AWP TOC money may be allocated by several or all Project Alliance Directors for this purpose, benefiting several or all AWPs. This top-level JCC coordinates, sponsors, and decides upon program training and development events. For example, program-wide internal seminars or professional development events to highlight and diffuse knowledge about rework mitigation strategies and practices described earlier in this chapter, or to explain how artificial intelligence (AI), AR and robotic devices can improve productivity.

The second JCC role is as a COP. Lave and Wenger (1991) identified and coined the term for groups of individuals who freely exchange ideas and insights about their particular interests. Seminal work on this concept of expertise exchange and development was attributed to Orr (1990) who undertook ethnographic research into a team of photocopy technicians and how they exchanged insights and converted their tacit problem-solving knowledge into explicit knowledge about the intricacies of photocopying machine problems and solutions.

Similarly, the JCC was formed to integrate APTs into the program and adopt a best-for-program program mindset while exchanging valuable knowledge about innovations each APT had developed and was willing to share. The KRA/KPI for innovation sharing outlined above provides an incentive to share ideas. The JCC is not restricted to project director levels but extends to various special interest groups within the LXRP across multiple disciplines and expertise/skills established for KRAs. The JCC also adds another support mechanism for the PAA behavioural requirements of respecting expertise and being open to assumptions being questioned. The JCC and PAA behavioural requirements support a natural tendency in many professionals to want to share insights. As IV01 stated:

Ultimately, we all work together; we've all worked around the industry a bit, sometimes we've worked in the same company, sometimes different companies; we're all friends and willing to share ideas with each other. And that's a funny thing because, in the contracts inside of our industry, we are all friends. We all work for the industry, no matter who pays our salary; we all work for the wider industry.

IV14 also observed how a JCC DE brought together AWP technical experts and enthusiastic users to explore and experiment with tools and processes, such as AR for visualising construction process planning and monitoring.

> We do have a what we call real-time camera ... it's a wearable piece of wearable tech.... Someone with the qualifications can go and do an inspection inside a pipe or confined space with this on, and that could be beamed into a Teams meeting ... we can hear them talking, and they can hear us conversing, and we can ask them to focus on areas and things like that.

Also, IV14 noted that many in the DE sub-JCC are passionate about how they can adapt and extend the use of AI for developing building information models to support time and cost planning and that they are working on an AWP testing its efficacy by cross-checking with a team of four cost estimators on the application's accuracy. They are using the feedback on cost estimating to improve AI reliability of. This experiment recognises the current limitation of AI and augments this with human intuition and expert feedback. The sub-JCC COP provides essential input and reflection to enable this radical innovation to develop further.

Another LXRP JCC support mechanism is the benchmarking and evaluation process. For example, the LXRP, on 4 November 2018, organised a 'Rework Symposium'[2] with representatives from each of the alliances to discuss how rework could be avoided in their projects. This symposium stimulated one alliance to initiate a research project to examine how rework could be benchmarked, as besides non-conformances, systems and processes did not exist to create a single point of truth to enable it to be undertaken (Matthews et al., 2022). Working is progressing to enable benchmarking of rework performance and other associated wastes to be conducted.

So, how does the JCC and its sub-JCC facilitate effective performance?

Figure 9-3 illustrates a simplified visualisation of how the JCC was supported. First, it was supported institutionally through governance arrangements. The top-level JCC comprising LXRP senior management had general oversight from each AWPs and the LXRP ownership corporate group, and it was part of their responsibility and accountability. As noted in Chapter 3, a key program strategy was enabling and encouraging innovation diffusion and continuous improvement. The PAA incentivised this through, for example, its KRAs, KPI 1.1, and 1.2. KPI 1.1 rewards submitting and gaining approval for innovations that could be demonstrated to improve

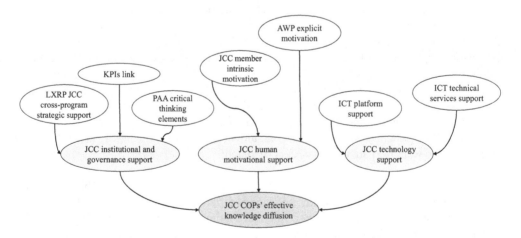

Figure 9-3 JCC support mechanisms

performance, and KPI 1.2 rewards taking up a previously registered innovation and adopting/adapting it. The PAA behavioural requirements specify critical thinking and an intellectually safe workplace. All APT participants are expected to challenge assumptions and speak freely about how work processes may be improved regardless of their place in the project hierarchy and based on their specific technical expertise. Thus, a systematic and institutionalised drive exists to support JCC COP activities.

The human element of a successful JCC COP entity cannot be overemphasised. Two motivational forces support the COPs. Drawing upon the Job Characteristics Model (JCM) developed by Hackman and Oldham (1976) and endorsed by Wegman et al. (2018). Walker and Lloyd-Walker (2020, pp. 230–236) explain how in alliancing, workers may be intrinsically motivated to collaborate and share insights and knowledge through work meaningfulness, interpretation of responsibility and interpreted feedback quality. The JCC COP groups support this motivational drive of those involved. Explicit motivation is marshalled through AWP incentivisation. Achieving KPI 1.1 and 1.2 psychologically rewards those who engage in JCC COP activities.

Third, the LXRP provides JCC technical support through its extranet platform and information communication technology (ICT) support staff team who develop, maintain, and support this facility.

In this holistic way, the JCC COPs become institutional components of the project governance arrangements and motivational cultural landscape that maintains enthusiasm and support for the JCC concept as a natural way to operate.

9.4.4 *Urban Design Advisory Panel*

A different workplace culture mechanism supporting innovation diffusion and continuous improvement is the role of community stakeholder engagement (Chapter 11). Specifically, engaging with the local community and working with local community members of the Urban Design Advisory Panel (UDAP) exposes LXRP APT members to innovative approaches to understanding what the local community values and how best to design and deliver projects that satisfy community expectations. These represent process and technology innovations. As IV09 stated:

> with the traders, we run what we call 'co-design', where we go to them and say, "We want to support you during the construction of this project; how can we best do that?" And so we partner with them to design a – usually a marketing/promotional type program, but how we can best support them? And then often there's elements of that that we can then hand over to them when we leave. So, if we build them a website, we build them social media channels, we build them a brand, for example, then we can often – we did this in Lilydale, where we handed over the 'Love Lilydale' brand at the end of the project. So there's sort of legacy items that we can hand over to the community at the end of that, and because we've partnered with them in the development of that, they've really got a commitment to continuing it, as well. And with all community engagement, it's really about how you – I think being really clear with the community upfront about what is and isn't negotiable, as part of the project.

This illustrates how the stakeholder engagement team have operated and continuously sought to learn from each episode of community engagement to calibrate their response and interactions with the community. This may represent an unexpected example of continuous improvement that can easily be overlooked on many infrastructure delivery projects. Yet, the LXRP have purposefully incorporated this project delivery continuous improvement aspect through its 'improving urban amenity' strategy (Chapter 3) and structuring the LXRP organisation (Figure 3–4) to support and facilitate innovation diffusion. Stakeholder engagement professionals are embedded in each project alliance team and regularly provide feedback and dialogue for feedforward through their JCC group providing the 4-Is and individual-group-organisation knowledge transfer theorised by Järvinen and Poikela (2001).

9.5   Tools and Support System Mechanisms

The LXRP have developed program-wide support systems for innovation diffusion and continuous improvement. Chapter 10 focuses on several DE tools that facilitate innovation and improvement. The use of UAVs, AR, visualisation technology, and robotics represent LXRP innovation and are part of what is happening but do not explain how or why this innovation diffusion is embedded to the extent that we observe on the LXRP.

The JCC provides a valuable face-to-face social knowledge exchange mechanism that can overcome many tacit knowledge-transfer problems because, in JCC forums, people can express themselves using more media than documentation and data; they can physically demonstrate and explain using emotions and all senses (including humour). However, the spoken word and physical demonstrations can be ephemeral. This is where a reliable and sophisticated electronic database system becomes invaluable to continuous improvement. The answer to the *why* question is that it has been embedded in the LXRP continuous improvement and innovation diffusion strategy in the belief that it will improve value for money and other performance criteria expressed in the KRAs. The question of *how* it has been achieved is answered in the above section. One key intriguing *how* question aspect is the mechanism and tools used to develop and support the *extranet* used as the main ubiquitous communication tool (for the LXRP entire team) that facilitates the enormous data repository and source of data and information underpinning the lessons-learned, knowledge and insights base, and a trove of future value that is impossible to fully appreciate.

We concentrate in this section on two system support mechanisms aspects. We reiterate the first aspect of strategy. How was the JCC enabling platform strategically envisaged and realised? Second, what examples can be provided to explain the scope and scale of the extranet facility so we can understand how this was developed and is operated?

At first, we were surprised that there was a paradox concerning the extranet and general ICT support facilities. The JCC's concept was drawn from the experience of the Rural Rail Link Authority (RRLA) program initiative to engage with external stakeholders, such as the local authorities along the rail line routes, to inform and collaborate with them to ensure that the RRLA program was well informed about potential disruptions and issues. The RRLA also formed an embryonic JCC to integrate the various forms of project delivery in that program (Chapter 2). It learned from that experience. Several senior LXRP staff had worked on the RRLA program and were able to carry that knowledge and those insights to further develop the LXRP's JCC. This became the top-down strategy to improve the JCC, but the means and methods were unclear when the LXRP was initiated.

The demand for the LXRP extranet requirement was elevated when the LXRP scope grew from its original 50 to the current 110 crossing removals and additional station renewals, and transformation. Additionally, several of the NOPs came to the LXRP with their expertise and experience in the sophisticated use of project/program-based extranet-type facilities. Several NOP organisations have used AR, BIM, UAV and other technologies. So, they drove a bottom-up organic strategy to maintain the extensive digital repository (Chapter 10). The paradox is that while the LXRP may be considered the global leader in innovation diffusion, DE and the like, it was achieved not from a top-down strategic approach or an organic ground-up approach but an iterative interaction of these forces.

We offer only a peek into the overall picture or story of the extranet aspect at this stage. LXRP developed customised templates and on-line administration tools to upload and communicate across the program that the JCC and LXRP used to monitor and enable achievement of continuous improvement. Templates and tools included: standardised agendas and JCC/ALT/AMT minute recording processes; dashboard reporting; and coordinated site occupation and rail service disruption event planning.

Another important mechanism is the improvement or innovation submission mechanism. Proponents from any alliance through their working group or COP committee (Chapter 3, Figure 3–6) submits an idea proposal that is assessed, evaluated and approved by the JCC executive level. This provides a highly effective integrative mechanism that ensures that all alliance project teams get access to the ideas and can built upon those ideas and innovations accessing the KPI1.1 or 1.2 incentives in doing so. It also improves collaboration and potential delivery performance. COPs, working groups, discipline champions and JCC technical sub-committees all play a role in this system. Figure 9-4 illustrates a 2019 example with the proponent's name blacked out.

The submission idea form illustrates the depth of approach sophistication. A concise explanation of the idea/innovation is required. Also, a justification is needed, based on what is proposed, the scale and extent of the current problem it addresses and the intended impact. Alignment with the 5-Greats and its strategic alignment with its supporters is specifically given. A visual

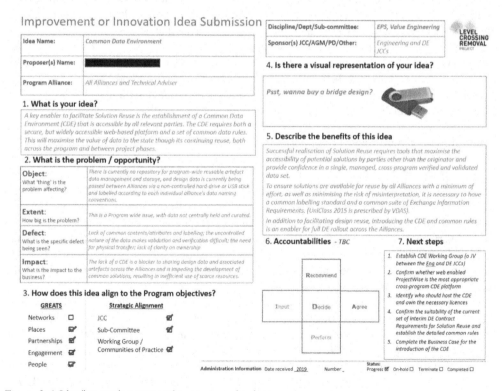

Figure 9-4 Idea/innovation proposal support mechanisms

(Source LXRP[3])

presentation explanation is given, often as a short video that can later be used for training and diffusion purposes. Its benefits are explained and proposed next steps outlined. It a highly parsimonious and well-crafted summary approach that succinctly provides a clear proposal. If/when approved it is developed further with detailed planning of its implementation with how-to adopt/adapt. New ideas are rewarded through this KPI-1.1 innovation incentive mechanism and demonstrated adoption/adaptation attracts KPI-1.2 incentives. Innovations may be technical, process-administrative, or related to people-skills development and so breeds a rich innovation landscape. Each alliance is required to demonstrate its continuous improvement ability through idea submissions and its performance is judged against criteria spectrum of fail (less than two proposals/year), minimum conditions of satisfaction (4/year), and exceptional (6/year). Proposals often are made between several project alliance teams and so this encourages collaboration.

The above process is managed through the LXRP extranet, its purpose being for information communication, collaboration and enabling innovation. All LXRP participants have access to the extranet which makes it a powerful integrative, collaborative and continuous improvement tool.

9.6   Global UK Innovation Strategy Example

9.6.1   *Crossrail project context*

London's Crossrail complex and challenging project comprised constructing a new railway – the Elizabeth line – from Reading and Heathrow in the west via 21 kilometres of twin-bore tunnels under central London to Shenfield and Abbey Wood in the east, including ten new and 30 upgraded stations. Many of Crossrail's tunnels were excavated and constructed close to existing Tube lines. For example, The Tottenham Court Road Station is underneath the junction between Oxford Street (Europe's busiest shopping street) and Charing Cross Road providing an interchange between the Elizabeth line and existing Northern and Central lines. The 250-metre platform and station located 24 metres below ground is about the length of Wembley Football Stadium. More than 200,000 passengers pass through the new station every day.

A separate company, Crossrail Limited (CRL), was established as the client organisation to ensure that project participants responsible for its delivery were entirely focused on a single task, with a clear remit and clarity of purpose. This integrated project delivery team of around 1,200 staff was formed including CRL and two delivery partners: a programme partner called Transcend (a joint venture between CH2M, AECOM and The Nichols Group) and a project delivery partner called Crossrail Central (a joint venture led by Bechtel and supported by Halcrow and Systra) to manage the construction of the complex central section of tunnels. CRL and its two delivery partners focused on the long-term goals of finishing the project on time, safely and within budget, dealing with external stakeholders and handing over a 'world-class railway' to the operator, whereas the more immediate priority of each contractor was to complete discrete piece of work and make a profit.

Crossrail had over 60 major contracts with individual firms and joint ventures to construct the railway, such as the £400 million western tunnel contract awarded to the BAM Nuttall, Ferrovial and Kier (BFK) joint venture. Contracts ranged in size from as little as £1 million for design services to over £500 million for the largest tunnelling and systems contracts. The client relied on collaborative contracts and various forms of persuasion and personal relationships to motivate the contractors (if necessary, talking directly with the contractor's CEO to wield influence) and keep the project on track towards completion. To overcome problems and resolve disputes, members of Crossrail's delivery partner also worked on site with contractors in co-located integrated project teams. As the project moved towards completion, the operator

(Rail for London) became a more active member of integrated project delivery team. After this project phase completion, the client and delivery partner organisations were finally disbanded.

9.6.2 *Crossrail project innovation strategy*

In 2012, Andrew Wolstenholme, Crossrail's CEO, engaged with Professor David Gann, Imperial College Business School, to develop an innovation strategy for the project (Davies et al., 2014). A core team of practitioners and researchers was established to create the innovation strategy (Crossrail, 2012) including Mark Thurston (John Pelton took over this role in 2014), Tim DeBarro, Prof. Andrew Davies (who joined University College London in 2012 but continued his involvement in the project), and Dr. Sam MacAulay. The Crossrail Innovation Forum was established to provide the support, resources, and advice required to steer the innovation strategy. The forum formed a biannual meeting established to monitor and steer the development and implement the innovation strategy involving senior members of Crossrail, Imperial College London, main contractors and suppliers.

In 2013 when the innovation programme became operational, individuals in contractors suppliers and external stakeholder organisations were encouraged to create, implement and share new ideas, technologies and practices and devise an in-house team to manage the innovation programme and established a database to capture all the innovative ideas, proposals and solutions submitted by project members (DeBarro et al., 2015).

Building on the success of Crossrail's innovation programme, Andrew Wolstenholme began working with others in the project, such as John Pelton, to build a bridge linking Crossrail's innovation and learning with other projects (Pelton et al., 2017). The "Infrastructure Industry Innovation Platform" (recently renamed the Industry Innovation Partnership or i3P (www.i3p.org.uk/) was established in October 2016 with the support of Crossrail, the Thames Tideway project, major contractors, government, and leading universities. To prevent innovation from occurring in isolation on each megaproject – as it had in the past – the i3P platform (which builds on database and collaborative approach originally created for Crossrail) was established to share new ideas, practices and technologies with other megaprojects in London and elsewhere in the UK. Many other infrastructure projects in the UK have followed Crossrail by establishing their own innovation programmes (e.g., the Thames Tideway sewer, Hinkley Point C nuclear power station, and HS2 railway). Together, they share the i3P pool of resources and collaborate on programmes of innovation aimed at improving the performance and outcomes of many projects. Future research should study the capabilities (Davies, Dodgson and Gann, 2016) required to effectively execute this learning and innovation required to improve the performance of megaprojects (Davies et al., 2017).

9.7 Conclusions

This chapter's focus is on innovation diffusion and continuous improvement within the IPD/Alliancing LXRP context supplemented with discussion of the approach taken in the UK triggered by Crossrail. There are key lessons to be learned from this chapter.

Effective innovation diffusion and continuous improvement requires removing sticky-knowledge challenges identified by Szulanski (1996), particulary the creation of a supporting fertile environment or ecology for innovation. Both LXRP and Crossrail purposefully developed an innovation strategy. As mentioned in Chapter 3, clear project strategy articulation is necessary but insufficient. What is required to institutionalise and embed a strategy, as outlined by Scott (2008; 2012), is a three-pronged approach. While a clearly articulated strategy explains the project's purpose, there needs to be governance measures in place as explained in this chapter to guide and steer the strategy's realisation. Additionally, an appropriate workplace culture environment (Chapter 8) provides a supportive mechanism for project participants who need to have the cultural-cognitive capabilities to interpret the regulative arrangements such as the PAA and the associated governance mechanisms in light of their norms and ability to take the initiative in how to diffuse knowledge. All this is best conducted by a well-integrated collaborative united team.

The LXRP leveraged its governance arrangements discussed above to create the novel JCC concept that evolved from experience gained from the RRLA, taking the JCC concept to another level of sophistication that went far beyond being an integration (Chapter 5) and collaboration (Chapter 6) mechanism to both provide a formal governance regime that embraced an formal COP and less formal working group model as illustrated in Figure 3–6 in Chapter 3.

LXRP's JCC shares similarities with the Crossrail's innovation strategy implementation mechanisms. For example both developed a digital platform (the LXRP extranet and Crossrail's i3P), both formally incentivised innovation, both developed means for the submission, peer-review and approval and dissemination of innovations that had been incentivised. The LXRP case mechanism articulated the 5-Greats and included specified workplace cultural values.[4] Crossrail developed their innovation forum (Pelton et al., 2017) as a similar mechanism to the JCC. What particularly stands out is the innovation and continuous improvement mindset and mechanisms and how that provides not only project innovation and improvement but also an industry innovation development and diffusion legacy.

The chapter illustrates two world-class examples of innovation and continuous improvement in projects and reports on the *what,* the *how* and *why* aspects. This offers deep insights for projects/program strategy developers to adapt and customise to their project/program contexts.

Notes

1 PowerPoint presentation LX PRESENTATION 20190710 JCC AGM PD Forum.pptx 23 July 2019, pp. 8–13.
2 Available at: https://vimeo.com/301757104/a32e3fdab0
3 PowerPoint presentation LX PRESENTATION 20190710 JCC AGM PD Forum.pptx 23 July 2019, p. 18.
4 The Cares values and culture: creativity, accountability, relationship, empowerment, and safety identified in the LXRP Blueprint 2023.

References

Ajzen, I. (1991). "The Theory of Planned Behavior." *Organizational Behavior and Human Decision Processes*. 50 (2): 179–211.

Australian Constructors Association (2017). *Changing the Game: How Australian Can Achieve Success in the New World of Mega-projects*, Australian Construction Association.

Ballard, G. and Howell, G. (1997). "Implementing lean construction: Reducing inflow variation." In *Lean Construction*, Alarcon, L. (ed.), pp. 93–100. Rotterdam, Netherlands, A.A. Balkema.

Brady, T. and Hobday, M. (2011). "Projects and innovation: Innovation and projects." In *The Oxford Handbook of Project Management*, Morris, P.W.G., J.K. Pinto and J. Söderlund (eds), pp. 273–294. Oxford, Oxford University Press.

Crossan, M.M., Lane, H.W. and White, R.E. (1999). "An organizational learning framework: From intuition to institution." *Academy of Management Review*. 24 (3): 522–537.

Crossrail. (2012). *Crossrail Innovation Strategy: Moving London Forward*, Crossrail London, UK.

Davies, A. and Brady, T. (2000). "Organisational capabilities and learning in complex product systems: towards repeatable solutions." *Research Policy*. 29 (7–8): 931–953.

Davies, A., Dodgson, M. and Gann, D. (2016). "Dynamic capabilities in complex projects: The case of London Heathrow Terminal 5." *Project Management Journal*. 47 (2): 26–46.

Davies, A., Dodgson, M., Gann, D. and MacAulay, S. (2017). "Five rules for managing large, complex projects." *MIT Sloan Management Review.* 59 (1): 73–78.

Davies, A., MacAulay, S., DeBarro, T. and Thurston, M. (2014). "Making innovation happen in a megaproject: London's Crossrail suburban railway system." *Project Management Journal.* 45 (6): 25–37.

DeBarro, T., MacAulay, S., Davies, A., Wolstenholme, A., Gann, D. and Pelton, J. (2015). "Mantra to method: Lessons from managing innovation on Crossrail, UK." *Proceedings of the Institution of Civil Engineers – Civil Engineering.* 168 (4): 171–178.

Ehrhart, M.G., Schneider, B. and Macey, W.H. (2014) *Organizational Climate and Culture : An Introduction to Theory, Research, and Practice*, London, Taylor & Francis Group.

Fireman, M.C.T., Saurin, T.A., Formoso, C.T., Koskela, L. and Tommelein, I.D. (2023). "Slack in production planning and control: A study in the construction industry." *Construction Management and Economics.* 41 (3): 256–276.

Hackman, J.R. and Oldham, G.R. (1976). "Motivation through the design of work: Test of a theory." *Organizational Behavior and Human Performance.* 16 (2): 250–279.

Imai, M. (1986) *Kaizen: The Key to Japan's Competitive Success*, New York, McGraw-Hill.

Järvinen, A. and Poikela, E. (2001). "Modelling reflective and contextual learning at work." *Learning, Working and Living: Mapping the Terrain of Working Life Learning.* 13 (7/8): 282–289.

Kolb, D.A. (1984) *Experiential Learning: Experience as the Source of Learning and Development*, Englewood Cliffs, NJ, Prentice-Hall.

Koskela, L. (2000). An Exploration Towards a Production Theory and its Application to Construction. Doctor of Technology, VTT Technical Research Centre of Finland. Helsinki, Finland, Helsinki University of Technology.

Latham, M. (1994). *Constructing the Team*, London, HMSO.

Lave, J. and Wenger, E.C. (1991) *Situated Learning – Legitimate Peripheral Participation*, Cambridge, Cambridge University Press.

Lawrence, T.B., Mauws, M.K., Dyck, B. and Kleysen, R.F. (2005). "The politics of organizational learning: Integrating power into the 4I framework." *Academy of Management Review.* 30 (1): 180–191.

Lehtovaara, J., Seppänen, O. and Peltokorpi, A. (2022). "Improving construction management with decentralised production planning and control: Exploring the production crew and manager perspectives through a multi-method approach." *Construction Management and Economics.* 40 (4): 254–277.

Li, C.-Y. and Hsieh, C.-T. (2009). "The impact of knowledge stickiness on knowledge transfer implementation, internalization, and satisfaction for multinational corporations." *International Journal of Information Management.* 29 (6): 425–435.

Lloyd-Walker, B.M., Walker, D.H.T. and Mills, A. (2014). "Enabling construction innovation: The role of a no-blame culture as a collaboration behavioural driver in project alliances." *Construction Management and Economics.* 32 (3): 229–245.

Love, P.D., Fong, P.S.W. and Irani, Z., Eds. (2005). *Management of Knowledge in Project Environments*, Series Management of Knowledge in Project Environments. Burlington, MA, Elsevier Butterworth-Heinemann.

Love, P.E.D. and Matthews, J. (2022a). "Error mastery in alliance transport megaprojects." *IEEE Transactions on Engineering Management.* 1–18.

Love, P.E.D. and Matthews, J. (2022b). "The social organization of errors and the manifestation of rework: Learning from narratives of practice." *Production Planning & Control.* 1–16.

Love, P.E.D. and Matthews, J. (2022c). "There is strength in numbers: Seven principles to contain and reduce error and mitigate rework in transport mega-projects." *IEEE Engineering Management Review.* 50 (1): 1–1.

Love, P.E.D., Matthews, J., Ika, L.A., Teo, P.T.T., Fang, W. and Morrison, J. (2021). "From Quality-I to Quality-II: Cultivating an error culture to support lean thinking and rework mitigation in infrastructure projects." *Production Planning and Control.*

Love, P.E.D., Smith, J., Ackermann, F. and Irani, Z. (2019). "Making sense of rework and its unintended consequence in projects: The emergence of uncomfortable knowledge." *International Journal of Project Management.* 37 (3): 501–516.

Love, P.E.D., Teo, P., Davidson, M., Cumming, S. and Morrison, J. (2016). "Building absorptive capacity in an alliance: Process improvement through lessons learned." *International Journal of Project Management*. 34 (7): 1123–1137.

Matthews, J., Love, P.E.D., Porter, S.R. and Fang, W. (2022). "Smart data and business analytics: A theoretical framework for managing rework risks in mega-projects." *International Journal of Information Management*. 65: 102495.

McCuen, R.H. (2023). "The mindset dimension of critical problem-solving." In *Critical Thinking, Idea Innovation, and Creativity*, pp. 195–216. Milton, Taylor & Francis Group.

Nonaka, I. and Takeuchi, H. (1995) *The Knowledge-Creating Company*, Oxford, Oxford University Press.

Orr, J. (1990). Talking About Machines: An Ethnography of a Modern Job. PhD Thesis. Ithaca, NY, Cornell University.

Othman, R. and Hashim, N.A. (2004). "Typologizing organizational amnesia." *The Learning Organization, MCB University Press*. 11 (3): 273–284.

Peansupap, V. (2004). An Exploratory Approach to the Diffusion of ICT Innovation a Project Environment. PhD Thesis, School of Property, Construction and Project Management. Melbourne, RMIT University.

Peansupap, V. and Walker, D.H.T. (2005a). "Exploratory factors influencing ICT diffusion and adoption within Australian construction organisations: A micro analysis." *Journal of Construction Innovation*. 5 (3): 135–157.

Peansupap, V. and Walker, D.H.T. (2005b). "Factors affecting ICT diffusion: A case study of three large Australian construction contractors." *Engineering Construction and Architectural Management*. 12 (1): 21–37.

Pelton, J., Brown, M., Reddaway, W., Gilmour, M., Phoon, S., Wolstenholme, A. and Gann, D. (2017). "Crossrail project: The evolution of an innovation ecosystem." *Proceedings of the Institution of Civil Engineers – Civil Engineering*. 170 (4): 181–190.

Rogers, E.M. (1983) *Diffusion of Innovations*, New York, Free Press.

Schneider, B., Ehrhart, M.G. and Macey, W.H. (2013). "Organizational climate and culture." *Annual Review of Psychology*. 64 (1): 361–388.

Scott, W.R. (2008) *Institutions and Organizations*, Thousand Oaks, CA and London, Sage.

Scott, W.R. (2012). "The institutional environment of global project organizations." *Engineering Project Organization Journal*. 2 (1–2): 27–35.

Scott, W.R. (2014) *Institutions and Organizations*, Thousand Oaks, CA and London, Sage.

Slaughter, E.S. (2000). "Implementation of construction innovations." *Building Research & Information*. 28 (1): 2–17.

Szulanski, G. (1996). "Exploring internal stickiness: Impediments to the transfer of best practice within the firm." *Strategic Management Journal*. 17 (Winter Special Issue): 27–43.

Szulanski, G. (2003) *Sticky Knowledge Barriers to Knowing in the Firm*, Thousand Oaks, CA, Sage Publications.

van Dyck, C., Frese, M., Baer, M. and Sonnentag, S. (2005). "Organizational error management culture and its impact on performance: A two-study replication." *Journal of Applied Psychology*. 90 (6): 1228–1240.

Walker, D.H.T. and Lloyd-Walker, B.M. (2015) *Collaborative Project Procurement Arrangements*, Newtown Square, PA, Project Management Institute.

Walker, D.H.T. and Lloyd-Walker, B.M. (2020). "Knowledge, Skills, Attributes and Experience (KSAE) for IPD Alliancing task motivation." In *The Routledge Handbook of Integrated Project Delivery*, Walker, D.H.T. and S. Rowlinson (eds), pp. 219–244. Abingdon, Routledge.

Walker, D.H.T., Love, P.E.D. and Matthews, J. (2023). "Generating value in program alliances: The value of dialogue in large-scale infrastructure projects." *Production Planning & Control*. 16pp.

Wegman, L.A., Hoffman, B.J., Carter, N.T., Twenge, J.M. and Guenole, N. (2018). "Placing job characteristics in context: Cross-temporal meta-analysis of changes in job characteristics since 1975." *Journal of Management*. 44 (1): 352–386.

Westrum, R. (2014). "The study of information flow: A personal journey." *Safety Science*. 67: 58–63.

Wiig, K.M. (1997). "Integrating intellectual capital and knowledge management." *Long Range Planning, Elsevier Science Limited*. 30 (3): 399–405.

10 LXRP, Digital Engineering and Technological Innovation

Jane Matthews, Peter E.D. Love, Derek H.T. Walker, Hadi Mahamivanan and Mark Betts

10.1 Introduction

How does the LXRP develop its digital innovation strategy across projects in the program?

Digital engineering (DE) is having a profound impact on the way major transport infrastructure projects are delivered worldwide. Significant projects that have benefited, or that are progressing with DE, include Jamaica Station, Rhode Island, in the United States (US) (LaShell and Goldman, 2021), the Wuhan Metro in China, Elizabeth Line (formerly known as Crossrail) in the United Kingdom (UK) (Peplow, 2016; Brown, 2023), Hong Kong's International Airport 3 Runway Concourse (Trimble Connect, 2021) and High-Speed 2 (UK) (Buddoo, 2021), to name a few.

DE is enabled by using collaborative and productive project delivery and management methods across an asset's life cycle (Transport for New South Wales, 2022). However, traditionally transport projects have not been delivered using collaborative project delivery (Regan, Smith and Love, 2017; Love et al., 2019) and thus have been unable to realise the benefits of digital technologies, such as Building Information Modelling (BIM) (Love et al., 2014). This has stymied the effective use of digital technologies to produce adequate as-built documentation, needed to manage, maintain, and operate assets over their lifecycle (Love et al., 2018; Love et al., 2021a; 2021b). As we have mentioned in Chapters 5 and 6, in each of LXRP's alliances, the Project Owner (PO) – Melbourne Transport Infrastructure Authority (MTIA[1]) – is actively involved with the delivery of each project and acts as a medium for conveying the information needs and requirements of their asset owners and operator, enabling them to be incorporated into the design process. In the case of traditionally procured projects, we seldom see the PO being embedded into the project team at its onset, having input into the project's design and asset information requirements, and being involved in formulating the Target Outturn Cost (TOC) and Target Adjustment Event (TAE) (Chapter 4). The collaborative structure of the alliance facilitated by the Project Alliance Agreement (PAA) provides a mechanism to support the adoption and implementation of DE as the organisational and cultural barriers that tend to inhibit information flow in projects are disassembled.

Simply put, DE uses an array of technologies (e.g., geographical information systems, GIS) – with BIM being critical – to create, capture, transfer and integrate data in a virtual space so that architects, designers, engineers, and construction organisations can effectively utilise it to design, plan, construct, maintain and operate physical infrastructure. Implementing technologies in a piecemeal fashion or shoehorning them into existing workflows and processes without considering the new skills required and the change to be enacted will result in their benefits going unrealised (Matthews et al., 2018).

DOI: 10.1201/9781003389170-10

A strategy must be established to provide a clear vision to harness the power of technology, data, and analytics to align with objectives, policies, directions, and priorities to ensure governments can acquire the benefits of DE. Equally, private sector organisations that deliver infrastructure assets need a DE strategy.

Figure 10-1 provides the structure for this chapter to address this question.

The LXRP and its alliances are using BIM to deliver their projects and are producing building information models [hereinafter models] to Level of Development (LOD) 350,[2] which are Issued for Construction (IFC) with LOD 400[3] for structural components.

As BIM is now a mainstream collaborative working methodology embedded in practice, we gloss over its use and role in daily work activities at the LXRP. Though, the level of development and detail required within a model by asset owners will vary depending on their requirements and needs.

The Department of Transport and Planning (DTP), an asset owner whom LXRP represents, requires a model to be handed over at completion, as well as drawings. Yet, it was not until 2019 that LXRP established a ProjectWise library to enable all staff to access the models, and the alliances were requested to provide federated Navisworks models (LOD 350) as a deliverable. A DE manager (IV11) from one of the alliances made the following remark "we hand over our [BIM] models to LOD 350 and drawings as they are a contractual requirement." However, because of LXRP's 'Digital Engineering Strategy,' approved in February 2021, they are currently undertaking a pilot study to deliver its first digital 'as-built' model for handover.

A challenge facing the LXRP alliances, which also generally confronts major infrastructure projects, is the production and handover of accurate 'as-built' documentation (Gallaher et al., 2004; Whyte, Lindkvist and Jaradat, 2016; Love et al., 2018). Errors and omissions can be contained in the 'Issue for Construction' (IFC) documentation and changes may also be required during construction. Thus, any amendments to the 'as planned' and documented design in the IFC drawings must be incorporated into the 'as-builts' produced.

The literature about how BIM is used in projects and integrates with other technologies such as Augmented Reality (AR), GIS, Internet of Things (IoT), scanning, sensors, Radio Frequency

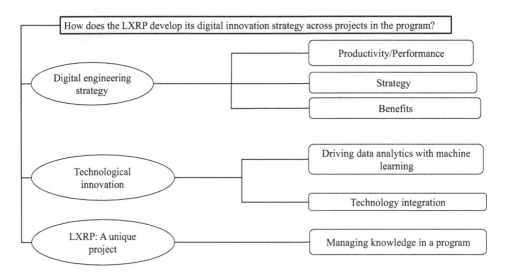

Figure 10-1 Chapter structure

Identification (RFID), Unmanned Arial Vehicles (UAVs) also known as drones, and virtual reality (VR) is too vast to discuss in this chapter. The LXRPs alliances are integrating BIM with an array of digital technologies such as VR in design and planning, RFID to track inventory, and UAVs to monitor progress, all widely used in the delivery of major infrastructure projects in Australia and around the globe.

So, instead of focusing on how BIM and associated technologies are utilised in the LXRP, we look at how one of its alliances strives to innovatively use DE, to enable and enhance its continuous improvement, by initiating research and development (R&D) projects. We examine how machine learning (ML) stimulates a data analytics agenda to help determine rework risks, and how better quality 'as-built' documentation can be produced by integrating several technologies. A DE strategy and the establishment of the Joint Coordination Committee (JCC) (Chapters 3 and 9) within the LXRP, alongside the drive of the alliances to improve their productivity, have enabled new ways of working using technology.

10.2   Digital Engineering Strategies

For decades the construction sector has been criticised for its low productivity and slow technology uptake (Barbosa et al., 2017). Tackling the problem of productivity requires the industry to engage in the process of change, which can be enabled by (Australian Constructors Association, 2017; Barbosa et al., 2017; Love et al., 2021a; Love et al., 2021b):

1. Reshaping regulations and improving transparency;
2. Collaborative project delivery (e.g., alliancing and integrated project delivery including constructability, value engineering and adoption of lean practices such as the Last Planner®):
3. DE (e.g., advanced analytics, BIM, and visualisation), new materials (e.g., prefabrication) and automation;
4. Integrating procurement and supply chains with digital technologies; and
5. Capability building and reskilling the workforce (e.g., apprentice schemes).

The expected productivity improvements of embracing the above areas have been forecasted to range between 48 to 60%, with collaborative project delivery and DE accounting for approximately 40% (Barbosa et al., 2017).

Several years ago, the McKinsey Global Institute digitisation index (Manyika et al., 2015) indicated that construction was one of the least digitised sectors, even though there had been calls for decades for it to embrace technology to address its productivity and performance issues (Latham, 1994; Egan, 1998). As software, methods, processes (e.g., understanding of workflows), standards (e.g., ISO 19650 'Organisation and Digitisation of Information about Building and Civil Engineering Works,' Industry Foundation Classes (IFC), and Information Delivery Manual) and server technologies (i.e., common data environments) to support BIM are continually evolving (Succar, 2009; Azhar, 2011; Sacks et al., 2018), its ability, in conjunction with other technologies, to drive efficiencies throughout a project's lifecycle has become evident (Young et al., 2014). As a result, many governments in countries such as Finland,[4] Germany,[5] Norway,[6] Singapore,[7] Spain[8] and the UK[9] have mandated its use to deliver their infrastructure assets.

We now focus on the UK as it has been instrumental in driving the use of DE throughout its construction sector. It has been a point of reference and source of knowledge for other countries to develop their strategies, roadmaps, and frameworks to facilitate the adoption of DE when delivering their projects.

10.2.1   *Developing a strategy for digitisation*

In a rapidly changing economic environment driven by innovation and digital transformation, governments must be able to adapt and respond to new developments by providing opportunities to promote economic growth, improve outcomes for citizens, tackle industry challenges, and help upskill the workforce. Within the context of infrastructure delivery, many governments have developed DE strategies, which describe 'how' an organisation will do things' and 'how' it will get there (Mintzberg, 1987). An organisation's mission, vision, goals, policies and plans to acquire the benefits of technology are typically identified within such a strategy (Bailey et al., 2018). The orientation of a DE strategy for governments is to drive efficiencies and enable the delivery of an asset so that it can be effectively operated and maintained.

In the UK, the government aspired to be a global leader in BIM and produced a policy paper entitled 'Government Construction Strategy' (Cabinet Office, 2011; 2012). This strategy mandated Level 2 BIM to be used on all public sector projects as a minimum by 2016. Level 2 refers to exchanging 3D models in a common format (such as IFC), and in a common data environment, where all stakeholders can view and access them. In 2016 the government announced they expected an industry shift from Level 2 to Level 3 BIM by 2020, where a single model, that includes scheduling, cost and project management is shared in a fully collaborative project environment (HM Government, 2016; Infrastructure and Projects Authority (IPA), 2016). Moreover, the UK BIM strategy meant that any firm wanting to tender for government contracts must use BIM.

The findings of the first National BIM Report undertaken since the introduction of the UK mandate, led Mark Bew, Chair of the HM Digital Britain, to state, "the successful delivery of the UK Government Construction Strategy Level 2 BIM programme now sees the UK take on a global leadership role and represents an internationally unparalleled achievement on the journey towards the digitisation on our built environment" (NBS, 2017, p. 4). While the report identified that Level 2 was well-established, it also pointed out that continual professional development and training were needed, standards and protocols were under-utilised, and clients needed to be educated about the process and benefits of BIM.

Acknowledging the need to raise awareness, educate, upskill, and help organisations with the adoption of BIM, the government, industry, and academia have worked collaboratively to achieve these goals. The establishment of the Centre for Digital Built Britain (CDBB)[10] in partnership with the Department of Business, and Energy, housed at the University of Cambridge (formally launched in 2018), as part of a transformative deal between the government and the construction industry (Clarke, 2017), played a critical role in disseminating knowledge and educating the industry about digital technology and its utilisation across an asset's life to create a digital twin. Put simply, a digital twin is a "virtual representation of physical objects across the lifecycle that can be understood, learned, and reasoned with real-time data or a simulation model that acquires data from the field and triggers the operation of physical devices" (Attaran and Celik, 2023, p. 2). The work of the CDBB came to an end in September 2022. Its legacy of knowledge will 'not ride off into the sunset.' Instead, it will provide the global construction industry with a critical resource to help transform and drive the breakthroughs in practice needed to improve the productivity and performance of projects.

Indeed, BIM mandates are essential – it has been successful in the UK, with many benefits realised – as adopting digital tools unlocks efficiencies that streamline communication, break down information silos, promote transparency, encourage cooperation, and improve productivity (Morkos, 2022). Still, simultaneously there must also be changes in government procurement policies to stimulate the much-needed structural, cultural, and behavioural changes needed to improve the productivity and performance of construction.

There is no question that the construction industry has been a laggard in its technology adoption. But with significant technological advances being made and aided by artificial intelligence (AI) and ML, we are seeing an increasing number of organisations in construction proactively engaging in DE. For example, construction organisations in Australia have developed DE strategies to establish the capability to adopt and implement varying technologies to drive innovation, reduce costs and time, improve safety, and increase their productivity in projects (Godley, 2023; Grogan, 2023; Nadel, 2023). It should be noted that Australian State governments have not mandated BIM. However, several States have developed DE strategies and frameworks to deliver their major infrastructure assets (e.g., Department of Planning Transport and Infrastructure, 2017; Department of State Development Infrastructure Local Government and Planning, 2018; Transport for New South Wales, 2018; Office of Projects Victoria, 2020; Department of State Development Infrastructure Local Government and Planning, 2022; Transport for New South Wales, 2022). NATSPEC (2022) has also produced a 'National BIM Guide' to facilitate a nationally consistent approach to adopting BIM.

The UK has, undoubtedly, been a primary source of innovation and a point of reference for developing various state governments' DE strategies in Australia, which we will briefly examine below. However, since mandating Level 2 BIM on all public sector projects in 2016, we have seen significant levels of adoption. In addition, juxtaposed with other technologies, BIM was central to developing, designing, engineering, and constructing the Elizabeth Line (Brown, 2023) and formed an integral part of the High Speed 2 delivery strategy to develop an operational digital twin (HS2, 2022).

The Level 3 BIM strategy created a vision to transform business models and processes in construction and integrate digital technologies to deliver assets that provide better value across their life cycle (HM Government, 2016). The 2016 strategic plan provided the impetus and established the case for providing the resources required to create the 'UK BIM Framework'[11] developed in collaboration between the British Standards Institute, CDBB and the UK BIM Alliance (UKBIMA).[12] The Framework embraces and assists with the implementation of the concepts and principles of ISO 19650 for managing information across an asset's life cycle and enabling the production of a digital twin, leveraged by technologies such as Industry 4.0[13] (e.g., Internet of Things, cloud computing, data analytics and artificial intelligence). A digital twin can take various guises (Love and Matthews, 2019; Remmert, 2022): (1) *status twin*, which is used for monitoring the physical condition of objects and equipment; (2) *operation twin*, which is to adjust operating parameters based on linked actions and/workflows; and (3) *simulation twin*, which is used to predict how an objective or device responds to operational conditions in the future, which is key for predictive maintenance.

While many governments aim to develop and operationalise a digital twin to manage their assets' performance, the quality of data and its structure in projects must be improved to ensure system interoperability (Love and Matthews, 2019). Moreover, the high fixed cost and complex infrastructure required to support a digital twin will need asset owners to justify the investment in technology and demonstrate its benefits and value (Love and Matthews, 2019; Technavio, 2022). The LXRP is not striving to develop a digital twin as part of its DE strategy for each crossing that is removed, as issues relating to integration and absorption into existing legacy systems still need to be solved. Commenting on this position, IV13 stated:

> these kinds of assets can be 100 years old or more. And so, there are a lot of legacy systems that also have to absorb existing processes, and how they can be merged in the near future, something that's still would be resolved by the DTP [Department of Transport and Planning]. So, for us, we're not going to go back and add to the system that was handed back at that

time. It's about moving forward to see how we transition; how do we advise the DTP to make the system work?

The inconsistency in data formats and structure and how it is collated during the construction of transport infrastructure assets impacts the ability to hand over a model to LOD 500, which is needed to create a digital twin. Reinforcing the difficulty in generating a digital twin, IV13 made the following remark:

> People talk about digital twins everywhere. But how people are capturing data through the construction phases is so inconsistent at the moment. I think, enviously, they've got a digital twin. But clearly, they just got their head in the clouds because the systems and processes to connect the asset owner from start to finish are not there.

The LXRP and its alliances' approach to DE is adapting and evolving as they learn from each completed project and share knowledge through JCC initiatives. The LXRP has several more years to run before its expected completion in 2030. With a learning ethos in place, and a willingness to stimulate technological and process innovation outcomes, we anticipate significant headway will have been made in establishing the knowledge needed to hand over 'only a digital model' and construct a digital twin of a new asset that can be integrated and utilised with existing legacy systems.

10.2.2   *LXRP digital engineering strategy*

The Victorian Digital Asset Policy (VDAS) (Office of Projects Victoria, 2020), built on and supported by the State's Asset Management Accountability Framework (AMAF) (Department of Treasury and Finance, 2016), provides Agencies and Departments with the digital requirements to support the planning, design, and construction of State Government assets. The Victorian Government developed the strategy in response to the need to deliver unprecedented levels of new infrastructure and to boost the capability of its Agencies and Departments. The strategy's underlying aim was to improve asset delivery's performance and productivity across its life cycle.

The VDAS recognises a need to preserve data generated from each stage of an asset's life cycle and that it must be fit for purpose. Such data is pivotal for decision-making, benchmarking, and continuous improvement, enabling assets to be productively managed. The VDAS provides a framing to apply DE to systematically manage and ensure the efficient flow of information across the asset lifecycle (Office of Projects Victoria, 2020). The VDAS is aligned with ISO 19650 and CDBB's Gemini Principles (Centre for Digital Built Britain (CDBB), 2018).

While the VDAS was established in 2020, we need to be cognisant that the LXRP, formerly referred to as the LXRA, commenced its program of work in 2015, originally to remove 50 level crossings by 2022 (Chapter 2). At this point, the LXRA did not have a formal DE strategy. After the incumbent Labour Government was re-elected in 2018 and announced an additional 85 crossings would be removed, there was a realisation that a strategy should be developed to help better operate, manage, and maintain the corpus of constructed new assets. Even though LXRP had not mandated BIM, the alliances openly embraced it, recognising its cost and productivity benefits. For example, applying it during the planning, design and engineering process and simulating the construction process through 4-Dimensional BIM (also known as 4D scheduling or 4D simulation). In brief, 4D BIM communicates to the project team and stakeholders exactly when (i.e., when a task will be installed and completed on a timeline), where, and how an asset will be constructed. Schedule data is synchronised to objects within the model, using bidirectional integration[14] in this case.

A good example of 4D BIM being used in one of LXRP's projects in the $1 billion Cranbourne Line Upgrade is presented in Figure 10–2 (Hawkes, 2021). The 4D modelling was undertaken collaboratively between organisations involved in the alliance delivering the project. A DE manager for the alliance was reported saying (Hawkes, 2021):

> Commonly, teams outsource the modelling to create point-in-time versions, but our team members created their own continuous feedback loop that allows everyone to respond and review actual engineering designs via an online platform.

The use of 4D BIM has enabled the alliance to identify and rectify problems such as clashes between temporary and permanent retaining walls and signalling cables and wall panels (Hawkes, 2021). It has also allowed the alliance to plan work virtually and rehearse the sequencing of activities, such as the bridge installation in Figure 10–2, to mitigate the risk of delays and safety instances, and minimise the inconvenience to the public of having road and rail closures.

While the alliances were creating designs in a digital model, they had to convert them into two-dimensional (2D) drawings in PDF format for the LXRP's reviewers and stakeholders to view and check. Moreover, 'as-builts' were also required to be submitted to the LXRP in a 2D PDF format. The conversion from digital models to 2D PDFs was a regressive step considering prevailing 'best practice,' with IV12 stating that:

> Everything has to be dumbed back down to analogue methods. So, to give you an example, you know they might do design in a digital model, but when it came to sharing that design with LXRP and the stakeholders and the reviewers, they would have to turn it into a 2D PDF drawing. The PDF drawing became the contract deliverable people would comment on.

Indeed, the LXRP is a complex program of work and is confronted with considerable production pressure. For one alliance, the adoption of DE technologies was not only viewed as tools to improve productivity but also explicitly as tools to enable work to be performed effectively, with IV11 maintaining that:

> Our digital engineering has been more focused on digital enablement. It's more about helping the construction team to deliver faster and more efficiently and not to be wasteful. Reducing waste is our ultimate goal. It is not about physical waste, but the time waste, the number of circles you run in, and the red tape you must go through. We are using it [digital engineering tools] to optimise our processes.

| Rail bridge over road | 4D Model |

Figure 10-2 4D BIM of a bridge on the Cranbourne Line

(With permission: Western Program Alliance)

With the alliances actively engaging in the digital space and LXRP operating in an analogue mindset, the potential to learn, innovate and provide a platform for transforming the construction industry was recognised. The JCC had in place a 'Value Engineering' working group, which they divided into 'Design and Engineering and 'Digital Engineering' subcommittees to garner more significant insights and acquire knowledge about the emerging developments and innovations that could be shared between the alliances and LXRP and used to drive continuous improvement. At this juncture, we point out that LXRP only receives digital design models for various stages of a project. Emphasising this position, IV12 noted:

> LXRP is not a generator of digital data assets. But it's increasingly becoming a consumer and, hopefully, passing that information on to the asset owner and operator soon.

Unquestionably, the alliances have been driving the use of BIM and associated digital technologies in the LXRP as they seek to improve their productivity and performance. The DE strategy developed by the LXRP has been influenced somewhat by the alliances, but more so by the VDAS (Office of Projects Victoria, 2020), which was developed at the same time, though, as mentioned above, it was only approved in February 2021. While DE was widely used to deliver projects, LXRP recognised that its potential was unrealised, particularly for the state's asset owners. As noted by IV13, the benefits of DE, particularly for asset owners, were 'untapped'; a strategy is needed to ensure that benefits are realised.

10.3   Benefits of Digital Engineering

The benefits of DE technologies in construction have been widely espoused by academics, bloggers, software vendors and the like, which include improved safety on-site, productivity, data capture, communication and reduced rework and project costs (Bryde, Broquetas and Volm, 2013; Matthews et al., 2018; XYZ, 2022; Aliu and Oke, 2023). There is no doubt that DE can transform practice in construction with evidence explicitly demonstrated with the Elizabeth Line (Brown, 2023). However, it cannot be assumed that adopting DE technologies automatically results in benefits being realised.

When technology is introduced to an organisation, processes must be redesigned to accommodate required new workflows to support its effective use and augmenting employees' capabilities and skills (Matthews et al., 2018). Yet little consideration by construction organisations has been typically given to embracing change and upskilling employees. However, the tide is changing as DE is becoming a critical part of infrastructure delivery, and organisations are developing strategies to take advantage of its benefits and add value to their operations.

A rigorous evaluation of an organisation's DE strategy and underpinning business case is needed to justify the resourcing required to deliver benefits. The LXRP were cognisant of this requirement and its role in influencing its asset owners to rethink how a digital model delivery at handover could better service their needs to manage their assets during operations and maintenance. The LXRP, in their business case to the MTIA executive, indicated that returns from requiring a digital model at handover would not be expected for three to four years; benefits were not quantified. Moreover, the business case provided a realistic view of what could be achieved within the purported time frame, with IV13 noting that a goal was to "drive consistency across MTIA" so that a digital model at handover would meet its requirements and expectations. The benefits of the DE strategy are materialising, though yet to be quantifiable for LXRP, with IV13 stating:

> We're seeing some evolution through construction, staging and planning. We see evolution in the design and in the issue of construction [drawings] and as-built stages. And now, we are

looking at the asset handover phases. And for the different parts of the project and piecing it together.

Quantifying the benefits of technology is imperative to determine its efficacy and value. This point has been made by Andrew Wolstenholme, Co-Chair of the Construction Leadership Council (UK), within the context of smart construction, who cogently stated (KPMG, 2016):

> for the industry to value off-site manufacture and the benefit it brings truly, we must actively do more to quantify its value and recognise the crucial part it will play in transforming our industry.
>
> (p. 6)

Equally, this applies to the use of DE technologies. The expected benefits of DE for the LXRP and its asset owners will differ from that of alliances.

The benefits of DE during construction, for example, are often exaggerated to justify the use of technologies. For example, the promotion of the benefits of Autodesk® BIM 360™ (now rebranded as Autodesk Construction Cloud) by (Bliss, 2017), suggests that this technology can reduce rework in construction. According to Bliss (2017), 24% of claims for rework that materialise from construction are due to insufficient detail and inaccurate specifications and logistics. Furthermore, Bliss (2017) suggests that delays in making information available to people in the field result in more requests for information (RFI) and can contribute to people making errors. Bliss (2017) claims have no scientific merit and are contrary to empirical studies showing that rework is a product of dysfunctional project culture, structure, and processes (Love et al., 2022b). Bliss (2017) then states that Skanska, a global construction group, adopted BIM 360™ and experienced a staggering 948% return on investment (ROI) from its implementation. Again, no evidence supports the claim that this software can provide such financial returns.

We do not discount that benefits can materialise from utilising Autodesk® BIM 360™; in fact, quite the contrary, new workflows must be re-designed to support its use (Matthews et al., 2015; Matthews et al., 2018). Still, questions need to be raised around the figures presented by Bliss (2107) to justify its adoption. In a similar vein to Bliss (2017), Brown (2008), Foreshaw (2014), Hou *et al.* (2015) and Huh et al. (2023), to name a few, have been zealous in their claims of benefits that can materialise from implementing disruptive technologies in construction.

Practical field evidence, underpinned by a robust method, demonstrating the benefits of digital technologies in construction are few and far between (Love and Matthews, 2019; Love et al., 2022a). Though Love et al. (2020), for example, demonstrated that when a digital systems information model (SIM) (i.e., object-oriented environment) is used to document the design of electrical, control and instrumentation systems instead of computer aided design (CAD) it results in a 90% cost reduction and a corresponding improvement in productivity.

The Perth Transport Authority has recognised the value of SIM in producing accurate and reliable 'as-builts' for their rail assets, as any amendments to a design during the construction can be documented in real-time and, if approved, effectively integrated with their asset management system, and used for operations and maintenance (Love et al., 2018). Similarly, XYZ (2022), through their AR-for-quality technology and workflow, has reduced the need for rework, and achieved a three to six percent reduction in the cost of producing as-builts, and an estimated 20-day saving at handover, equating to an ROI of 9%.

With any new technology, benefits cannot be delivered without change, and change without benefits cannot be sustained (Matthews et al., 2018; Love and Matthews, 2019). In the case of construction, this change centres on organisations preparing themselves for a new way of working (e.g., education, upskilling, and new processes) where cyber-physical systems are viewed as key enablers to improved and sustained business performance.

For projects delivered using traditional procurement strategies, the benefits arising from using BIM, for example, are often marginal. A major issue contributing to this situation is the absence of integration and collaboration between organisations involved with design and construction processes. Furthermore, in the case of traditional procurement strategies, asset owners' requirements are seldom, if ever, considered and embedded into a project's design process.

Collaboration within projects and throughout their supply chains is needed to harness the benefits of digital technologies. The LXRP, aided by its PAA with its alliances and the established JCC, provide an environment where experiences and knowledge of implementing digital technologies can be shared, enabling their benefits to be realised. As mentioned above, the journey to developing a digital model for handover has begun, and a pilot study is underway.

10.4   Technological Innovation

As revealed in Chapters 3 and 9, the PAA provides a governance provision and incentive scheme to encourage cross-project knowledge sharing. Consequently, the LXRP, through its PAA, fosters an environment for the alliance's *desorptive capacity* to flourish. Desorptive capacity is an additional dynamic capability that enables the safe transfer of knowledge from one alliance to the other four (Bravo et al., 2020). It complements an organisation's *absorptive capacity,* which is its ability to identify, assimilate, transform, and use external knowledge, research, and practice (Cohen and Levinthal, 1990). Put differently, absorptive capacity measures the rate at which an organisation can learn and use scientific, technological, or other knowledge from its external environment – it measures the ability to learn. The transference of knowledge from one alliance to another does not develop absorptive capacity *per se.* An alliance cannot build its absorptive capacity if it only acquires and assimilates external knowledge, as it needs to transform and exploit it somehow (Zahra and George, 2002). So, if an alliance can transfer its newly acquired knowledge and integrate it within its project's learning routines, its absorptive capacity can be enhanced.

We now explore two DE R&D projects being undertaken by one of the alliances – engendered by the incentive to develop technological innovation – which are aiming to:

1. Use a knowledge-based engineering system (KBES) that utilises *descriptive* (insights into the past), *predictive* (understanding the future) and *prescriptive* (handling similar situations in the future) analytics to determine the risk paths of rework; and
2. Integrating technologies (e.g., BIM, drones, GIS, and photogrammetry) to identify deviations between 'as-designed' and 'as-built' items to help improve the accuracy of 'as-builts' at handover.

Both R&D projects are 'in-progress' and are innovative in their intent, as there have been no solutions to the problems being addressed previously reported in the extant literature.

10.4.1   *Driving data analytics with machine learning*

As part of an alliance's approach to continuous improvement, it implemented a War on Waste (WoW) strategy. Representatives from each of the alliance's participating organisations and key subcontractors formed a working committee to identify wastes and determine how they can be mitigated. The WoW committee initiated a pilot study to examine the wastes impacting productivity and performance. It emerged that rework, unnecessary transportation, and motion (i.e., unproductive time) were major wastes of concern.

Addressing rework became the committee's lodestone as it significantly contributed to the former wastes but also had the potential to affect safety adversely (Love et al., 2015). Yet as mentioned in Chapter 9, rework is a problem that often confronts construction projects, being considered uncomfortable knowledge or a *zemblanity* by organisations. As a result, rework risks are seldom considered and accounted for before and during construction, yet they are a stark reality of practice.

Acquiring knowledge about rework would allow the alliance to 'anticipate what might go wrong,' affording it adeptly to adapt and respond to events that may occur quickly and mitigate their negative consequences accordingly. The development of KBES provides the alliance not only knowledge about rework, which can be shared and used for continuous improvement but can also be used to estimate project costs and schedule better, improve planning and safety, and contribute to the accurate production of 'as-built' documentation.

Approaches for determining rework risks are absent from the literature (Love et al., 2022b). Furthermore, there is limited knowledge about *why* and *how* rework manifests and its costs beyond those contained in non-conformances reports (NCRs). Before a problem can be addressed, it must be acknowledged that it exists to begin the journey to amend the issue. The alliance openly recognised the problem and realised that rework was confined to being documented in NCRs and i/RFIs (internal), punch lists and change orders.

Figure 10–3 maps the locations and formats where rework-related data is stored within the alliance's information system. Indeed, acquiring data for decision-making is a complex task as it is often duplicated and incomplete. The data is noisy as no standard definitions exist, and the vocabulary and formats for capturing rework are inconsistent. With so much noise, a *smart data* approach was adopted, driven by determining what data are required for prediction rather than what is available, enabling it to be processed and turned into actionable information for effective decision-making.

The framework for developing a KBES is presented in Figure 10–4. As mentioned above, the research commenced with identifying the locations of rework. It started with analysing NCRs to garner an understanding of their links to other documents and the nature of vocabulary used to describe rework. Additionally, there was less noise in the NCRs, making their analysis much easier – they were deemed a higher-quality dataset to commence the research.

To reduce the dimensionality of the NCR data, topic analysis modelling – a text classification and ML technique and basic activity of Natural Language Processing (NLP) – was employed to discover latent topics in documents and determine relationships between words, topics, and documents (Matthews et al., 2022).

Traditional or shallow ML classifiers such as Support Vector Machine, Hidden Markov model or Conditional Random Fields have been widely used to classify text in construction (Zhong et al., 2020a; Zhong et al., 2020b). While traditional ML approaches have become popular for classifying text from documents in construction and can yield good classification performance, they are "time-consuming and inefficient to use due to their reliance on manual-handcrafted features" (Zhong et al., 2020b, p. 2; Matthews et al., 2023).

The limitations of traditional ML have resulted in the increasing use of a generative (i.e., unsupervised learning) statistical model for topic modelling, the Latent Dirichlet Allocation (LDA) algorithm. This algorithm can automatically extract features from text descriptions and identify the topics within documents. However, LDA requires "detailed assumptions and careful specification of hyperparameters" to ensure valuable results (Gallagher et al., 2017, p. 529). It also cannot model relationships between topics and performs poorly with short sentences (Lin et al., 2017).

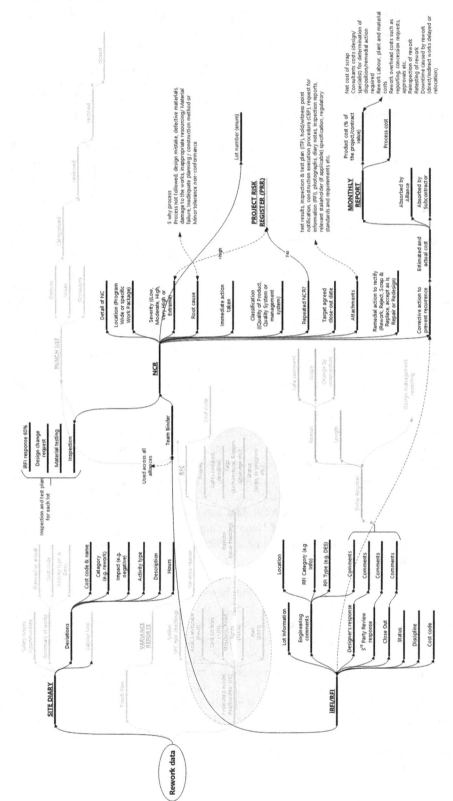

Figure 10-3 Mapping the sources of rework data in an alliance

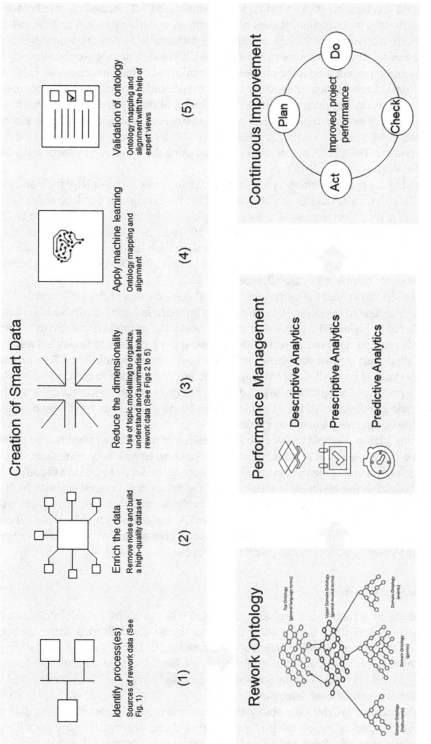

Figure 10-4 Framework for developing a KBES for assessing rework risks in an alliance

Correlation Explanation (CorEX) (i.e., a semi-supervised ML method) is an alternative to LDA and addresses its limitation. It does not assume an underlying model and instead learns maximally informative topics through an information-theoretic framework word-level domain knowledge can be flexibly incorporated within CorEX through anchor words, allowing topic separability and representation to be promoted with minimal human intervention. Thus, topics emerging from documentation echo actual rework events and the development of an ontology that mirrors the reality of practice. A detailed description of the methodology adopted to create the ontology and deploy the topic modelling, particularly the training of the CorEX algorithm, can be found in Matthews et al. (2022; 2023). Figure 10–5a provides an overview of the topic modelling process. The CorEX algorithm was trained using the VicRoad Standards and a dataset of 19503 NCRs (Love et al., 2018).

Determining the optimal number is a matter of trial and error for most NLP topic analysis algorithms. However, an advantage of CorEx over LDA, for example, is that it provides feedback in the form of a Total Correlation value for each topic – common issues arising in the NCRs related to 28-day compressive concrete tests. Figures 10–5b to 10–5d illustrate the types of analysis undertaken to understand nuances and idiosyncrasies associated with the topics manifesting from the 28-day compressive concrete strength tests across all NCRs obtained from eight-level crossing removals completed by the alliance.

Figure 10–5b shows that the output for 'Topic 8' contains two problematic words, 'Kevin' and 'Steph,' highlighted in yellow. We also note that the material supplier (whose name has been named redacted in Figure 10–5c) has been used to match the example document to 'Topic 23.' These words must be removed as part of the data-wrangling process to fine-tune topics. A binary correlation chart is produced (shown in Figure 10–5c) to determine the NCRs where topics consistently match (Figure 10–5c). Here Topics 5, 9, and 23 are found to exist in four NCRs and additionally performed a topic correlation, denoted in Figure 10–5d, to ascertain the number of topics appearing in documents. For example, Figure 10–5d shows that 'Topic 9' and 'Topic 23' appear in 300 documents (denoted by circles).

Performing such an analysis enables the identification of patterns and combinations of words to determine collective issues that lead to an NCR requiring rework to be completed. Thus, the determination of recurring themes assists with developing the scope to create a classification of causes embedded in the reality of practice. While the research is progressing with NCRs and other documents are examined, the varying use of vocabulary and data formats requires increasing attention to pre-processing and training the CorEX algorithm. New knowledge about the nature of rework is unfolding, informing the alliance's continuous improvement strategy and ability to implement practices to mitigate its occurrence.

10.4.2   *Integrating technologies*

The production of accurate asset registers juxtaposed with 'as-builts' for rail infrastructure assets is challenging as there is no standardised data format for delivering them, such as the Construction Operations Building Information Exchange (CoBie), available for buildings. In essence, CoBie is a subset of the Industry Foundation Class – the open standard for sharing and exchanging models between various software applications – that can be communicated using Excel worksheets or relational databases.

The CoBie format provides the ability to capture and record important project data at the point of origin, including equipment lists, product data sheets, warranties, spare parts lists, preventive maintenance schedules and so on. Once an asset is constructed and in service, this information is critical for supporting operations, maintenance, and asset management. Without

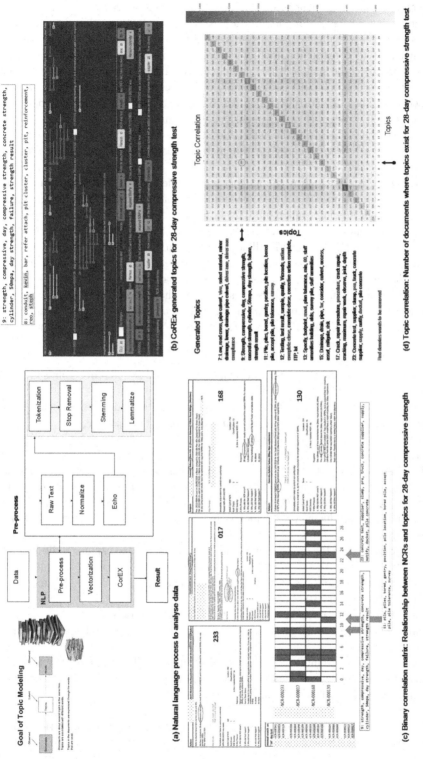

Figure 10-5 Topic modelling process and examples of outputs for 28-day concrete compressive strength test

a standard to capture and record from the design through to asset handover for infrastructure assets, producing a meaningful and accurate asset register that complements and mirrors the 'as-builts' becomes a challenge. Equally, this applies to creating a usable digital twin.

During an asset's construction, quality control (QC) plays a crucial role in ensuring that asset deliverables meet the standards and guidelines established in the contract by the PO. A QC procedure ensures that 'as-built' assets comply with specified requirements enabling deviations to be identified. Such deviations can then be rectified and the accuracy of 'as-builts' assured. The prevailing QC process is labour-intensive, time-consuming, and error-prone – it is unproductive and costly! Recognising the need to address this problem, an alliance initiated an R&D project to examine how technology can automatically detect deviations between 'as-designed' and 'as-built' objects.

The R&D project initiated by the alliance aims to develop and implement a new QC procedure using photogrammetry and GIS techniques to detect and tag objects automatically. The new QC procedure is being trialled on a $234 million rail-over-road project comprising a 2km rail bridge, a new station, an upgraded car park, bicycle facilities and a pedestrian bridge. Works commenced in December 2021 and will be completed in 2024.

The new QC that takes advantage of an array of integrated technologies is presented in Figure 10–6. Here photogrammetry techniques are used to capture images of the 'as-built' asset and produce a 3D model (Phase 1), objects are identified (Phase 2), and tagged and labelled (Phase 3) with relevant information extracted from the BIM, such as their location, size, and type. The workflow, as shown in Figure 10–6, for detecting and labelling objects and determining deviations comprises eight steps:

1. Produce coded targets in Agisoft Metashape[15] to be used as reference points to provide accurate orientation and scale.
2. Install the coded targets on the physical asset. Measure the distance between targets to determine the scale of the digital model constructed in Agisoft Metashape.
3. Calibrate the drone camera to ensure the photogrammetry produces high-quality, clear and accurate images.
4. Define the drone's flight path (imaging network) based on photogrammetry principles. The image network design for the drone's photogrammetry ensures high-quality data collection, which is critical for accurate 3D mapping and modeling and also reduces the number of images required for producing the 3D model (in this instance a reduction from 3000 to 250 was achieved).
5. Capture images using the drone. Hand-held cameras are used to capture additional images of key assets.
6. Import the drone images into Agisoft for processing and generate a 3D textured scene of the 'as-built' assets (Figure 10–7).
7. Import images of key assets into ArcMap to extract geolocations.
8. Export the key assets' attributes from the design [BIM] model, and map to corresponding ArcMap geolocations
9. Import both the Agisoft model and the geolocated asset attributes into ArcScene to detect and tag (i.e., using GIS) key assets/objects. The result is an 'as-built' GIS model, enhanced with semantic data extracted from the design (BIM) model.

The alliance is testing and validating the new QC procedure for several scenarios. Simply tagging and labelling objects with relevant information (such as location, size, and type) using GIS is helping site supervisors 'find the right information in the right location' promptly compared to previously having to sift through countless drawings. Identifying deviations between 'as-designed' and 'as-built' may require rework. Still, discovering and addressing the problem is more cost-effective before an asset is handed over to the operator. The newly QC procedure enabled by several technologies, which are interoperable with a model, provides much-needed knowledge that can be utilised to help create a digital 'as-built' model for handover, which the LXRP is striving to achieve.

Phase 1: Data collection: Collect images using photogrammetry techniques from site using UAV and hand-held cameras.

Phase 2: Object detection: Using the GIS system for object detection in the 3D [BIM] model.

Phase 3: Object labelling: Using the GIS system to tag the detected objects with relevant information.

Phase 4: Testing and validation: Testing the system's performance in various scenarios and validating its accuracy.

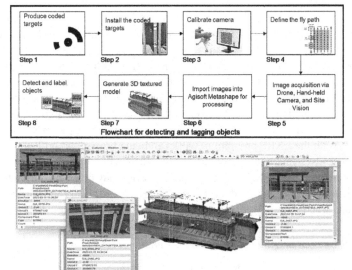

Figure 10-6 The proposed new QC procedure for detecting deviations

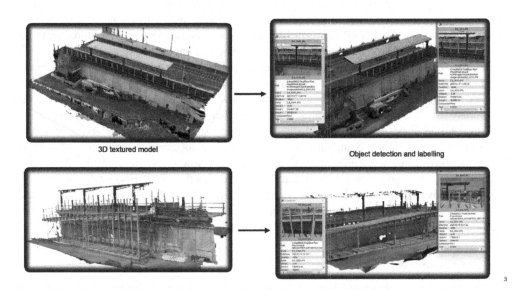

Figure 10-7 Examples of 3D textured models and object detection and labelling

10.5   LXRP: A Unique Project

The LXRP and its alliances acknowledge the critical role that DE plays in removing level cross-ings throughout the Melbourne Metropolitan area and constructing new assets such as stations, bridges, roads, car parks and community-based facilities. While the UK took a hierarchical approach to DE through its BIM mandate, within the LXRP, it has emerged bottom-up and top-down. The alliances possess an enthusiasm and willingness to adopt digital technologies and embrace change to reduce costs, improve the nature of work, workflows, quality, safety, and productivity and add value throughout an asset's life.

The level of DE deployed throughout the LXRP aligns with international best practices for rail infrastructure procurement (e.g., LOD 350 model). Indeed, there are areas where LXRP needs to catch up, such as the requirement of a digital model at handover for asset management, but this is being addressed and will no doubt become a requirement in future projects. Equally, it is leading the way as the PAA incentivises the alliances to experiment and develop innovative technological solutions – through the established JCC – to problems confronting practice, which are jointly shared. The R&D projects we have identified above are just two examples that the alliances are carrying out in their pursuit of continuous improvement.

With the LXRP program of works all being brownfield projects, integrating existing assets and legacy systems with the digital models for newly constructed assets on the same rail line poses challenges for the MTIA and operator. Lessons may be gleaned from the delivery of other rail infra-structure projects around the globe, such as the Elizabeth Line (Brown, 2023). But such lessons are not directly transferred as the context and conditions facing LXRP and its asset owners are unique.

10.5.1   *Managing knowledge in a program*

Even though the LXRP is a unique and complex program of work, lessons can be learned from its experiences with engendering a DE strategy and its implementation by alliances in their pro-jects. The LXRP's collaborative contracting model through its PAA and JCC fosters knowledge sharing, and while some aspects of work undertaken by alliances are unique, many elements are repeated. Learning from experiences garnered, particularly mistakes made, provide the basis for stimulating innovation. The capture, storage, transfer and sharing of knowledge within and between alliances and the LXRP and its utilisation has been and continues to be fundamental to successfully removing level crossings and installing new and upgraded assets.

The Micro Focus IDOL technology[16] has been implemented (Crozier, 2022) because knowl-edge is dispersed in the LXRPs' numerous systems and across those of the alliances and there is a need to capture and connect all organisations involved with delivering the program of work to stimulate learning (e.g., accessing and reusing knowledge acquired from previous projects),

With IDOL in place, access to knowledge repositories, such as the Intra-and-extranet, Engi-neering Hub, and InEight's Teambinder (housing technical documents, drawings, and official documentation), which were previously unconnected, are now searchable. The LXRP recog-nises the importance and power of information and the benefits of the IDOL technology as it provides improved decision-making, risk management and stimulates learning. It can "unlock value" and transform practices as silos in projects are dismantled and information becomes readily available, bolstering the potential to use technology effectively (Crozier, 2022).

10.6   Conclusions

This chapter examines the emergence of the LXRP's DE strategy to stimulate technological in-novation across its program of projects. The DE strategy, approved in February 2021, emerged out of a need to drive change and transform how assets are managed, maintained, and operated. Consequently, the LXRP is conducting a pilot study to create a digital model at handover instead of relying on both a partial model (i.e., LOD 350) and drawings. By the same token, the alliances engaged with DE, specifically with BIM, at the onset of the program in 2015. Additionally, the alliances have actively pursued – incentivised by the PAA to engage in R&D – a strategy to determine how digital technology can effectively support their continuous improvement initiatives and engender innovation.

The JCC acts as a catalyst for the alliances to share and transfer their knowledge and innovations between them. The introduction of a knowledge-based sharing platform by the LXRP, leveraged by the IDOL technology, provides alliances access to information that can be used to plan, design, construct, and hand over assets more efficiently and effectively.

Notes

1 Restructured to become the Victorian Infrastructure Delivery Authority (VIDA) in April 2024, see Chapter 3.
2 LOD 300 'Accurate Modelling and Detailed Design' involves accurate modelling and detailed shop drawings, where elements are defined with specific assemblies, precise quantity, size, shape, location, and orientation. Non-geometric information can also be embedded within model elements at LOD 300. At LOD 350 'Great Detail and Construction Documentation' includes additional detail and elements to LOD 300 representing how they interface with various systems. It also provides clear graphics and written definitions.
3 LOD 400 'Fabrication and Assembly' represented as specific assemblies, complete with fabrication, assembly, and detailed information, in addition to precise quantity, size, shape, location, and orientation. Non-geometric information can be embedded within model elements at LOD 400.
4 Details of the Finnish BIM requirements for infrastructure can be found at: www.infrakit.com/finnish-bim-requirements-for-infrastructure-2015-released/
5 Details about BIM and public procurement in Germany can be found at: https://public-buyers-community.ec.europa.eu/communities/bim-and-public-procurement/resources/german-bim-implementation-strategy-federal. Also details about Germany's governmental BIM initiative, BIM4INFRA2020 can be found in Borrman, A., Konig, M. and Liebich, T. (2021). Germany's government BIM initiative – The BIM4INFRA2020 project implementing the roadmap. *8th International Conference on Computing in Civil and Building Engineering (ICCCBE)* Santos E. T. and S. Sheer. São Paulo, Brazil, Brazil, pp. 425–465 .
6 Details about the Norwegian (Statsbygg) BIM Manual Version 1.2.1 can be found at: https://dok.statsbygg.no/wp-content/uploads/2020/06/statsbyggs-bim-manual-1–2-1_en_20131217.pdf
7 Details about the Singaporean BIM Guide Version 2 can be found at: www.corenet.gov.sg/media/586132/Singapore-BIM-Guide_V2.pdf
8 Details of the BIM Plan for procurement in Spain can be found at: https://cibim.mitma.es/
9 Details of the UK BIM Framework can be found at: www.ukbimframework.org/. The BIM Level 3 implementation strategic plan can be found at: https://assets.publishing.service.gov.uk/government/uploads/system/uploads/attachment_data/file/410096/bis-15–155-digital-built-britain-level-3-strategy.pdf
10 The Centre for Digital Britain details can be found at: www.cdbb.cam.ac.uk/. The overarching mission of CDBB was "'to develop and demonstrate policy and practical insights that will enable the exploitation of new and emerging technologies, data and analytics to enhance the natural and built environment, thereby driving up commercial competitiveness and productivity, as well as citizen quality of life and well-being". Case studies, tools and guidance documents can be found on this website and information about CDBB's involvement with the 'Construction Innovation Hub' and 'National Digital Twin Programme.'
11 The UK BIM Framework Guidance can be viewed at: https://ukbimframeworkguidance.notion.site/UK-BIM-Framework-Guidance-20a045d01cfb42fea2fef35a7b988dbc
12 The UKBIMA was launched in 2016 as a cross-industry alliance reaching the whole built environment to advocate the adoption of BIM in the UK and wider digital transformation. It is now called 'nima' and details can be found at: https://wearenima.im/

13 Industry 4.0 refers to the fourth industrial revolution. Industry 1 was mechanisation, 2 was electrification and 3 was computerisation, 4 includes a range of digital technologies.
14 Bidirectional means transferring data between systems A and B in both directions.
15 Agisoft Metashape is a software product that performs photogrammetric processing of digital images and generates 3D spatial data for GIS applications, visual effects production, and to produce indirect measurements of objects of various scales.
16 IDOL is explained at www.microfocus.com/en-us/products/information-data-analytics-idol/overview

References

Aliu, J. and Oke, A.E. (2023). "Construction in the digital age: Exploring the benefits of digital technologies." *Built Environment Project and Asset Management*. 13 (3): 412–429.

Attaran, M. and Celik, B.G. (2023). "Digital Twin: Benefits, use cases, challenges, and opportunities." *Decision Analytics Journal*. 6: 100165.

Australian Constructors Association (2017). *Changing the Game: How Australian Can Achieve Success in the New World of Mega-projects*. on-line, Australian Construction Association.

Azhar, S. (2011). "Building Information Modeling (BIM): Trends, benefits, risks, and challenges for the AEC industry." *Leadership and Management in Engineering*. 11 (3): 241–252.

Bailey, C., Mankin, D., Kelliher, C. and Garavan, T. (2018) *Human Resource Management*, Oxford, Oxford University Press.

Barbosa, F., Woetzel, J., Ribeirnho, M., Sridhar, M., Parsons, M., Bertram, N. and Brown, S. (2017). "Reinventing construction: A route to higher productivity." www.mckinsey.com/capabilities/operations/our-insights/reinventing-construction-through-a-productivity-revolution

Bliss, A. (2017). "5 ways BIM 360 Docs reduces construction rework." *B Autodesk 360*, Autodesk. https://connect.bim360.autodesk.com/bim-360-docs-reduces-rework

Borrman, A., Konig, M. and Liebich, T. (2021). "Germany's government BIM initiative – The BIM4INFRA2020 project implementing the roadmap." In *8th International Conference on Computing in Civil and Building Engineering (ICCCBE)*, Santos, E.T. and S. Sheer (eds), pp. 425–465. São Paulo, Brazil.

Bravo, M.I.R., Stevenson, M., Moreno, A.R. and Montes, F.J.L. (2020). "Absorptive and desorptive capacity configurations in supply chains: An inverted U-shaped relationship." *International Journal of Production Research*. 58 (7): 2036–2053.

Brown, C. (2023). "Grasping the nettle: Integrating the first UK's digital railway." In *Crossrail Learning Legacy*. London, Crossrail. https://learninglegacy.crossrail.co.uk/documents/grasping-the-nettle-integrating-the-uks-first-digital-railway/

Brown, K. (2008). *BIM-implications for Government*, Brisbane, Cooperative Research Centre for Construction Innovation. www.construction-innovation.info/images/pdfs/2004–032-A_BIM_Case_Study_Final_(20081030).pdf

Bryde, D., Broquetas, M. and Volm, J.M. (2013). "The project benefits of Building Information Modelling (BIM)." *International Journal of Project Management*. 31 (7): 971–980.

Buddoo, N. (2021). "How digital tools are transforming the delivery of HS2." In *New Civil Engineer*, London, Institute of Civil Engineers www.newcivilengineer.com/innovative-thinking/how-digital-tools-can-help-with-major-project-delivery-04–08–2021/

Cabinet Office (2011). *Government Construction Strategy*, London, UK Government Cabint Office. https://assets.publishing.service.gov.uk/government/uploads/system/uploads/attachment_data/file/61152/Government-Construction-Strategy_0.pdf

Cabinet Office (2012). *Government Construction Strategy One Year On Report and Action Plan Update*, London, UK Government Cabinet Office. https://assets.publishing.service.gov.uk/government/uploads/system/uploads/attachment_data/file/61151/GCS-One-Year-On-Report-and-Action-Plan-Update-FINAL_0.pdf

Centre for Digital Built Britain (CDBB) (2018). *Gemini Principles*, Cambridge, University of Cambridge. www.cdbb.cam.ac.uk/DFTG/GeminiPrinciples

Clarke, G. (2017). "Government and industry cement deal to give UK construction the edge." London, Department for Business, Energy and Industrial Strategy. www.gov.uk/government/news/government-and-industry-cement-deal-to-give-uk-construction-the-edge

Cohen, W.M. and Levinthal, D. (1990). "Absorptive capacity: A new perspective on learning and innovation." *Administrative Science Quarterly*. 35 (1): 128–152.

Crozier, R. (2022). "Vic level crossing removal project to digitally 'change' construction sector – With knowledge capture, sharing and smart search." *itnews*. www.itnews.com.au/news/vic-level-crossing-removal-project-to-digitally-change-construction-sector-578664

Department of Planning Transport and Infrastructure (2017). "Building information modelling implementation." Adelaide, Australia, Government of South Australia..www.bpims.sa.gov.au.

Department of State Development Infrastructure Local Government and Planning (2018). "Digital enablement for Queensland infrastructure: Principles for BIM implementation." Brisbane, Queensland Government. www.statedevelopment.qld.gov.au/__data/assets/pdf_file/0020/32915/bim-principles.pdf

Department of State Development Infrastructure Local Government and Planning (2022). "State infrastructure strategy." Brisbane, Queensland Government. www.statedevelopment.qld.gov.au/industry/infrastructure/state-infrastructure-strategy

Department of Treasury and Finance, V. (2016). "Asset management accountability framework." Melbourne, Victoria State Government. www.dtf.vic.gov.au/infrastructure-investment/asset-management-accountability-framework

Foreshaw, J. (2014). "Woodside tags a smart saving idea." *The Australian*. www.theaustralian.com.au/technology/woodside-tags-asmart-savings-idea/newstory/c9b2268d11f5f202fca98b6ce52337cf

Gallagher, R.J., Reing, K., Kale, D. and Ver Steeg, G. (2017). "Anchored correlation explanation: Topic modeling with minimal domain knowledge." *Transactions of the Association for Computational Linguistics*. 5: 529–542.

Gallaher, M.P., O'Connor, A.C., Dettbarn Jr., J.L. and Gilday, L.T. (2004). *Cost Analysis of Inadequate Interoperability in the U.S. Capital Facilities Industry*, Gaithersburg, Maryland.

Godley, M. (2023). "Digital engineering saves time, improves safety and reduces costs." *CPB Supply Chain*. www.cpbcon.com.au/en/news-and-media/2023/digital-engineering-saves-time-improves-safety-and-reduces-costs

Grogan, A. (2023). "McConnell Dowell builds on its digital engineering capabilities." Melbourne, Inside Construction. www.insideconstruction.com.au/news/features/mcconnell-dowell-builds-on-its-digital-engineering-capabilities/

Hawkes, H. (2021). "How digital engineering tools helped this major infrastructure project to life." Create, Engineers Australia. https://createdigital.org.au/digital-engineering-tools-level-crossing-removal-project/

HM Government (2016). Level 3 *Building Information Modelling – Strategic Plan*. Digital Built Britain. https://assets.publishing.service.gov.uk/government/uploads/system/uploads/attachment_data/file/410096/bis-15–155-digital-built-britain-level-3-strategy.pdf

Hou, L., Wang, X. and Truijens, M. (2015). "Using augmented reality to facilitate piping assembly: An experiment-based evaluation." *Journal of Computing in Civil Engineering*. 29 (1): 05014007.

HS2 (2022). *Digital Engineering and BIM*. www.hs2.org.uk/building-hs2/innovation/digital-engineering-and-bim/

Huh, S.-H., Ham, N., Kim, J.-H. and Kim, J.-J. (2023). "Quantitative impact analysis of priority policy applied to BIM-based design validation." *Automation in Construction*. 154: 105031.

Infrastructure and Projects Authority (IPA) (2016). *Government Construction Strategy 2016–2020, Report of HM Treasury and Cabinet Office*. London, Infrastructure and Projects Authority. https://assets.publishing.service.gov.uk/government/uploads/system/uploads/attachment_data/file/510354/Government_Construction_Strategy_2016–20.pdf

KPMG (2016). "Smart construction: How offsite manufacturing can transform our industry." https://assets.kpmg.com/content/dam/kpmg/pdf/2016/04/smart-construction-report-2016.pdf

LaShell, D. and Goldman, M. (2021). "5D: The new frontier for digital twins online." *WhereNext Magazine*. www.esri.com/about/newsroom/publications/wherenext/5d-the-new-frontier-for-digital-twins/

Lin, K.P., Shen, C.Y., Chang, T.L. and Chang, T.M. (2017). "A consumer review of driven recommender service for web e-commerce." In *2017 IEEE Conference on Service-oriented Computing and Applications (SOCA)*, Benatallah, B., C. Huemer and T. Ito (eds), pp. 206–210. Kanazawa, Japan, IEEE.

Love, P.E.D., Ika, L., Matthews, J. and Fang, W. (2021a). "Shared leadership, value and risks in large scale transport projects: Re-calibrating procurement policy for post COVID-19." *Research in Transportation Economics*. 90 (100999.

Love, P.E.D., Ika, L.A., Matthews, J., Li, X. and Fang, W. (2021b). "A procurement policy-making pathway to future-proof large-scale transport infrastructure assets." *Research in Transportation Economics.* 90: 101069.

Love, P.E.D. and Matthews, J. (2019). "The 'how' of benefits management for digital technology: From engineering to asset management." *Automation in Construction.* 107: 102930.

Love, P.E.D., Matthews, J., Fang, W. and Luo, H. (2022a). "Benefits realization management of computer vision in construction: a missed, yet not lost, opportunity." *Engineering.*

Love, P.E.D., Matthews, J., Simpson, I., Hill, A. and Olatunji, O.A. (2014). "A benefits realization management building information modeling framework for asset owners." *Automation in Construction.* 37: 1–10.

Love, P.E.D., Matthews, J., Sing, M.C.P., Porter, S.R. and Fang, W. (2022b). "State of science: Why does rework occur in construction? what are its consequences? and what can be done to mitigate its occurrence?" *Engineering.* 18: 246–258.

Love, P.E.D., Matthews, J. and Zhou, J. (2020). "Is it just too good to be true? Unearthing the benefits of disruptive technology." *International Journal of Information Management.* 52: 102096.

Love, P.E.D., Sing, M.C.P., Ika, L.A. and Newton, S. (2019). "The cost performance of transportation projects: The fallacy of the Planning Fallacy account." *Transportation Research Part A: Policy and Practice.* 122: 1–20.

Love, P.E.D., Teo, P., Carey, B., Sing, C.-P. and Ackermann, F. (2015). "The symbiotic nature of safety and quality in construction: Incidents and rework non-conformances." *Safety Science.* 79: 55–62.

Love, P.E.D., Zhou, J., Matthews, J., Lavender, M. and Morse, T. (2018). "Managing rail infrastructure for a digital future: Future-proofing of asset information." *Transportation Research Part A: Policy and Practice.* 110: 161–176.

Manyika, J., Ramaswamy, S., Khanna, S., Sarrazin, H., Pinkus, G., Sethupathy, G. and Yaffe, A. (2015). "Digital America: A tale of have and have-mores." McKinsey & Company. www.mckinsey.com/industries/technology-media-and-telecommunications/our-insights/digital-america-a-tale-of-the-haves-and-have-mores

Matthews, J., Love, P.E.D., Heinemann, S., Chandler, R., Rumsey, C. and Olatunj, O. (2015). "Real time progress management: Re-engineering processes for cloud-based BIM in construction." *Automation in Construction.* 58: 38–47.

Matthews, J., Love, P.E.D., Mewburn, J., Stobaus, C. and Ramanayaka, C. (2018). "Building information modelling in construction: Insights from collaboration and change management perspectives." *Production Planning & Control.* 29 (3): 202–216.

Matthews, J., Love, P.E.D., Porter, S. and Fang, W. (2023). "Curating a domain ontology for rework in construction: Challenges and learnings from practice." *Production Planning & Control.* 1–16.

Matthews, J., Love, P.E.D., Porter, S.R. and Fang, W. (2022). "Smart data and business analytics: A theoretical framework for managing rework risks in mega-projects." *International Journal of Information Management.* 65: 102495.

Mintzberg, H. (1987). "The strategy concept I: Five Ps for strategy." *California Management Review.* 31 (1): 11–24.

Morkos, R. (2022). "Britain & BIM: Digital adoption drives construction breakthrough." *Forbes.* www.forbes.com/sites/forbestechcouncil/2022/09/02/britain--bim-digital-adoption-drives-construction-breakthrough/?sh=594fee1417a8

Nadel, J. (2023). "John Holland works to digital construction strategy 2025." *itnews.* www.itnews.com.au/news/john-holland-works-to-digital-construction-strategy-2025-594309

NATSPEC. (2022). *NATSPEC National BIM Guide.* Construction Information Systems Limited. https://bim.natspec.org/documents/natspec-national-bim-guide

NBS (2017). *National BIM Report 2017.* Newcastle upon Tyne, UK, RIBA Enterprises. www.thenbs.com/knowledge/nbs-national-bim-report-2017

Office of Projects Victoria (2020). *Victorian Digital Asset Strategy.* Melbourne, State of Victoria (Department of Treasury and Finance. www.vic.gov.au/victorian-digital-asset-strategy

Peplow, M. (2016). "London's Crossrail is a $21 billion test of virtual modelling." London, IEEE Spectrum. https://spectrum.ieee.org/londons-crossrail-is-a-21-billion-test-of-virtual-modeling

Regan, M., Smith, J. and Love, P.E.D. (2017). "Financing of public private partnerships: Transactional evidence from Australian toll roads." *Case Studies on Transport Policy.* 5 (2): 267–278.

Remmert, H. (2022). "What is a digital twin?" DIGI. www.digi.com/blog/post/what-is-a-digital-twin

Sacks, R., Eastman, C., Lee, G. and Teicholz, P. (2018) *BIM Handbook: A Guide to Building Information Modelling for Owners, Contractors and Facility Managers*, Hoboken, N.J., John Wiley & Sons.

Succar, B. (2009). "Building information modelling framework: A research and delivery foundation for industry stakeholders." *Automation in Construction.* 18 (3): 357–375.

Technavio (2022). "Digital twin market by end-user, deployment, and geography – Forecast and analysis 2022–2026." www.technavio.com/report/digital-twin-market-size-industry-analysis#:~:text=The%20digital%20twin%20market%20share,at%20a%20CAGR%20of%2039.48%25

Transport for New South Wales (2018). *The Digital Engineering Framework.* Sydney, Transport for New South Wales. www.transport.nsw.gov.au/digital-engineering/digital-engineering-framework-0

Transport for New South Wales (2022). *Digital Engineering Framework. DMS-ST-208.* Sydney, Transport for New South Wales. www.transport.nsw.gov.au/system/files/media/documents/2022/Digital-Engineering-Framework-v4.0.pdf

Trimble Connect (2021). "Hong Kong deploys connected construction for country's largest-ever infrastructure project." DBM Vircon Use Case, Trimble Connect. https://connect.trimble.com/dbm-vircon-use-case

Whyte, J., Lindkvist, C. and Jaradat, S. (2016). "Passing the baton? Handing over digital data from the project to operations." *Engineering Project Organization Journal.* 6 (1): 2–14.

XYZ (2022)." Investing in innovation: ROI of AR in construction." *XYZ Realty.* www.xyzreality.com/resources/9-x-roi

Young, N.W.J., Jones, S.A., Bernstein, H.M. and Gudgel, J.E. (2014). *The Business Value of BIM: Getting Building Information Modelling to the Bottom Line. Smart Market Report.* McGraw-Hill Construction. https://images.autodesk.com/adsk/files/final_2009_bim_smartmarket_report.pdf

Zahra, S.A. and George, G. (2002). "Absorptive capacity: A review, reconceptualization, and extension." *Academy Of Management Review.* 27 (2): 185–203.

Zhong, B., He, W., Huang, Z., Love, P.E.D., Tang, J. and Luo, H. (2020a). "A building regulation question answering system: A deep learning methodology." *Advanced Engineering Informatics.* 46: 101195.

Zhong, B., Pan, X., Love, P.E.D., Sun, J. and Tao, C. (2020b). "Hazard analysis: A deep learning and text mining framework for accident prevention." *Advanced Engineering Informatics.* 46: 101152.

11 LXRP and Community Engagement

*Derek H.T. Walker, Kirsi Aaltonen, Peter E.D. Love
and Mark Betts*

11.1   Introduction

How does the LXRP manage its community engagement and legacy transformation strategy?

This chapter aims to provide an appreciation of appropriate strategic community engagement practices and the rationale for their adoption and discuss how internal project stakeholders (the alliance project team (APT) and subcontractor and supplier engagement relationships were undertaken were integrated (Chapter 5) and how they collaborated (Chapter 6). This chapter focuses on the LXRP as an alliancing stakeholder engagement case study.

We begin with a short theoretical introduction to stakeholder engagement and a discussion of the LXRP case study example. To fully appreciate the LXRP legacy stakeholder engagement strategy, we interviewed over 20 senior LXRP staff, three of whom had intimate and overview knowledge about how the strategy was developed and operated. We then widen the discussion with global stakeholder engagement project examples to compare and contrast the LXRP example.

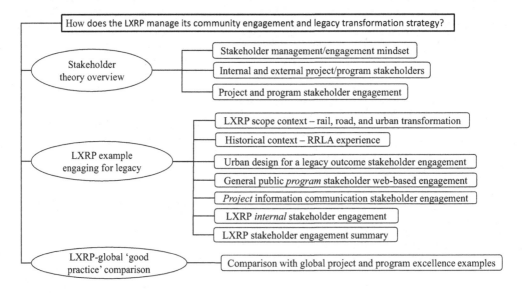

Figure 11-1 Chapter structure

DOI: 10.1201/9781003389170-11

11.2 Stakeholder Engagement Theory

Interest in stakeholder engagement has recently grown in the project management (PM) literature. Many literature sources discuss stakeholder identification theory from numerous perspectives. Early project stakeholder theory stretches back to the 1980s (Littau, Jujagiri and Adlbrecht, 2010) focusing on defining the stakeholder and how they may impact project initiation and delivery success. Littau et al. (2020) reviewed the project management stakeholder literature over 25 years and segregated trends into stakeholders having *an interest or stake in* a project (passive), about them being able to *influence and impact* (potentially active) and having an interest and influence and impact (active) needing to be addressed. The general *stakeholder management* concept proposed that stakeholder characteristics included: power (to or over); legitimacy (whether stakeholders should be managed or considered); urgency (how their impact or influence may impact project performance); and salience (the extent to which they 'count' or not, and how they perceive their identity and the identity of the project delivery organisation and the project owner) (Mitchell, Agle and Wood, 1997).

More recent research has expanded our understanding of project stakeholders from a *legitimacy* (Derakhshan, Mancini and Turner, 2019) and *identity* (Cornelissen, Haslam and Balmer, 2007) perspective. Legitimacy is important to *internal stakeholders,* project delivery participants and government departments associated/liaising with project delivery teams. Project delivery stakeholders need to be recognised as *relevant, valuable*, and *valued* – i.e., being legitimate recipients of the confidence level required for effective project delivery. It is also important for *external stakeholders* (affected communities, local businesses, facility users, affected general public etc.) to accept the project delivery entity as legitimately doing so, and that the project has a legitimate purpose. Legitimacy facilitates support for the relevance of a project's strategic aims and validity, and confidence in decisions made by the delivery entity during project delivery.

Infrastructure project identity may be crafted at three levels of shared meaning (Cornelissen, Haslam and Balmer, 2007). *Corporate-level* identity is reflected by how the project owner creates, maintains, and communicates its raison d'être and encourages shared meaning of its true purpose that often extends beyond the delivery of a 'thing' to creating social value that may vary in its composition (Gil, 2021, p. 15). Its mission, vision and strategic aims influence how project delivery employees perceive their project delivery role. Identity is shaped by what the project owner is *perceived* to stand for, its values and expected performance (Hollensbe et al., 2014). This can influence inter-team and interpersonal relationships and motivation to collaborate (Hackman and Oldham, 1976).

Organisational identity informs how the project delivery entity perceives its members' shared meaning about the nature of their project delivery role/task. This may be influenced through effective developmental team coaching (Hackman and Wageman, 2005, p. 271) and internal stakeholder engagement. The main aim of many infrastructure projects is social value delivery (Drouin and Turner, 2022, Chapter 1). However, this may be unclearly articulated. Participating team entities may assume a 'business-as-usual' (BAU) expectation, based on shared iron triangle[1] performance meaning. The above influences the individual's *perceived* identity as part of the project organisation.

Individual project delivery participants often align their identity with the espoused and enacted project organisations' values. At this level, people within the delivery entity, and external stakeholders, may adopt an identity of being either a part of their home-based organisation or the project delivery organisational entity. This may profoundly impact how they share meaning and are motivated. Hackman and Oldham's (1976) model of task perception explains how three psychological states can trigger high-level motivation to be committed to identified behaviours, such as the extent of commitment to excellence or mediocrity. Individuals associated the task with their identity, their values and aspirations. Three psychological states govern how tasks may be seen as meaningful, the individual's role as being responsible and their being provided high-quality feedback on their performance contribution. An overwhelming number of LXRP interviewee unsolicited comments revolved around their pride and delight at being associated with this program, they all seemed to fully endorse and appreciate the urban improvements and industry and cultural transformation legacy.

Social alliancing and IPD infrastructure projects call for team integration and collaboration to achieve best-for-project outcomes (Department of Infrastructure and Transport, 2011). Therefore, corporate and project organisational level identity need to be characterised by an identity that encourages individual identification with clear project values that support a best-for-project mindset and high levels of collaboration (Chapter 6). If infrastructure projects' aims rely upon IPD/Alliancing team integration and collaboration, then a concerted effort is required to ensure alignment of corporate, organisational, and individual *internal stakeholder* identity with values and goals consistent with social infrastructure outcomes.

Similarly, the Hackman and Oldham (1976) motivation model may be applied to *external stakeholder* identity. These stakeholders may be persuaded to adopt positions and act consistently with their perceived identity. Individually or in groups, their task of supporting or opposing a project may be influenced by the above-described three psychological states forming their identification with the program's values. Therefore, a concerted effort is also needed to align corporate, organisational, and individual *external stakeholder* identity with values and goals consistent with social infrastructure outcomes. Therefore, stakeholder engagement elicits different responses to stakeholder management. Ali (2023, p. 70) makes a cogent point about stakeholder engagement's purpose.

> [alliance participants] organisations define the project concept (master plan) and shared project goals. This would help the project alliance to understand the context of the hospital construction project, related challenges, complexities, dynamics, and requirements that the project should meet. This would also help the project alliance to develop a fully informed master plan that meets the needs of the entire network of healthcare stakeholders. When a project alliance understands the business side of the healthcare process, then they are in a better position to plan, design, and construct a state-of-the-art healthcare facility – a hospital – that meets the care needs. This would develop cooperation among project stakeholders which is a prerequisite for developing collaborative relationships and can be viewed as a starting point for project stakeholders to align their interests to the shared project goals.

The hospital alliance project related quote applies equally to other project types. The central concept is that participants gain a fuller and deeper understanding of the project's purpose through stakeholder engagement, enabling them to plan and act better. They are often more highly motivated for a best-for-project outcome because they see the relevance of their efforts.

11.2.1   *Stakeholder management/engagement mindset*

Stakeholder engagement theory perspectives include: the business perspective of managing business stakeholders (Freeman, 1984; Donaldson and Preston, 1995); a project management perspective in which stakeholders' relationships are managed (Winch, 2004; Bourne, 2005;

Aaltonen, 2010); and, more recently, including from an integrated project delivery (IPD) and alliancing literature perspective, where stakeholders are *engaged* with rather than *managed* (Aaltonen et al., 2020).

Two different conceptual project-owner stakeholder relationship mindsets exist. For decades, stakeholder *management* has been associated with risk management, minimising adverse project opponent impact, or taking an opportunity to encourage project support (Eskerod and Huemann, 2013; Huemann, Eskerod and Ringhofer, 2016). Instead of stakeholders being a negative or manipulative influence, emphasis has shifted towards their value co-generation potential (Porter and Kramer, 2011; Tjahja and Yee, 2022). Therefore, the empathic design of products/services and their delivery with strong user input often involves prototyping and modelling (Leonard and Rayport, 1997) resulting in co-value production continually delivering superior outcomes. This 'new stakeholder theory (NST)' co-value perspective extends co-value generation opportunities. NST holds that co-value generation may occur after a project's stage-gate process approval to accommodate engagement opportunities with external stakeholders to jointly produce additional value (value creep) – often relying upon additional resource expenditure (Gil, 2023). Therefore, as Gil argues, project owners and project delivery teams should be prepared for this situation.

Gil and Fu (2022) take a value co-generation and value distribution perspective of stake-holder relations and argue that external stakeholders may attempt to *manage* the interaction process opportunistically to extract maximum value in return for minimum interference through hold-up blocking strategies or they may choose to *engage* with the project team in a positive collaborative manner. The engagement process involves negotiation between actors to decide how actors should behave to achieve *reasonable* win-win outcomes that achieve the project purpose while creating and distributing project-specific value that the participating stakeholders value. The project owner may assume that direct project delivery participants and non-market stakeholders (project product users and other externally impacted stakeholders) are beneficiar-ies of project value. Still, the degree of effort-resource contribution benefit distribution may be contested or poorly understood.

Stakeholder perceptions about a project delivery entity's legitimacy, their relationship with the entity, and the extent that they can trust that entity (to be honest, ethical, and professional) impact stakeholder engagement willingness. The project delivery organisation must nurture that sense of legitimacy to engage in meaningful interaction and communication with stakehold-ers. This affects how trust is formed and maintained. Trust, as discussed in Davis and Walker (2020) building upon seminal work by Mayer et al. (1995), is based on the perceived ability, benevolence and integrity of the entity and the trustee's willingness to risk trusting that entity. Legitimacy positively or negatively shapes that impression of ability, benevolence, and integ-rity. For example, suppose the project delivery organisation promises X, Y, or Z, and it does. In that case, its legitimacy is credible, and its identity tends towards being able, benevolent and/ or having sound integrity. If it fails to deliver a promise, then the identity suffers from illegiti-macy. Each organisational public announcement or internal staff communication, triggers a test of trust to build, maintain or degrade that organisation's stakeholder trust. Therefore, having capable and professional stakeholder relationship staff is vital in managing the mindset and perceptions stakeholders may hold about the delivery organisation.

11.2.2 *Internal and external project/program stakeholders*

Alliance stakeholders include numerous direct internal members, such as alliance participant organisations and the project supply chain, their families and the individuals and organisations to which the alliance project team (APT) are accountable. Much of this book deals with how

APT participants, as vital *internal* stakeholders, conduct themselves and engage in an integrated and collaborative manner. Internal stakeholders are discussed in Chapter 6 from collaboration and Chapter 5 from an integration perspective. However, there remains a gap in IPD/Alliancing research into *how* the APT engages with impacts the non-market stakeholder community, before design or post-design, during project delivery, and into the infrastructure's operational phase. Few studies focus on how an APT engages with communities impacted by a project. We are aware of only one alliance project study being published (Smith, Anglin and Harrisson, 2010; AAA, 2012) or explicitly analysed from a community engagement perspective (Lloyd-Walker and Walker, 2017). Drouin and Turner (2022, Chapter 3) discuss of internal stakeholder (project participants) consideration and engagement in external stakeholders and how megaprojects are 'marketed' to stakeholders (2022, Chapter 5).

Gil and Pinto (2018) argue that programs of projects and large megaprojects (such as Cross-rail, Heathrow Terminal 2, the 2012 London Olympics, and High Speed 2 (HS2)) require a strategic organisational design issue to be solved that allows a polycentric focus on stakeholder engagement. Each project within a program, or component of a megaproject, will have different stakeholder groups, so their sense of value will vary. While an overall protocol may be applied, individual project nuances should be recognised and addressed. Gil and Pinto's case studies (2018) illustrate how project scope and value may be increased through sponsors of ancillary projects contributing resources and support to incorporate their projects into an extended program of projects.

11.2.3   *Project and program stakeholder engagement*

Programs of work may be undertaken as a sequence of projects, or a set of projects being undertaken in parallel, that deliver a coherent strategic objective. As explained in Chapter 5, these may be well integrated, but taking of polycentric stakeholder engagement view highlights challenges presented by a widening program or project scope, and how benefits should be explained to impacted stakeholders and the public. This view presents stakeholder communication challenges for system and project-to-program interface management. Program versus single-project stakeholder engagement strategies offer both synergy and conflicting characteristics.

Project stakeholder management performance has been operationalised by visualising stakeholder characteristics as a framework for evaluating and managing stakeholders (Bourne, 2005). However, with its manipulative inference, the restricted focus on managing stakeholders was soon overtaken by developing a theory to better engage with stakeholders (Bourne, 2011; Eskerod, Huemann and Ringhofer, 2015). Understanding stakeholders, particularly those with strong opinions and possible influence, has been an emerging field of stakeholder theory. Aaltonen and Kujala (2016) undertook a systematic review of the stakeholder literature to explore the concept of stakeholder landscapes, developing a four key project stakeholder landscape dimension framework and their various sub-factors: "complexity (element and relationship complexity), uncertainty, dynamism and the institutional context" (p. 1537). These help improve our understanding and identifying better ways to engage with them. However, a few examples explain how stakeholder engagement operates within a program IPD/Alliancing context. Aaltonen et al. (2020) focus on project alliancing stakeholder engagement, but only briefly, as an integrated program-wide stakeholder engagement strategy.

Research on understanding how to be aware of project-team external stakeholders has taken several forms. One study, analysed 250,253 Twitter tweets to gather qualitative data to develop network models from insights about stakeholder opinions and intentions on the United Kingdom (UK) High Speed 2 project (HS2) (Williams, Ferdinand and Pasian, 2015). This research found that: stakeholders could be clustered into temporally stable groups and that larger group clusters could directly contact the project team to influence them. Clusters undergo coordination and identity development processes, with supporters highlighting project advantages from their perspective and opponents using political, environmental, and other broader issues in their arguments (p. 101). Williams et al. (2015, p. 102) also state that:

> the stability of groups … over a year indicates that users engaged in these conversations have strengthened their intra-cluster relationships, suggesting the development of a coherent identity. As a result, they can now engage in sustained action and be resistant to information that does not match their pre-existing beliefs…. This indicates that the project team will need to consider alternative approaches to explaining benefits of HS2 before construction or face opposition. However … a targeted strategy for communicating may be able to create an open dialogue between participants.

These findings provide valuable lessons for stakeholder engagement.

Salient literature places an authoritative focus on a more inclusive concept of stakeholder engagement (Eskerod, Huemann and Ringhofer, 2015). Aaltonen and Sivonen (2009, p. 139) conclude from analysis of *how* to engage with stakeholders on four global case studies in emerging markets that five basic strategies offer an appropriate response using:

1. *An adaptation strategy* – obeying the demands presented by stakeholders to cope with the demands and achieve the project objectives by adjusting to external stakeholder pressures.
2. *A compromising strategy* – negotiating with the stakeholders, listening to their project claims, offering dialogue possibilities or making reconciliations and offering compensation, generally engaging openly with project stakeholders.
3. *Avoidance strategy* – loosening stakeholder claim attachments and protecting against claims through responsibility transfer to another actor in the project network.
4. *Dismissal strategy* – ignoring presented stakeholder demands and dismissing related pressures and their requirements during project execution; and
5. *Influence strategy* – proactively shaping the values and demands of stakeholders through information sharing and stakeholder relationship-building".

Strategy 2 appears somewhat similar to a value co-generation mindset because it explicitly mentions dialogue which could be seen as a solid strategy to find a win-win best for project solutions rather than being subservient.

Aaltonen et al. (2015) concludes from a comparative hermeneutical study of two nuclear waste management projects, one in Finland and another in the United States of America (USA), that it is important to realise that project-external stakeholder behaviours and actions are dynamic with the extent of salience (relevance) waxing and waning in response to their degree of support or opposition. Therefore, it is essential to develop a stakeholder engagement strategy and monitor and adjust it over time to fit with changing stakeholder salience perceptions.

According to AccountAbility (2015, p. 11), core stakeholder engagement principles include: inclusivity – people should have a say in the decisions that impact them; materiality – decision makers should identify and be clear about the issues that matter; and responsiveness – organisations should act transparently on material issues.

Stakeholder considerations are crucial to project delivery, and definitions of how success is perceived by project-program stakeholders. Similarly, for a program of integrated infrastructure projects, stakeholder engagement must address the project perspective and how it links to the program in delivering integrated value. The corporate identity is broadened to a program set of deliverables, and legitimacy extends to the project component of the program logic.

In the following sections, we now focus our analysis on external stakeholders.

11.3   The Level Crossing Removal Project

11.3.1   *LXRP scope context*

The LXRP context is comprehensively discussed in Chapter 2. The program of works is a 15-year megaproject of $19.8 billion, removing 110 level crossings with associated rail station rebuilds, signalling works and urban transformation along the train lines to improve rail precincts and multimodal transport access.

For a program of megaproject scope and scale, several opportunities arise that are particular to the LXRP. First, as a long (15-year) program of works, there are enormous opportunities for learning how to develop a best-practice strategy to deliver meaningful and valuable stakeholder engagement. Second, few programs of work (globally) that we have observed and researched have opportunity to coherently engage with all stakeholders, motivating internal stakeholders by illuminating their contribution and role in achieving a noble purpose, and managing and achieving co-value generation with external stakeholders. As one interviewee (IV09) commented, often project contractors arrive on site and start working, only to often deal with angry residents and business owners faced with a *fait accompli*. The LXRP had the opportunity to avoid this and be proactive, ensure that they maximise co-value generation with residents and businesses, leave a legacy, fulfil its stated purpose, and motivate all participants to gain a sense of achievement from the program.

The program was long overdue, with successive governments of all political persuasions having avoided the issue of dangerous level crossings, increasing traffic delay time at crossings with the expansion of Melbourne's population from two million in the 1970s to over five million currently and projected to almost eight million by 2050. Train services are adversely impacted as trains need to slow at intersections with road crossings and are stopped when 'accidents' occur on the crossings. The price tag of many billions of dollars had deterred most governments until Labor, in the 2014 state government election, committed to 50 rail crossing removals by 2018. There was a palpable sense of purpose to rectify an intolerable situation. The 2014 election helped to legitimise the program's aims.

Figure 11–2 illustrates the overall strategic considerations. The outcome needed, program's purpose had four main strands: fixing the problem of dangerous crossings; fixing the transport network to free up road congestion at boom-gates and free up the train downtime in slowing at crossings; developing a vibrant community to regenerate areas 'liberated' by road-rail grade separations and; improve rail station precinct to create vibrant community hubs; and remove barriers separating communities being from one side or the other of rail tracks, create horizontal parklands and provide a legacy of an aesthetically pleasing environment that better serves community needs.

The choice to achieve four broad strategic aims was to deliver the program via alliance projects. Project alliancing was selected to overcome the high levels of risk, uncertainty, and complexity of working in a brownfield environment with live electric lines and operating trains and road traffic interfaces and the reputational need for the project owner to take a 'hands-on' role to ensure that social benefits, environmental impact, and rail/road traffic disruption were minimised.

The LXRP organisational structure was set at the project to start as a high-level program support mechanism for stakeholder engagement and across projects to integrate internal stakeholders

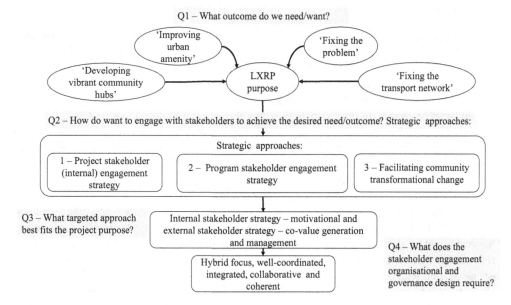

Figure 11-2 Stakeholder engagement strategy

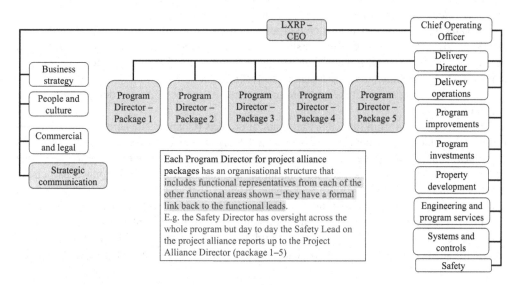

Figure 11-3 Strategic communications within the LXRP organisational chart

to reinforce the program purpose and enable them to engage cross-project within the program. It has embedded team members that report and work with their project team and their 'corporate' entity within the program, as illustrated and highlighted in yellow in Figure 11–3.

The Strategic Communication function carried out the stakeholder engagement activity, both internal and external. It also liaised with the MTIA website (https://bigbuild.vic.gov.au/) which engages with external stakeholders across all MTIA infrastructure activities.[2]

11.3.2   *LXRP historical context*

The LXRP had an early strategy for engaging with stakeholders enhanced by learned lessons from the Regional Rail Link Authority (RRLA) program of mixed new rail and rail upgrades to improve Victoria's regional access to Melbourne undertaken (Chapter 2). Why would LXRP specifically engage with local community stakeholders? Might that be setting up a situation for being coerced into spending more on projects that VfM cost-benefit considerations could justify? Two perspectives can be taken on this strategic direction, seeing this as a risk and/or opportunity.

First, taking a risk management perspective is a belief that community engagement may encourage unreasonable demands that suit minority-group selfish agendas rather than the greater community good and that this may lead to disputation and even hold up through demonstrations. Thus, creating (implicit or explicit) community benefits agreements (CBAs) may make sense to incentivise community stakeholders to support a project by negotiating an acceptable outcome with a project owner. CBAs have been used in North America by miners, and other resource developers, with indigenous peoples who hold land rights that may hinder resource extraction (Dorobantu and Odziemkowska, 2017). The same principle may apply more broadly. This takes a risk management approach and considers collaborative engagement as reducing transaction costs associated with community opposition (Odziemkowska and Dorobantu, 2021).

Second, taking an opportunity perspective is the belief that harnessing community social and intellectual capital can improve the outcome value and/or reduce negative impacts through disputation. Locals know their proximal context well and what they value most. They may identify value propositions that an LXRP APT may not have contemplated and could forge genuine win-win outcomes. Here, there is a focus on value contribution and sharing (Gil and Fu, 2022) where value can be construed to include transaction cost minimisation. As noted earlier, Ali (2023, p. 70) reminds us that effective stakeholder engagement between an integrated project delivery team and external stakeholders who have valuable local knowledge helps achieve all concerned to better understand the essential project purpose.

Thus, considering and addressing community concerns can be argued to deliver VfM and or best value (BV).

The LXRP learned from the RRLA program of projects (Chapter 2) where an innovative series of stakeholder engagement meetings between RRLA senior executives and senior executives from local councils along the rail corridors was held to discuss plans and project progress. One of the RRLA participants relayed a story about how, at one meeting, a local council participant mentioned in response to a 6-month look-ahead briefing, that an oil/gas pipeline laying permit had been recently granted and that there would be a likely clash of works between the rail works and pipe works. This potential problem was readily mitigated and re-planned to avoid delays, additional costs, or claims for 'extras' that would otherwise happen. This illustrates the value of the RRLA stakeholder engagement approach and this RRLA model were adapted by the LXRP.

11.3.3   *Urban design for a legacy outcome stakeholder engagement*

Chapter 3 discusses the LXRP strategy. Strategic goals set for the LXRP by the Government of Victorian were illustrated in Chapter 3, Table 3-1. Strategic Goal 1, separating road and rail networks at crucial junctions using one of the set options (road under, road over or hybrid), was developed with extensive and intensive APT stakeholder team consideration (the rail and road facility operators). It outlined feasible options for engagement with the affected local community, ensuring that they had every opportunity to influence the finalised design option that suited their preference. Urban design options were developed taking account of 'developing a vibrant community' and 'improving urban amenity' goals. Thus, urban design outcomes and a community communication engagement strategy was developed.

The MTIA's Big Build initiative has a specific urban design strategy guided by the LXRP's Urban Design Framework (UDF).[3] The framework states, "the UDF sets out what is to be achieved in terms of urban design quality and performance" (Victorian State Government, 2020, p. 6). It considers existing urban design guides and other relevant documents to maintain coherence with local needs and aims. It is a transformational strategy stating, "The design for each place and precinct affected by the project must consider the unique characteristics, issues and opportunities in its location and community. Consideration must also be given to the dynamism of communities and the needs of those who may live in and use these areas today and in the future" (p. 10). The UDF strategy sets out its vision, aspirations, principles, objectives, measures, and qualitative benchmarks. Principles clarify the direction of the program being both transformative and focused on value generation rather than just fixing a rail/road interface problem. Principles include: a sense of identity and place; urban integration; connectivity and wayfinding to multi-modal transport; accessibility and inclusivity, prioritising walking and cycling with easy access to multi-modal transportation; a safe environment; amenity to deliver an experience of a great public place that contributes to a thriving, equitable and prosperous community; vibrancy; and resilience and environmental sustainability (pp. 14–17).

To ensure that the framework is not considered purely aspirational, Key Result Areas (KRAs) and associated Key Performance Indicators (KPIs) were developed as part of alliance project governance and the incentivisation gain/pain sharing Project Alliance Agreement (PAA) provisions/requirements. These have evolved into a 'legacy' KRA that requires an improved project facilities outcome for the local community. Interestingly, these relate to three levels, the immediate site area, the precinct, and the affected rail corridor. The guide provides numerous illustrations to articulate its expectations. Figure 11–4 is a photograph of the space beneath the Skyrail section at Carnegie Station. It illustrates the level of amenity, integration and other espoused principles above. It presents a view towards the station, bike path (left), paved walkway, and amenities, including painted concrete table tennis facilities, a basketball court, and exercise

Figure 11-4 Photograph of space below the Carnegie Station Skyrail area

(Source: Walker, 2022)

stands where the rail line previously cut off public access areas. It is now a longitudinal park precinct.

The composition of the Urban Design Advisory Panel (UDAP) "includes members working within the government who have expertise in architecture, urban design, strategic planning, transport planning and landscape architecture. A representative from the OVGA [Office of the Victorian Government Architect] is the Chair of UDAP and drives high-quality outcomes and integrated design for the projects" (Victorian State Government, 2020, p. 39). UDAP's role is both advisory and providing feedback on the design as it progresses through the project Target Outturn Cost (TOC) process. The TOC process is a 15–20 weeks integrated team design and planning process involving the APT (as internal stakeholders) that results in the TOC fixed price/time target development that forms the basis of the project alliance gain/pain sharing incentive arrangements (see, Walker and McCann, 2020; Walker, Vaz Serra and Love, 2022). The UDAP provides an example of how external stakeholders can gain a voice and help influence a project's outcome by participating in co-generating value.

A specific example of engagement with local experts on a Revitalisation Reference Group for the Carrum station $50 million revitalisation precinct design development and refinement. This entailed three community engagement rounds to craft designs for the precinct comprising community representatives and industry experts who met monthly to discuss salient design aspects important to the community as feedback information for the design team's consideration. This is one of many such engagement activities.

11.3.4   Public communication strategy

From the program's commencement, there has been a purposeful general information strategy to ensure that people are informed about proposed projects, initial community response to proposed project design and implementation approaches, and day-to-day information about traffic disruptions and alternative bussing arrangements where track sections are closed for project works. Initially, this was an LXRP-only website. However, when the Melbourne Transport Infrastructure Authority (MTIA) was formed in 2019 (see Chapter 2 for context details), the Victorian Government badged the MTIA works as part of its 'Big Build.'[4] The LXRP shares a more extensive website.[5]

The use of this platform for communicating effectively and dynamically conveys a range of information. At the most stakeholder-passive level of engagement, it provides corporate information updates on progress, what has been achieved and what the aims and objectives of the LXRP are. It also provides dynamic information on disruptions, and through its calendar link[6] activities, dates and time information is downloaded to a user's phone/computer etc., device. The LXRP home page also has an 'explore the level crossing removal projects' map showing all removed and in-progress projects so that stakeholders can click on their desired project, and if removed, it gives the month and year removed and has a link to click on about that removal project. News items and details are presented, showing the LXRP an opportunity to educate readers about what was done and highlight innovations (such as the *straddle carrier*[7]) or other positive features that form a sophisticated public relations tool while performing a public education role. This investment in the web platform is not insignificant in cost, skill resources, leadership, and management energy. Politically, it is highly effective in reinforcing a policy message about the value of public transport and the government that has not only developed a strategy for the LXRP but also has overseen its actions and outcomes. The website also reinforces positive messages about the community transformation (legacy) policy, delivering social value rather than restricting the message about the program in cost/time terms only as a quick and cheap fix.

11.3.5   Project information communication strategy

The LXRP has a strategy for community engagement to discuss options with affected stakeholders when initiating a design for a crossing removal (Chapter 2):

- Road over the rail – The rail line remains at its existing level, and a new road bridge is constructed over the rail line.
- Road under rail – The rail line remains at its existing level, and the road is lowered to pass underneath the rail.
- Hybrid options – Hybrid options are variants of road over/under rail and rail over/under road options where both the road and rail are raised/lowered. More recently, this option includes an elevated rail over the road – the Skyrail option with a stretch of elevated track with a horizontal parkland occupying the existing ground-level rail track.

Each option raises community concerns about visual impact, access, and other urban planning aspects. The LXRP makes the overall design option decision based on technical data and information of a specialised nature. However, fine-tuning the project strategy involves community engagement. This engagement strategy provides for a comprehensive range of online and face-to-face 'town hall' meetings with presentations and Q&A sessions, social media platform engagement, and other smaller group and more focused mechanisms, particularly to engage with value co-generation opportunities and practical issues such as temporary access during disruptions when the LXRP is 'occupying' space for direct works and logistics access to the site. It is interesting to note that as the program progressed, residents' attitudes changed with a better understanding of the options. One of the authors recalls the at-home debates about the three options for the Carnegie station line solution, which has a long Skyrail section that released considerable rail-line land to be converted to a long parkland asset (Figure 11–4).

The LXRP has, according to IV09, IV10, and IV11, a remit to ensure minimum disruption to traffic (road and pedestrian) and commercial activities and to engage with the community and trade owners to find out what they value most when the LXRP develop the disruption strategy to enable works to proceed. Engagement at this level is about how to compensate best and minimise disruption. For example, avoiding cutting off shop(s) from passing traffic, and as IV09 and IV11 noted, distributing vouchers to potential pedestrian passing trade for affected businesses. In this way, trader owners are not compensated *per se,* but this community engagement mechanism provides some compensation.

The LXRP regularly arranges community events and advertises these through the Big Build web platform,[8] where community members can drop in to see progress on projects and be briefed by LXRP engagement people. The web page also shows pictures and text to allow virtual drop-ins. Another initiative was the 'Super-sized Machines on Show at Scienceworks' at the Scienceworks museum in Melbourne held between 31 March and 15 April 2018. It showcased the big blue 'monster machines' used to remove level crossings on a narrow part of the Cranbourne/Pakenham rail line in Melbourne's southeast. A replica of specialist equipment comprising two gantry cranes, a straddle carrier and 90-metre support beams made up of 5,000 Lego bricks showed how they combine to lift, shift and install the 'building blocks' for a 2.4-kilometre section of elevated rail. Project engineers gave short talks for children between the 10 and 12 April 2018. Such events illustrate the nature of community engagement.

The engagement process follows three broad phases – an engagement phase, design engagement, and development of engagement objectives and activities. Each alliance project package has its quirks, so while activities follow the general LXRP protocol, each project's context is individually considered. The general protocol includes stakeholder mapping; market research to understand community sentiment and key specific issues; developing engagement strategies and communication materials; and informing and engaging with the community to explain, adapt and report how the construction will impact them and through dialogue with the community, to refine the delivery plan. For example, in 2022, the LXRP had: embedded International Association for Public Participation (IAP2) trained communication and engagement in each project team; undertaken 40 engagement activities; received 440,000 website visitors and almost 2 million page views; issued more than 500,000 newsletters; provided nearly 250 e-updates to 33,104 subscribers and received more than 4300 contributions from the community; and had almost 75,000 followers across their social media channels.

11.3.6   *Internal engagement strategy*

The LXRP has reinforced its internal stakeholder strategic communication approach by not only embedding communications specialists within each alliance project (Figure 11–3) to gather useful *external* stakeholder information and insights to be communicated but also to gather and distribute stories and information that can be shared across projects to enhance a sense of identity, purpose and legitimisation of their role within a project, and program, delivering social benefit. As noted earlier, Hackman and Oldham (1976), as mentioned earlier, identify job relevance, effective feedback and a sense of responsibility as being task motivators. The project-embedded strategic communication staff and the corporate level staff that help distribute and formulate *external* stakeholder engagement content play an important role in project team integration, willingness to collaborate and motivation. One of the authors who has undertaken research with the LXRP over several years in which project alliance participant staff and subcontractors, and suppliers were interviewed and studied had overwhelmingly stated that the LXRP project alliances were psychologically and motivationally rewarding projects to work on.

Another vital internal stakeholder communication strategy developed by the LXRP, based on RRLA experience, was the Joint Coordination Committee (JCC). The JCC operates at several levels. At one level, it integrates projects across the program to reflect on and exchange ideas, insights and information on the effectiveness of stakeholder engagement to develop a lessons-learned database that informs continuous improvement in (both internal and external) stakeholder engagement. It also fosters collaboration across projects, reinforced by governance and key project performance measures (KPIs) relating to and rewarding innovation diffusion and legacy (satisfying external stakeholder aspirations by leaving the project with a vibrant community hub and urban amenity). Several KPIs are entrenched into each project alliance agreement (PAA) that measures and incentivises the vibrant community hub and urban amenity key result areas. Each APT also can benefit from other projects in the program adopting and/or adapting innovations that APTs register with the JCC. The JCC also operates as a set of forums for alliance managers, program directors, and specialists in design and delivery. Each forum acts as a community of practice (Lave and Wenger, 1991) to exchange ideas, promote continuous improvement and enhance their sense of identity with their project and the program. For example, in 2022, there were 70 lessons learned case studies in the JCC portal.

11.3.7   *LXRP engagement strategy summary*

Interviewees IV9, IV10 and IV11 were particularly vocal about three important mindset aspects. First, they stressed the importance of *purpose* and ensuring it is front-and-centre of all engagement action. Second was the realisation that the LXRP is a social benefit program, not

just a quick fix to remove dangerous and congested road-rail crossings. It has vital urban transformational and renewal aspects and the need to improve road and rail transport network effectiveness. Third was the issue of gaining and maintaining trust with both internal and external
stakeholders. They stressed the importance that trust is fragile and dynamic, that what was
promised needs to be delivered and that when disruption is inevitable, as it is, honest and informative communication about the disruptions and suggestions about how to cope are clear
and reasonable.

Stakeholder engagement was understood to be paramount in ensuring that internal stakeholders
realised the nature of the impact of their actions and influence on project performance and how
they interact with other project participants within an integrated program of projects, particularly
on complex brownfield projects. Maintaining a sense of purpose was seen as vital, and the purpose
accorded with the program's intended benefits and outcomes. Potential problems with external
stakeholder opposition impacts were also fully understood, with the LXRP being determined to
ensure that external stakeholders were fully engaged to be 'controlled or managed' and to be positive co-generators of value, being part of a collaborative dialogue to enhance benefits and value.

An important observation we make is that the LXRP was a highly strategised megaproject
with a few clear and well-articulated strategic goals (Figure 11–2) and that this strategy's operationalisation and development of its design and delivery (Figure 11–3) designed-in mechanisms
for the Strategic Communications entity to enhance program integration and collaboration as
well as serve as a stakeholder engagement conduit to external stakeholders. Table 11-1 illustrates our overall insights.

We can make specific observations when we consider the Aaltonen and Sivonen (2009)
Engagement strategy categories. The LXRP stakeholder strategy is centred on adaption and codesign. LXRP, through its integrated design, delivery, project owner representatives and road and
rail operator teams, devise the overall technical design solution but cross-referenced to consider
specific local community sentiment. This adheres to the collaboration core value in Alliancing.
There is also an element of compromise through the collaborative dialogue process, but dialogue aims to find optimum win-win solutions rather than a win-lose attitude (Senge, 1990).
There is no evidence of a dismissal or avoidance strategy being adopted. Critics may highlight
the overall strategy of Skyrail, rail-under-road and the like, decisions made by the LXRP team.
Still, they ultimately have the greatest combined technical expertise, and the UDAP influences
their decision. The only avoidance strategy that may be perceived is to leave detailed design and
delivery decisions to the last most responsible moment to ensure minimising rework based on

Table 11-1 Engagement strategy observations

Strategy insights	*Comments*
Response to overall engagement strategy	The Strategic Communications entity was highly effective and won an international award, the International Association for Public Participation (IAP2) *2021 Australasia Organisation of the Year Award,* for its industry-leading approach to stakeholder engagement. This is an innovation that may prove globally ground-breaking for megaprojects.
Response to urban design engagement strategy	From our observations, this has been successful for many of the station precincts. Carnegie provides one example Figure 11–4.
Response to website engagement strategy	This appears highly successful, and the above 2021 award recognises that performance.
Response to project information engagement and JCC strategy	This provides another innovation, and we note that it seems particularly effective in integrating projects within the program and helping create a sense of LXRP identity.

community push-back. Much of the web-based communications were intended to inform and influence many stakeholder groups, and social media platforms and *town hall* meetings were perceived by some stakeholder groups as influence and advocacy attempts. However, any influencing strategy was more based on gaining informed support or reasoned opposition to plans.

11.4   LXRP Global Best-Practice Comparison Discussion

Global best practice of engaging stakeholders to create value has rapidly replaced the traditional "management of stakeholders" approach in project and program management theory and practice, as showcased by the LXRP example. In advanced project-based organisations, positively involving internal and external stakeholders has become the primary approach to organising activities, as opposed to the outdated "us and them" mindset that previously dominated strategic stakeholder management approaches. This engagement ideology aligns closely with the core values and identity of project and program alliancing, emphasising the development of inclusive and collaborative ways of working across a broad range of stakeholders as a central pillar.

The LXRP stakeholder demonstrated engagement approach is characterised by a systematic, authentic and continuous process, employing multiple communication channels and organising solutions. These practices align strongly with stakeholder engagement approaches observed in two global leader examples, which also involved highly complex stakeholder landscapes: The Future Hospital Programme and the Tampere Tunnel project in Finland, executed as part of a larger national alliance development program. Both cases showcased extensive internal and external stakeholder engagement by applying innovative digital practices and diverse communication channels. Notably, the Tampere Tunnel project received national and international acclaim for its exceptional community stakeholder engagement, earning the first prize in the 2018 International Project Management Association (IPMA) Mega-Sized Projects category. Several shared stakeholder engagement themes in these exemplary cases and the LXRP have discussed: (1) Collaborative values, transparency and facilitation of trust; (2) Digitalisation as a driver of stakeholder engagement; (3) Novel coordination roles and governance practices facilitating stakeholder engagement; and (4) Multi-channel engagement: community engagement in social media and co-locational space.

11.4.1   *Collaborative values, transparency and facilitation of trust*

Developing collaborative stakeholder relationships, characterised by high levels of relational capital, including mutual trust and commitment, has yielded improved project performance outcomes. The Future Hospital Programme and Tampere Tunnel project share a common feature, having established stakeholder collaboration and cooperation as fundamental values upon which the projects are built. Promoting community spirit was essential in fostering stakeholders' cooperation, extending the mindset of working together towards a shared goal with external stakeholders. As with the LXRP, the strong collaborative project culture among internal stakeholders influenced their approach to external, external stakeholders. Moreover, these collaborations gave project actors significant intrinsic non-monetary benefits, such as social reputation and identification.

Rather than perceiving stakeholders and their claims as potential risks to the project, diverse stakeholders were regarded as potential co-creators of value. Engaging with them was seen as a means of reducing resistance to change. The Tampere Tunnel project focused on working collaboratively with external stakeholders instead of engaging in conflicts, aiming to swiftly and efficiently address challenging issues through problem-solving. As one manager explained, "stakeholder collaboration is crucial as it enables the development of ideas that can be incorporated into the project". Similarly, within the Future Hospital Programme, respect, and a dialogical approach towards stakeholders provided drivers for generating novel ideas: "You can have constructive arguments and have a debate in a nice way. And when we involve these people such as users early in the design process, it becomes our collective design, together."

However, establishing a shared value basis did not occur spontaneously; instead, it required regular workshops and training sessions throughout the program's lifecycle. Additionally, it was considered highly important for managers to lead by example, actively embodying stakeholder-oriented collaborative values.

11.4.2 *Novel coordination roles and governance practices facilitating stakeholder engagement*

The Future Hospital Programme, in particular, has demonstrated an innovative approach toward its organisational arrangements and interface roles. It is specifically designed to facilitate stakeholder engagement across the boundaries of temporary projects and the permanent organisation. A good example is the introduction of service designers who are tasked with advancing the end users' perspective. They act as boundary spanners, interpreters, and facilitators, ensuring a comprehensive understanding of the diverse stakeholder and user needs, effectively translating them for the hospital project organisation. These interface roles and coordination bodies are crucial for successful stakeholder engagement, as observed in the LXRP, as they foster dialogue between stakeholders and program personnel. Furthermore, emphasising the identification of the appropriate stakeholders to be engaged in the dialogical process is extremely important. Individuals in coordinator roles representing their respective groups, such as specialised medical areas, were also considered a critical factor.

In addition to coordination roles, establishing appropriate governance arrangements that support stakeholder engagement played a significant role in both cases, similar to the LXRP example. The project plans of both cases highlight user and community engagement as key project objectives. Schemes, engagement indicators, and incentive structures were developed to incentivise and reward stakeholder engagement, including those related to favourable publicity. Consequently, the importance of stakeholder engagement should be reflected in the governance structures and KRAs. Furthermore, stakeholder engagement practices were aligned with industry-level objectives aiming for a more transparent and trustworthy modus operandi within the construction sector. Communication systems and structures were established to collect feedback from external stakeholders, and lessons learned from stakeholder experiences were systematically gathered. It was also crucial to provide stakeholders with regular reports and communication about decisions based on their input.

11.4.3 *Digitalisation as a driver of stakeholder engagement*

The external stakeholder engagement landscape has recently transformed rapidly, primarily due to digitalisation and the emergence of novel tools and engagement platforms. The Future Hospital Programme employed an array of digital objects and channels to engage with and involve diverse stakeholders and representatives in the decision-making process, notably utilising virtual space tools and architect-facilitated workshops, known as the VALO™ method. These workshops aimed to gather insights and opinions on design choices from stakeholders such as future end users, doctors, nurses, and professional staff.

This approach implemented interactive 3D virtual modelling techniques, including Computer Aided Virtual Environment (CAVE), to present alternative space solutions to the users. These sessions allowed users to experience the proposed spaces firsthand and provide feedback on the proposed solutions. Facilitators also posed diverse questions to the participants during these workshops. Such innovative digital experiences and visualisations offer advanced means of engaging community stakeholders in decision-making. By moving beyond two-dimensional drawings, virtual representations provide more tangible avenues for community stakeholders to envision how the project will transform their environment in the future and contribute their development ideas. Furthermore, these workshops serve as valuable tools for fostering future-oriented dialogue with stakeholders, ensuring their comprehensive understanding of forthcoming changes.

Similarly, digital tools also played a pivotal role in facilitating stakeholder engagement within the LXRP. These tools offered concrete means of advancing stakeholder engagement practices, enhancing communication, and enabling effective collaboration.

11.4.4   *Multi-channel engagement: Community engagement in social media and co-locational space*

Both cases underscore the significance of employing multiple channels for effective stakeholder engagement. Social media platforms have gained increasing prominence in facilitating stakeholder engagement processes. For instance, in the Tampere Tunnel case, Facebook served as the primary channel for informing local community members about the project, amassing over 2,000 followers. A dedicated Facebook community was established, enabling collaboration, shared decision-making and stakeholder empowerment. The project organisation maintained a specialised team responsible for managing stakeholder engagement, ensuring daily communication and interaction with stakeholders on the Facebook platform. The key objectives related to stakeholder engagement included providing essential and timely information about the project's progress and impacts to key community stakeholders, enhancing the project's public image, facilitating project progress through effective communication, preparing for communication regarding potential disruptions, and collecting and disseminating information about environmental changes and their effects on people and the landscape. Notably, Facebook was also utilised to foster stakeholder dialogue, enabling their participation in polls and decision-making processes. Traditional stakeholder engagement methods complemented these digital channels, such as public events, town hall meetings, presentations, conventional media communication, and joint workshops in the project's co-locational space. The integration of analogue and digital engagement channels was also emphasised in the Future Hospital Programme, where workshops involving users, including patient group representatives, incorporated digital virtual environments and physical simulations of hospital spaces.

Like the LXRP, adopting multi-channel strategies is vital, as it provides diverse stakeholder groups with meaningful and timely engagement opportunities tailored to their specific needs. However, it is important to note that while social media has been heralded as a promising tool for local community engagement, it is often constrained by limited resources and may not necessarily be the most effective avenue for fostering genuine and authentic dialogue. For this purpose, a range of different media and setups is necessary.

11.5   Conclusions

We interviewed staff with relevant LXRP engagement strategy experience to seek their insights. Our frame of reference is based on a series of initiatives, stakeholder engagement processes, and how each fits with the Aaltonen and Sivonen (2009, p. 139) five strategy designations and interview data relating to how interviewees perceived that stakeholder salience and support/ opposition level experienced over the project planning and delivery phases. We also comment on identity and legitimacy reflections on the LXRP. The LXRP approach was compared to two leading alliance project examples from Finland, demonstrating global stakeholder engagement best practices and highlighting similarities in the techniques deployed.

The chapter illustrates *how* stakeholder engagement on these complex megaprojects is necessary, and we provided explicit examples of these practices, their impact and how they were strategised as part of legitimate VfM/BV benefits. Many leading global infrastructure project delivery teams deploy these practices to a greater or lesser extent. What becomes clear in this chapter is that stakeholder engagement is not merely a box-ticking exercise to appease potential project critics but an integral and vital component of VfM/BV project delivery.

In summary, best global practice suggests projects should adopt the following:

1. A proactive internal *and* external stakeholder engagement strategy based on value co-generation with stakeholders.
2. Ensuring collaborative values, transparency, and facilitation of trust as the strategic foundation.
3. Adopting novel coordination roles and governance practices to facilitate stakeholder engagement, such as the RP's JCC and the development of the Urban Design Advisory Panel.
4. Developing effective digital technology tools and stakeholder engagement platforms as a driver of stakeholder engagement combined with traditional personalised human-to-human approaches.
5. Focussing on multi-channel engagement: continuous and authentic community engagement in social media and in co-locational spaces.

Notes

1 Time, cost and quality (fitness for purpose).
2 Note that from April 2024 MTIA was restructured to for the Victoria Infrastructure Delivery Authority – see Chapter 2.
3 https://bigbuild.vic.gov.au/__data/assets/pdf_file/0011/635771/LXRP-Urban-Design-Framework-v5-October-2020.pdf
4 https://bigbuild.vic.gov.au/
5 https://bigbuild.vic.gov.au/projects/level-crossing-removal-project
6 https://bigbuild.vic.gov.au/disruptions/calendar
7 https://bigbuild.vic.gov.au/news/level-crossing-removal-project/springing-into-action
8 Refer to https://bigbuild.vic.gov.au/projects/level-crossing-removal-project/community/events#-UnionStation

References

AAA (2012). "Lessons learnt on Sugarloaf." *The Australian Pipeliner*. Melbourne, The Great Southern Press.
Aaltonen, K. (2010). Stakeholder Management in International Projects. PhD Thesis, Department of Industrial Engineering and Management. Espoo, Helsinki University of Technology.
Aaltonen, K., Huemann, M., Kier, C., Eskerod, P. and Walker, D.H.T. (2020). "IPD from a stakeholder perspective." In *The Routledge Handbook of Integrated Project Delivery*, Walker, D.H.T. and S. Rowlinson (eds), pp. 288–314. Abingdon, Oxon, Routledge.
Aaltonen, K. and Kujala, J. (2016). "Towards an improved understanding of project stakeholder landscapes." *International Journal of Project Management*. 34 (8): 1537–1552.

Aaltonen, K., Kujala, J., Havela, L. and Savage, G. (2015). "Stakeholder dynamics during the project front-end: The case of nuclear waste repository projects." *Project Management Journal*. 46 (6): 15–41.

Aaltonen, K. and Sivonen, R. (2009). "Response strategies to stakeholder pressures in global projects." *International Journal of Project Management*. 27 (2): 131–141.

AccountAbility (2015). *AA1000 Stakeholder Engagement Standard*, London.

Ali, F. (2023). Framework for analyzing, developing, and managing stakeholder network relationships in collaborative hospital construction projects. Thesis by publication, Faculty of Technology, Industrial Engineering and Management Research Unit, University of Oulu Graduate School. Oulu, Finland, University of Oulu.

Bourne, L. (2005). Project Relationship Management and the Stakeholder Circle. Doctor of Project Management, Graduate School of Business. Melbourne, RMIT University.

Bourne, L.M., Ed. (2011). *Advising Upwards – A Framework for Understanding and Engaging Senior Management Stakeholders*. Series Advising Upwards – A Framework for Understanding and Engaging Senior Management Stakeholders. Farnham, Gower.

Cornelissen, J.P., Haslam, S.A. and Balmer, J.M.T. (2007). "Social identity, organizational identity and corporate identity: Towards an integrated understanding of processes, patternings and products." *British Journal of Management*. 18 (s1): S1-S16.

Davis, P.R. and Walker, D.H.T. (2020). "IPD from a participant trust and commitment perspective." In *The Routledge Handbook of Integrated Project Delivery*, Walker, D.H.T. and S. Rowlinson (eds), pp. 264–287. Abingdon, Oxon, Routledge.

Department of Infrastructure and Transport (2011). *National Alliance Contracting Guidelines Guide to Alliance Contracting*. Department of Infrastructure and Transport A.C.G. Canberra, Commonwealth of Australia. www.infrastructure.gov.au/infrastructure/nacg/files/National_Guide_to_Alliance_Contracting04July.pdf

Derakhshan, R., Mancini, M. and Turner, J.R. (2019). "Community's evaluation of organizational legitimacy: Formation and reconsideration." *International Journal of Project Management*. 37 (1): 73–86.

Donaldson, T. and Preston, L.E. (1995). "The stakeholder theory of the corporation: Concepts, evidence, and implications." *Academy of Management Review*. 20 (1): 65–91.

Dorobantu, S. and Odziemkowska, K. (2017). "Valuing stakeholder governance: Property rights, community mobilization, and firm value." *Strategic Management Journal*. 38 (13): 2682–2703.

Drouin, N. and Turner, J.R. (2022) *Advanced Introduction to Megaprojects*, Northampton, UK, Edward Edgar Publishing

Eskerod, P. and Huemann, M. (2013). "Sustainable development and project stakeholder management: What standards say." *International Journal of Managing Projects in Business*. 6 (1): 36–50.

Eskerod, P., Huemann, M. and Ringhofer, C. (2015). "Stakeholder inclusiveness: Enriching project management with general stakeholder theory." *Project Management Journal*. 46 (6): 42–53.

Freeman, R.E. (1984) *Strategic Management: A Stakeholder Approach*, Boston, Pitman.

Gil, N. (2021). "Megaprojects: A meandering journey towards a theory of purpose, value creation and value distribution." *Construction Management and Economics*. 1–23.

Gil, N. and Fu, Y. (2022). "Megaproject performance, value creation, and value distribution: An organizational governance perspective." *Academy of Management Discoveries*. 8 (2): 1–27.

Gil, N. and Pinto, J.K. (2018). "Polycentric organizing and performance: A contingency model and evidence from megaproject planning in the UK." *Research Policy*. 47 (4): 717–734.

Gil, N.A. (2023). "Cracking the megaproject puzzle: A stakeholder perspective?" *International Journal of Project Management*. 41 (3): 102455.

Hackman, J.R. and Oldham, G.R. (1976). "Motivation through the design of work: Test of a theory." *Organizational Behavior and Human Performance*. 16 (2): 250–279.

Hackman, J.R. and Wageman, R. (2005). "A theory of team coaching." *Academy of Management Review*. 30 (2): 269–287.

Hollensbe, E., Wookey, C., Hickey, L. and George, G. (2014). "From the Editors: Organizations with purpose." *The Academy of Management Journal*. 57 (5): 1227–1234.

Huemann, M., Eskerod, P. and Ringhofer, C. (2016) *Rethink! Project Stakeholder Management*, Newtown Square, PA, Project Management Institute.

Lave, J. and Wenger, E.C. (1991) *Situated Learning – Legitimate Peripheral Participation*, Cambridge, Cambridge University Press.

Leonard, D. and Rayport, J.F. (1997). "Spark innovation through empathic design." *Harvard Business Review*. 75 (6): 102–113.

Littau, P., Jujagiri, N.J. and Adlbrecht, G. (2010). "25 years of stakeholder theory in project management literature (1984–2009)." *Project Management Journal*. 41 (4): 17–29.

Lloyd-Walker, B.M. and Walker, D.H.T. (2017). "The Sugar Loaf Water alliance: An ethical governance perspective." In *Governance & Governmentality for Projects – Enablers, Practices and Consequences*, Muller, R. (ed.), pp. 197–220. Abingdon, Oxon, Routledge.

Mayer, R.C., Davis, J.H. and Schoorman, F.D. (1995). "An Integrated model of organizational trust." *Academy of Management Review*. 20 (3): 709–735.

Mitchell, R.K., Agle, B.R. and Wood, D.J. (1997). "Toward a theory of stakeholder identification and salience: Defining the principle of who and what really counts." *Academy of Management Review*. 22 (4): 853–886.

Odziemkowska, K. and Dorobantu, S. (2021). "Contracting beyond the market." *Organization Science*. 32 (3): 776–803.

Porter, M.E. and Kramer, M.R. (2011). "Creating shared value." *Harvard Business Review*. 89 (1/2): 62–77.

Senge, P.M. (1990) *The Fifth Discipline – The Art & Practice of the Learning Organization*, Sydney, Australia, Random House.

Smith, S., Anglin, T. and Harrisson, K. (2010) *Sugarloaf Pipeline A Pipe in Time*, Melbourne, Sugarloaf Pipeline Alliance, Melbourne Water.

Tjahja, C. and Yee, J. (2022). "Being a sociable designer: Reimagining the role of designers in social innovation." *CoDesign*. 18 (1): 135–150.

Victorian State Government. (2020). *Urban Design Framework: Version 5 – Principles & Objectives, Measures & Qualitative Benchmarks*, Government V. Melbourne.

Walker, D.H.T. and McCann, A. (2020). "IPD and TOC development." In *The Routledge Handbook of Integrated Project Delivery*, Walker, D.H.T. and S. Rowlinson (eds), pp. 581–604. Abingdon, Oxon, Routledge.

Walker, D.H.T., Vaz Serra, P. and Love, P.E.D. (2022). "Improved reliability in planning large-scale infrastructure project delivery through Alliancing." *International Journal of Managing Projects in Business*. 15 (8): 721–741.

Williams, N.L., Ferdinand, N. and Pasian, B. (2015). "Online stakeholder interactions in the early stage of a megaproject." *Project Management Journal*. 46 (6): 92–110.

Winch, G.M. (2004). "Managing project stakeholders." In *The Wiley Guide to Managing Projects*, Morris, P.W.G. and J.K. Pinto (eds), pp. 321–339. New York, Wiley.

12 LXRP and Program Value Generation Performance

Derek H.T. Walker, Nathalie Drouin, Mark Betts and Peter E.D. Love

12.1 Introduction

How does the LXRP measure and communicate its best value delivery performance across projects in the program?

Our chapter objective is to explain how the LXRP strategy performed in terms of achieving its strategic goals and how it delivered VfM/BV in those terms. VfM is traditionally limited to treating costs and benefits in monetary terms while BV considers difficult to monetise aspects (MacDonald, 2011). We use the term VfM/BM to infer total cost/value. It is important to understand that the LXRP is a *program* of alliance projects, so ultimate program success and its performance cannot be thoroughly evaluated until well after the completion of the project, which is expected to be in 2030.

Chapter 2 discusses the program's context and scope. Chapter 3 outlines the program's four strategic core goals (Victorian State Government, 2017, p. 6) as illustrated in Table 3-1 and Figure 3-1. The four strategic goals are: (1) separating road and rail networks at critical junctions; (2) implementing a Metropolitan Network Modernisation Program; (3) improving the urban amenity and physical integration of activity precincts and communities along rail corridors; and (4) improving integrated land use along rail corridors to create vibrant community hubs. While performance is often explained in VfM terms, this chapter describes *how* the firm documented goals were assessed in VfM/BV terms because, as a social infrastructure project, some benefits cannot be accurately monetised (Raiden et al., 2019; Drouin and Turner, 2022, Chapter 1 p. 7–8). Thus, BV has been argued to be a more reliable measure than VfM.

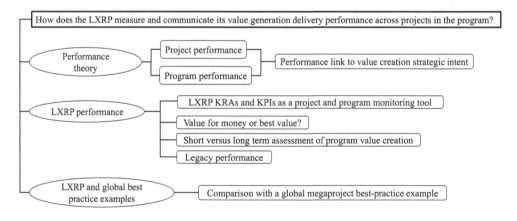

Figure 12-1 Chapter structure

DOI: 10.1201/9781003389170-12

Figure 12-1 illustrates the chapter's structure. We introduce project and program performance theory to explain these concepts and followed by a discussion of LXRP specific performance and how value was conceptualised. We set the LXRP case study within a global context with a discussion of a similar Canadian megaproject case study.

12.2   Project/Program Performance Theory

What do we mean by project or program performance and success? This debate emerged from Atkinson's (1999) critique of the traditional time, cost, and quality triumvirate, which has been considered to be a factor of a project's success (Pinto and Slevin, 1987), and benefits obtained from processes and established measures (Zwikael and Globerson, 2006) (Zwikael, 2018). Project performance has evolved into a more nuanced holistic performance success concept, focusing on performance as achieving strategic objectives considering different project types (Shenhar et al., 2007). Shenhar et al. (2001) investigated temporal success perspectives – from short-term efficiency, short-term customer impact, long-term business success, and long-term preparation for the future. This view resonates with infrastructure projects (and programs) primarily about creating social value in supporting community activity effectiveness and creating a legacy (Raiden et al., 2019). Value creation benefits and delivery performance may thus be measured from an economic, environmental, and social perspective, together with sustainability goals identified by the United Nations (2016).

Often, high-profile government projects are targeted by opportunistic politicians to criticise a political administration by accusing them of incompetence in their project/program's delivery. This tendency for strategic misrepresentation and obfuscation in project delivery for short-term political gain has been referred to as a 'dark side' of project management (Locatelli et al., 2022, see their Table 1). Strategic misrepresentation and optimism bias are frequently cited as root causes of cost or time 'blowouts' in large-scale transport projects (Flyvbjerg, Holm and Buhl, 2002; Flyvbjerg, 2014). However, it has been argued by many scholars that this is a misleading causal claim (Ika, Love and Pinto, 2021) and a distraction as many other factors impact a project's cost and time estimation and performance (Ahiaga-Dagbui and Smith, 2014). Walker et al. (2022) argue that even when focusing on cost/time performance, we need to ensure that the initial benchmark for these measures is realistic with a given procurement framework. Many megaprojects are affected by concurrent cost and value 'creep' that could be viewed as logical, desirable and represent best practice (Gil and Pinto, 2018; Gil, 2021). For example, linked project creep affords synergistic value-creep advantages. Therefore, an integrated program of projects mindset builds a clearer benefits performance picture. Consequently, we ignore the emotive literature that supports opportunistic politician arguments about cost/time blowouts and instead focus on performance from a strategic intent and multi-faceted benefit realisation perspective. In this sense, the focus is on a project's strategic purpose.

12.2.1   *Project performance*

Projects are initiated for a purpose (Gil, 2021, p. 15). The benefits realisation literature discusses performance in output and outcome terms where the specific (often tangible and measurable) output delivers what was promised (Bradley, 2010; Zwikael, 2016). This may have dimensions such as a maximum cost/time but more often includes other key result areas (KRAs).

Walker et al. (2015) studied 60 Australian alliance projects, finding 15 KRAs (besides cost/time) cited by project alliancing experts. Many KRAs were environmental and social, with some expected (safety) KRAs, but others that were difficult to quantify and measure accurately. Traffic disruption was an identified KRA measuring how detours, partial road closure and other congestion-inducing factors may inconvenience road users. Legacy was another social KRA, measuring the extent to which the post-project impact improves its pre-project condition.

Attempts may be made to measure many less tangible KRAs, but they are all subject to debate and controversy. For example, how do we realistically measure the monetary cost/value of 'life and limb' safety? Cost data on life and limb was developed by insurance actuarial analysts decades ago. Are such calculations realistic and defendable? Difficulty in measuring costs and benefits compromises many cost-benefit models, business cases often fail to measure what they should and include what should not. Thus, their usefulness may be questionable (Baden-Fuller and Morgan, 2010; Volden, 2019). We do not suggest abandoning business case development or cost-benefit analysis; however, performance measures and assessment should reflect a project's purpose and be considered indicative, not 'gospel.'

If performance is the outcome of project/program value generation, then we need to be clear about the nature of value. Drevland (2019) wrote a PhD thesis about projects being a value-delivery system and, with his colleagues, offered the following insights on value:

> Value is the result of an evaluative judgment. This judgment is guided by values and based on the evaluator's knowledge at hand. It is always based on comparing two or more alternatives in a given context. This context envelops all get and give consequences for a particular party from a decision made on the basis of the value judgment. The get and give consequences are always in the form of gained or lost experiences, or expressed in monetary terms as a placeholder for experiences. The consequences are not summative; the value judgment is done by considering them all at once.
>
> (Drevland, Lohne and Klakegg, 2018, p. 38)

Therefore, project performance assessment for many KRAs should be based on stakeholder experience. Additionally relevant to this perspective is how stakeholder engagement leads to the co-generation of mutually desirable 'shared value.' Porter and Kramer (2011, p. 65) argue that shared value performance should concentrate on supporting local groupings of commercial and social suppliers to grow opportunities as part of a more sustainable value-performance objective. Thus, LXRP's performance in enhancing community hub revitalisation is valid and vital KRAs. Chapter 11 discussed LXRP community engagement activities.

12.2.2   *Program performance*

No project is an island (Engwall, 2003) and program performance is linked to the summative impact of project performance for integrated programs projects. Program performance, measuring the effectiveness in delivering the program's overall strategy, cannot be appreciated or finalised until well after completion.

For example, improving the rail network's effectiveness through reduced rail trip journey time, increased capacity in rail service numbers, and coping with increased demand, cannot be fully judged until the final rail crossing in the program is removed. Similarly, Melbourne-wide rebuilt station precinct revitalisation is challenging to assess. However, early signs of public feedback and community support are encouraging. Two points are noteworthy:

1. The integrated project within the program strategy and its operation strongly indicates a high-performance level. In the next section, we discuss interviewee responses.

2. The program is also assessed through the Victorian Auditor General's Office (VAGO), with two reports published when writing this chapter (Victorian Auditor-General's Office, 2017; 2020). These independent bodies provide reasonable assessments as there are conducted without 'fear or favour' and allow for the audited entity (in this case, the LXRP) to respond to questions and correct factual errors before publication.

Ultimately, the public and local community can realistically judge if the program meets its strategic social aims. For example, are the completed projects easing congestion at the former crossing choke points? Are the new intersections safer? Have the rebuilt rail station precincts improved the inter-modal transport experiences of rail passengers? Have local traders benefited from increased commercial returns due to increased demand for the products/services? Are communities located around completed crossing-removal areas more vibrant, and do the local community and visitors appreciate the urban transformation results?

This program was transformational, so the intended transformation can only be effectively judged decades after the event. The Sydney Opera House presents an example of a facility, initially proposed and considered a performance culture venue, but which became a tourism icon drawing millions of visitors to Australia. On that basis, it is no doubt a highly successful project (Murray, 2004).

From a VfM/BV *program* perspective, it is important to consider the values held by LXRP participants and what is valued at project and program levels. Interviewee 10 (IV10) made internal LXRP documents available to us showing 12 elements of VfM reported for each project to inform program performance through specific KRAs that include and exceed the Barrett (2005) 5Es. The following headings list the 12 elements. Each item is required to be considered and addressed, explaining how VfM/BV will be achieved when developing each Alliance Work Package (AWP) Proposal:

1. Design solution
2. Proposed approach and program (plan) for delivery
3. Continuous improvement
4. Risk management
5. Team potential and commitment
6. Urban design
7. Sustainability concepts
8. Stakeholder management (engagement)
9. Price
10. Team structure potential
11. Commercial arrangements
12. Integrated development opportunities

Nine of these criteria are encompassed by the LXRP vision, termed their '5-Greats'[1]: great network that strives for a great transport network – strive for a great transport network by reducing congestion, improving safety and unlocking capacity to run more trains; great places that create safe, vibrant, attractive and connected places for communities that enhance the travelling experience; great partnerships that leverage great partnerships to drive the safest and most efficient delivery for our works; great engagement that harness the power of community feedback and collective wisdom to deliver great outcomes; and great people who unleash the potential of people by investing in their skills, capabilities and wellbeing (see Chapter 3, Table 3-2). These 5-Greats visions and VfM/BV criteria are supported by stated shared LXRP values that inform the cultural transformation and people development strategy (see Chapter 7) and senior team leadership development (see Chapter 8) and are fundamental to recruiting LXRP people and alliance syndicate teams (see Chapter 5, Section 5.4). The five values are:

1. Creativity (through initiative, learning and sharing knowledge and insights)
2. Accountability (through passion, promise and delivery)
3. Relationships (through connectivity, listening and engaging in dialogue)
4. Empowerment (through leadership, decision-making and actions)
5. Safety (through considering everyone, everywhere and every day)

Analysis of interview transcripts and LXRP internal documents indicate the project performance agenda and LXRP performance approach being strategically and purposefully drilled into the workplace culture (Chapter 7). The 12 criteria used to report performance based on the 5-Greats reflect LXRP participants shared vision and values. KRAs and KPIs that measure performance reinforce the workplace culture. Its continuity from 2015 entrenches a performance perspective that affects the LXRP itself and spreads wider, influencing the five project alliance teams, their participants (including sub-contractors and suppliers) and those the LXRP engages with. The strategic aim includes positively influencing an improved industry culture, being an industry improvement catalyst (productivity and human satisfaction) and providing a legacy.

While the LXRP 2015–2030 program is expected to experience some fluctuations in its project delivery performance experience, experiential learning is an ingrained part of each project's delivery. It is likely that some projects, or phases of a project, will perform at sub-optimal levels – but with valuable lessons learned. The objective is for the *overall program* to deliver societal benefits.

12.3   LXRP Value Generation Performance

How does the LXRP manage performance communication to its internal and external stakeholders about its performance and achieving its strategic goals? It does this through formal and less formal community engagement governance mechanisms.

Chapter 11 discussed community engagement providing examples of how the LXRP manages its communications strategy. It won the International Association for Public Participation *Australasia Organisation of the Year Award* for its 2021 industry-leading approach to stakeholder engagement. The industry award recognised community engagement excellence in the program, including undertaking 40 engagement activities; achieving almost two million website views; delivering more than 500,000 newsletters; receiving more than 4300 contributions from the community; and building a following of more than 75,000 followers across all social media channels.[2]

General performance communication to the government and the public is delivered through annual reports such as the Victorian Government Department of Transport 2021–22 annual report. However, some detail of KRA performance is often less noticeable. Performance on some specialist LXRP areas, such as Training for the Future,[3] is communicated through online annual reports. Other specialist parts of the LXRP similarly publicly report performance online, but searching for these is not straightforward. Dynamic programs of projects of this scale and scope have challenges maintaining accurate central public-accessible data and reports. There is always a balance to be met between the cost of transparency and its value.

More detailed and accurate performance data and analysis can be found in LXRP internal communications, such as presentations compiled by each Alliance Project Team (APT) for its Alliance Management Team (AMT) and Alliance Leadership Team (ALT) meetings. Interviewee IV01 explained the performance data gathering and reporting process:

> We haven't got ALT champions for each KRA because there are so many of them. But each of those names [he points to a whiteboard in his office] is an AMT champion.

The APT assembles the relevant performance data and presents it at AMT and ALT meetings. KPI performance is also transparently available to all LXRP team participants. Anyone authorised with LXRP extranet access, this includes any alliance or program level participant or stakeholder with the extranet access account, can view currently reported performance in real-time. The extranet is an LXRP innovation (Chapter 9) that makes a range of reporting, digitised data processing and project/program-level communications. Subsection 12.3.1 is devoted to that aspect.

The ALT's scrutiny of the project's performance is an important component of the alliance's governance (Ross, 2003). The LXRP has a highly formalised and rigorous ALT performance review role that includes reflecting on and monitoring KPI reporting, resolving issues and decisions referred by project AMTs requiring higher-level authority, influence, and consideration to the management emerging or opportunities.

As discussed more fully in Chapter 3, the AMT may be equated to a board of corporate enterprise directors. Their role in reviewing performance is vital, as is their access to resources within their home organisations to be called upon when necessary. We offer the following observations from an internal AMT effectiveness report. First, the Project Alliance Agreement (PAA) that guides project governance states that the ALT's primary duty is to provide strategic leadership and direction, ensure transparent governance and accountability, and monitor and guide project performance. KRAs and KPIs facilitate understanding and monitoring key benefit performance, focusing on what delivers VfM/BV. Positive, trustful, but independently reflective relationships are developed between the ALT and AMT through mutually understanding their roles, capacity to exert influence (internally within the alliance and towards external stakeholders) and facilitating necessary actions to address emerging problems or seize potential opportunities. Mutual trust and commitment are needed, maintaining a balance between control and restraint, naivety and vigilance.

Like any steering committee style group, ALTs need to have committed members who consistently participate in ALT meetings and are well-prepared but not overloaded with relevant discussion materials. COVID-19 presented challenges in co-locating the ALT and for members recently joining an ALT. IV10 said that LXRP has a policy to train all ALT members based on the Australian Institute of Company Directors' course because their roles and obligations are similar. IV10 also stated that while each ALT varies in its culture due to membership differences and project-specific aspects, it is consistently chaired by the project delivery director representing the LXRP project owner (PO), and its typical agenda comprises approximately 30% development so that each meeting participant reflects on and considers what they are doing and how they function and how that may be improved and knowledge about this aspect transferred to future ALT project members. About 30% of the time is devoted to current project performance reviewing KPIs etc., to understanding how performance may affect project milestones. About 40% of the time is spent on strategic matters and what needs to be done to meet lead-time constraints, cost issues and other stakeholder engagement and project personnel aspects. Each ALT meets monthly, and each meeting takes a full day (12 noon to 6 pm).

The above LXRP ALT process, although not innovative or new on most alliancing projects, has its main difference in how performance is treated and presented. In conventional project delivery, there would be a tendency to report 'good news' and try to hide 'bad news.' In the LXRP alliancing context, participants are expected to flag unexpected performance outcomes so that these are *understood* and addressed. Favourable outcomes may be leveraged, and lessons learned to amplify positive impact. Unfavourable outcomes can trigger investigation and discussion about how adverse trends may be remedied and assumptions being questioned about causes so that any re-thinking of plans can be undertaken before lasting damage is done for example with the 37-day blitz (Chapter 6, Section 6.3.1) when the APT realised that delayed access to the site required by the POs. The APT had to re-think project planning and logistics requirements. The PAA's no-blame behaviours about performance are helpful when the APT's resilience mentality is deployed to probe and respond to challenges.

Performance measurement, appreciation, reporting and perceived accountability include less-formal tools and mechanisms with formal governance mechanisms. Strategic plans are operationalised through KRAs, but how are they defined, what are they designed to measure? Performance is also about VfM/BV delivery through short and long-term assessment but community legacy remains long after the project's completion.

12.3.1   *LXRP KRAs and KPIs as a project and program monitoring tool*

The Australian Commonwealth Government Department of Infrastructure and Transport (2011b) National Alliance Contracting Guidelines specifies that the PO will draft the PAA at the tender target outturn cost (TOC) development stage and that the LXRP's PAA, according to interviewee IV01, specifies that:

> at the program level – innovation, continuous improvement and efficiency, industry capability and inclusion, and there's some government policy that we've got to maintain there like major project skills guarantee and 2.5% Aboriginal engagement, we've got to maintain the key resources (because we're here for five years so we've got to maintain key resources). And then we've got effective engagement, so that's different from normal construction-related community engagement.

Key performance indicators (KPIs) are developed as KRA measures at the TOC stage. The rationale for the alliance team (and not the PO) developing them is that the APT then 'owns' KPIs. The APT becomes more attentive to being responsible and accountable for KPIs. Interestingly, one interviewee commented on how KPIs impact performance motivation and how at the TOC stage trade-offs between profit margin and anticipated gain-sharing through performance excellence could be finessed. IV05 observed that:

> package four decided to put part of their margin on the line and said, well, we will take a lesser margin and put some of that into a KRA pool and then measure us against the KRA pool, and we'll win back our margin by our performance in the KRA pool. So instead of getting an X% margin, they got effectively an X-1% margin, and the extra 1% went into a KRA pool to try and incentivise the performance there. Which we quite liked; we thought that was a neat way of trying to put something in the KRA pool, which was meaningful.

Some KPIs can have more easily allocated meaningful measures, such as safety standards, that were expected to be met, e.g., recordable injury, frequency rate etc. Other KPIs, such as the station precinct urban design, needed a more nuanced assessment approach that maintained the integrity of purpose with fairness in its application. Safety KPIs incur a penalty (pain) for failure to meet them but no gain for over-achieving. In this case, this avoids rewarding what should be automatically delivered.

Cost and time KPIs are easy to measure and act on. The TOC process ends with fixed cost and time numbers, ensuring fairness and encouraging innovation and resilience (Chapter 3). The PAA specifies that all project direct costs are reimbursable up to the TOC value with alliance participants' tendered profit margin taken out of the TOC figure and held as an at-risk escrow account to be adjusted by the gain/pain sharing incentive arrangement. If the actual end cost exceeds the TOC, then the difference is met by the 'pain-sharing' reduction of the profit margin and is capped at that margin (Department of Infrastructure and Transport, 2011a). The PO takes on all costs over that capped figure. Profit gain/loss provides a viable incentive for the APT to maintain its performance motivation. If the actual cost is under the TOC figure, then the APT (including the PO representative) participants share that' gain.' The PAA also treats performance holistically, with all APT participants being measured on the *project performance* outcome, not *discipline-team effectiveness*. Accordingly, the APT is incentivised and motivated to act as a united team. The impact of this integrated team concept is that it is in each participant's interest to ensure that: nobody exercises opportunistic behaviours; assumptions are questioned when necessary; and decision-making is consensus led. This reduces transaction costs associated with potential disputes (Haaskjold, 2021). The PAA also encourages collaboration within the APT and nurtures resilience to unexpected events because the TOC is fixed. The 15–20-week LXRP TOC process ensures that risk and uncertainty are also effectively analysed. Fairness is maintained through a Target Cost Event adjustment process (Walker, Vaz Serra and Love, 2022) that classifies unknown, unanticipated event categories justifiably 'owned' by the PO and events to be allocated in the APT TOC contingency. This approach provides a more reliable cost/time TOC figure and a more effective performance assessment of these KPIs.

Each LXRP strategy relating to the 5-Greats has KRAs with operationalised KPIs developed at the TOC stage and refined by project alliance experience. Continual effort is made with each project package to fine-tune KPIs to ensure 'raising the bar' through experience while maintaining relevance.

Figure 12-2 illustrates the KRA dashboard that helps identify and monitor project and program performance. It is linked to KPI 1.1 innovation and continuous improvement initiation submissions for approval by the ALT and Joint Coordination Committee (JCC) related technical sub-committee (see Chapters 3 and 9 for more details about the JCC) and KPI 1.2 innovation and continuous improvement initiative adoption. Part of the TOC cost/time, design and delivery plan process (see Chapters 3 and 5) to produce a more reliable TOC (Walker, Vaz Serra and Love, 2022) is predicated on a requirement to demonstrate continuous improvement initiatives and to adopt innovation initiatives developed on previous alliance projects within the program. Innovation and continuous improvement are central to LXRP's value (creativity) and strategy.

Figure 12-2 is blurred but provides a screengrab image from the LXRP extranet system. KRA 1 is linked to the 5-Greats and the LXRP cultural value 1 of 'creativity.' The screenshot indicates at that time (circa 2022, it is a real-time system), 365 KPI 1.1 innovation submissions were recorded (1.1 – adopting and approved 1.2 innovation from another alliance project), and 140 KPI 1.2 submissions (submitting an innovation for others to adopt/adapt) had been approved across the filters selected (below the KPI 1.2 box). These innovations are spread across various disciplines, each having JCC sub-committees that discuss these proposals and ultimately approve them as valid innovations and appropriate to be accepted into the system. The 'Proposal discipline' panel in Figure 12-2 also illustrates the extent of cross-disciplinary and cross-project collaboration. Submission status is also tracked in another panel in the figure, where it is accepted or rejected (by the LXRP and project ALTs). Other panels enable drilling down by alliance

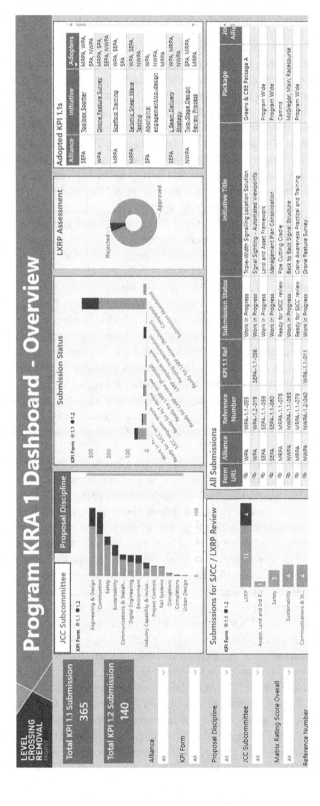

Figure 12-2 Screen capture image from the extranet for the KPA dashboard

project and submission reference code. This highly sophisticated dashboard monitoring and control mechanism indicates an advanced level of strategised and operationalised performance monitoring and control. There is also a series of videos produced and available on this system that explains the submission ideas as a set of best practice briefing resources that each alliance participant can 'graze through' to apply or adapt for their projects and, in doing so, be eligible for a KPI 1.2 outcome. This level of detail and summary, available to the ALT, suggests that this JCC approach may be a rare globally sophisticated performance management mechanism. It also provides valuable continuous improvement within-across the project system.

12.3.2 *Value for money or the best value?*

Governments require and expect projects to be delivered efficiently and effectively, delivering best value benefits. At the business case stage-gate process, the PO must demonstrate that the proposed project's approach delivers VfM/BV but it is an easily misunderstood term because it comprises elements of cost and value, and value is perceived. Thus, specifying *values,* that is, what is *valued* by stakeholders, and what can be costed, is required. Based on research studies of stakeholders, Barrett (2005), based on research studies of stakeholders, has derived the 5Es of VfM.

1. Efficacy – did the process produce the required output?
2. Efficiency – was the transformation carried out with minimum resources?
3. Effectiveness – were long-term requirements met?
4. Ethicality – was the transformation acceptable from a value judgement perspective?
5. Elegance – was the solution over or under-engineered, aesthetically pleasing, well designed?

Unfortunately, most people and institutional clients only focus on efficacy and efficiency and perhaps effectiveness while measuring VfM as required. They may find ethicality and elegance considerations too difficult to specify or monitor as these are often impossible to monetise as a benefit to complete a cost-benefit analysis. The above 5Es may be better termed best value (BV) because of confusion about what VfM embraces (MacDonald, 2011). Therefore, we refer to VfM/BV being used to emphasise the full 5Es.

Consider the difficulty in monetising for a VfM report on the aesthetic value of a railway station precinct re-development. Many of the old-fashioned stations built in the early/mid last century are unfit for today's passenger experience, sense of personal safety, and intermodal transfer links. Rail line public access was purposely fenced off and made inaccessible, present-ing urban barriers. Combining a Skyrail road/rail grade separation, a new station, and turning rail line land into parklands with recreational appurtenances transformed many station area pre-cincts (Chapter 11, Figure 11–2). How could a KRA creating urban transformation and a vital community hub, be operationalised into KPIs encompassing VfM/BV benefits that cannot be easily monetised? An LXRP interviewee explained the Urban Design KRA operationalisation into KPIs that can legitimately reflect VfM/BV.

IV05 elaborated on the Urban Design KRA for Project Alliance package 1:

Urban design was based on a score that we had an independent panel assess their tender design at, and then we said you've got to try and achieve what you promised with your tender submission. So, we've assessed the functionality of your station design and the ar-chitectural features and the shared paths and bike paths, and all that sort of stuff is – we had that scored by an urban design panel, which involved members of the Office of Victoria

Government Architect and VicTrack, PTV [Public Transport Victoria], VicRoads, council were all member organisations of the urban design panel. And so, we got them to score various components of their solution against urban design principles, which we put out as part of the tender document that we wanted them to try to achieve. And then we said you've now promised that you can deliver this; that's now your baseline. And during the course of the work, things changed. Trees had to be removed from there to there to make something else fit, etcetera or whatever the case might be. So, it was about working through all those and still trying to get as good an outcome as what was originally intended.

The above example illustrates how VfM/BV *could* be made accountable and responsible for its development maintained while not being limited to a *cost-benefit* numbers game.'

12.3.3   *Short versus long-term assessment of program value creation*

While we are yet to assess the whole program performance impact, with many crossing removals taking the program into 2030, we can offer tentative conclusions on projects within the program that have been completed.

One of the projects, the Carnegie Station, was completely rebuilt and opened in June 2018. One of this chapter's co-authors has direct experience of this station from a before-to-after perspective. With grandchildren living near the station, when visiting them he experienced lengthy traffic delays at the Grange Road rail crossing. He was also aware of pollution from the then at-ground-level rail line and rail traffic noise, and he considered the existing station shabby.

When the LXRP announced the project that would transform that area, the community debate was set *against* a Skyrail solution. However, despite significant local support for the 'trench' road/rail grade separation option, the project proceeded as a Skyrail. Chapter 11, Figure 11–2 illustrates the 'after' situation with an attractive rail station, paved bike/pedestrian trail, and parkland with recreational facilities that his grandson enjoys where the rail track had previously been situated.

The Skyrail reduces ground-level rail noise. Residents now appreciate that the Skyrail was a superior solution to the trench rail-road grade separation. It is impossible to calculate the realised monetary benefit value of the bike/pedestrian path, though it is easy to cost infrastructure work. Benefits include aesthetics and health value by facilitating exercise through walking and bike riding. Anecdotally, it appears that there has been a positive legacy left for that station that most likely represents VfM/BV.

Interestingly, the National Museum of Australia in Canberra was delivered as a project alliance with several quality measure KRAs for that project, including a visitor's museum experience quality measurement and other quality aspects (Keniger and Walker, 2003) suggesting that aesthetic and post-project quality perception KPIs can, with due consideration, be measured.

12.3.4   *Legacy performance*

Legacy may be viewed from many perspectives. One vital legacy strategised by the LXRP was to enhance and develop the skills base of the infrastructure construction industry sector, specifically the more general construction industry, through a concerted educational and training program (see Chapters 7 and 8). There is a critical skills shortage in Australia and a recent government report highlights the problem and suggests mitigating actions (Infrastructure Australia, 2021). The LXRP is part of The Victorian Transport Delivery Authority (VIDA) that also operates a range of up-skilling programs for graduates destined as APT participants and trade apprentices for facility operators and project-participating organisations. Victoria's Big

Build website[4] directs potential employees to its comprehensive website that details what training will be provided, what trainees may expect, what is expected of them, and how to apply.

Linked to the above is the LXRP response to the government strategy for improving social inclusion and employment diversity[5] as documented in the Victorian Government Social Procurement Framework Annual Report 2020–21. That report declares the Department of Transport is performing well by embodying the framework's aims in its workplace culture. It cites an LXRP case study example:

> The Women in Transport mentoring program is a six-month, industry-wide initiative run twice a year by the Level Crossing Removal Project on behalf of the Department of Transport.
>
> It provides professional development and networking opportunities for women mentees to encourage more women to enter and stay in the transport industry.
>
> The aim is to give women a fresh avenue for development as they progress in their careers, and to foster relationships that can help them identify and realise their professional goals.
>
> In 2020, the program saw 535 people matched to form 261 pairs of mentors and mentees (with some mentors taking up to two or three mentees). Mentees and mentors attend three professional development sessions staggered across six months.
>
> (Victorian Government, 2021, p. 22)

This example of KRA performance illustrates how KPIs may be designed from the KRA, Thus, operationalising difficult-to-measure aspects of a strategy. This performance example shows a particular activity (the mentoring program), explains its purpose and provides indicative performance metrics (535 people matched, forming 261 mentor-mentee pairs) and how that was achieved. This form of KPI expresses inferred performance and legacy value, the opportunity provided for career and personal development but cannot articulate the impact. That would require a series of potentially debatable legacy outcomes such as the extent of career advancement, participant satisfaction with the experience and many other BV outcomes that cannot be easily monetised and hence recorded as VfM measures.

Other legacy performance examples could be discussed, for example, the joy of experiencing recreational facilities in a parkland setting where barren rail lines once dominated the urban landscape, possible mental health well-being derived by rail passengers using an upgraded station and surrounding precinct, or the legacy of creating local industry supply-chain opportunities through mandated social procurement policies that are more widely being adopted globally as demonstrated in Sweden (Troje and Gluch, 2020) and the UK and North America (Loosemore, Keast and Alkilani, 2023).

12.4 Global Megaproject IPD-Alliancing Type Project/Program Performance

How do LXRP's performance and wider transformational philosophy compare with global best-practice? We offer a Canadian example that adopts a similar holistic performance perspective.

Recently, researchers Drouin and Brunet (2023) examined the public value generated by replacing the old Samuel De Champlain Bridge with a new, more modern bridge. Thus, we highlight in this section how public value, VfM/BV, for Canadians was achieved through three key dimensions: *legitimacy, accountability* and *efficiency* (Brunet and Aubry, 2016). The bridge is a key economic structure for the city of Montreal and its surroundings for the transport of goods and the mobility of people between the south shore suburbs, the city centre, and the north shore suburbs. The old bridge, dating from the 1950s, had been badly damaged by the harsh Canadian winters. Under the responsibility of Infrastructure Canada, the Government of Canada and its federal agency, The Jacques Cartier and Champlain Bridges Incorporated,[6] a private-public partnership was set up to build the new structure. The new bridge is 3.4 km long across the St. Lawrence River. It has a three-corridor design, including two three-lane corridors for vehicle traffic and a two-lane transit corridor for a light rail transit system. It includes a multi-use path for pedestrians and cyclists. The new bridge corridor project was designed to achieve four objectives (for more detailed information, see Infrastructure Canada (2019)[7]):

1. *Ensure continued safety and service*: Remove traffic from the existing bridge, maintain the corridor's safety throughout the process and deliver a long-term solution that meets all the government's requirements.
2. *Promote economic growth*: Improve system connectivity, strengthen the local, regional and national economy, and promote economic growth.
3. *Provide VfM/BV for Canadians*: Provide a stable, high-quality structure at the lowest possible life cycle cost consistent with the government's vision for the project.
4. *Foster sustainable development and urban integration*: Deliver a project that considers its environmental and social context, protects the natural environment, showcases Montreal's landscape and promotes sustainable transportation.

Construction of the new bridge followed a 42-month timeline, representing a considerable technical challenge. Cooperation and coordination among the project team was essential to meet this tight deadline. The new bridge was designed and constructed in 48 months (6 months delayed) because of many barriers and challenges, such as: the deterioration of the existing bridge, which impacted the City and the Province of Québec economically; limited construction periods; and, severe weather constraints. An accelerated bridge construction approach was favoured to meet the 48-month schedule (Nader, 2020). Basically, it was like constructing a Lego project with various prefabricated parts in collaboration with multiple subcontractors and orchestrated by a consortium leader called Signature on the St. Lawrence Group (SSL). This approach was characterised as an important innovation in the erection scheme (Nader, 2020). The project measured sustainability required by the project owner Infrastructure Canada, Government of Canada. Based on *Envision status*, the project must demonstrate that it delivers a range of environmental, social and economic benefits through five categories: *Quality of Life, Leadership, Resource Allocation, Natural World, and Climate* (Institute for Sustainable Infrastructure, 2018a)[8]. The project met the following key factors that contributed to the new bridge project earning an Envision Platinum award (Institute for Sustainable Infrastructure, 2018b)[9]; (Drouin and Brunet, 2023) evidenced by:

1. *Sustainable growth and development*: Improved mobility of people and goods that benefit local and provincial economies.
2. *Community quality of life*: Provisions for sustainable mobility, increased safety and improved traffic flows benefit the surrounding communities.
3. *Leadership in sustainability*: Sustainability was a crucial part of the project from the outset, and SSL developed many policies and tools to ensure and improve sustainable performance. Stakeholders from all affected communities and organisations were engaged in dialogue with the government and SSL, and their concerns were addressed.
4. *Innovation and exceptional performance*: According to the Institute for Sustainable Infrastructure (2018b) 'The project exceeded the highest levels of achievement within the Envision system for several credits in the Quality of Life category, earning the project bonus points for exceptional performance.' The project applied an innovative construction approach to avoid interfering with traffic on the St. Lawrence Seaway. In addition, it used a new de-icing system to prevent the problem of ice falling from the cable stays onto the bridge deck.

According to Drouin and Brunet (2023) and Brunet and Aubry (2016), the project achieved public value or VfM by meeting three dimensions: legitimacy, accountability and efficiency. First, the Government of Canada move beyond the lower bidder scheme to aim at VfM over the long run. Then, a centralised decision-making process, a regulatory system based on a rigorous analytical process including sustainability issues, and developing the client's (owner's) capability enhanced legitimacy (Brunet, 2021). In addition, the project owner (Infrastructure Canada) was responsible for proactively addressing the needs of multiple stakeholders for consultation and representativeness. The decision by the Government to adopt the Envision framework reflected its sensitivity to up-date impact assessment tools by integrating social, environmental and governance facets to legitimate the project and allow accountability.

More specifically to accountability, the Government played its role as a "strong owner" by selecting a delivery method that allows strict compliance regarding quality and respect of the schedule (Winch and Leiringer, 2016). A comprehensive process was established that includes a transparent procurement process (Drouin and Brunet, 2023). Finally, efficiency is the realm of VfM/BV. The Samuel De Champlain new bridge showed a rational basis to support decisions such as the delivery method, sustainability certification, enhanced control over SSL, the partner to ensure efficient delivery and quality, and improved front-end development with efficient decision-making (Brunet, 2021). Summarising, this megaproject led by a public entity, Infrastructure Canada, Government of Canada, attained a high level of VfM/BV with the establishment of key success factors such as: a transparent procurement process; the selection of the reliable, well-structured consortium led by SSL which was able to demonstrate technical, managerial and operational capacities as well as innovation to carry out the project and; to foster sustainable project development (Drouin and Brunet, 2023).

There is a commonality in the transformational and holistic performance views across both case studies. The LXRP is a program of alliance projects, whereas the Samuel De Champlain Bridge was a single project relationship-based collaborative partnership organisation. However, the concept of a "strong owner" selecting a delivery method that allows strict governance (Winch and Leiringer, 2016) was evident in both instances. The primacy of VfM/BV being defined by economic, environmental, and social performance is common, particularly their emphasis on using projects as an opportunity for leaving a broader social and environmental legacy. Both projects' vision focuses on project delivery value through co-created integrated collaborative teamwork performance and generated project outcome performance through a legacy of social value. Both cases are recognised as having set strategies for and delivered VfM in its broadest sense.

12.5   Conclusions

Infrastructure megaproject and program performance cannot be reasonably assessed solely on 'iron-triangle' measures. This chapter moves beyond a restricted view of 'good' project delivery performance. As Drouin and Turner (2022) note, performance on government infrastructure megaprojects should be measured on social benefits delivered through the project strategy.

This chapter followed a broad VfM/BV perspective focusing on how the LXRP measure and communicates its performance across projects in the program. The Canadian case study example illustrates commonalities in LXRP philosophical orientation and seizing the opportunity for delivering a lasting positive transformational legacy. The project delivery performance discussion was widened through the Canadian megaproject case study discussion, comparing how the LXRP delivery value generation performance might relate to global best/good practice exemplars.

This chapter's question is: *How does the LXRP measure and communicate its best value delivery performance across projects in the program?* The answer links to defining a project's purpose and how that is interpreted by project owners and stakeholders that benefit from the project's delivery. The way in which it is communicated is vital to perceived project delivery or project outcome success.

Figure 12-1 mapped out the chapter's content and logic. We briefly introduced project performance theory and sought to focus on delivering value rather than a restricted iron triangle perspective. Also, this chapter's major project management theoretical contribution is that we have discussed infrastructure *programs* and *project* performance management strategies. We showed how these strategies were aimed at VfM/BV co-generation through the integrated collaborative APT approach.

This also links closely with Chapter 3, which discusses how the LXRP program performance strategy was operationalised through its choice of governance mechanisms. In the following section, we focused specifically on the LXRP's performance management approach. Key to understanding how the LXRP measure and communicate its performance criteria is the development of governance mechanisms such as KRAs and their associated KPIs that shape how KRA's successful performance 'looks like' and how that is enacted through the LXRP's organisational design. Communicating and demonstrating VfM/BV is a complicated process. Clarity is king. We saw that the PO determines the program strategy.

Chapter 2 explains the LXRP context that led to it articulating its purpose and what benefits it sought to deliver. The KRAs operationalise what the vision of success looks like, the 5-Greats. We then explained how these were operationalised into specific, clearly articulated measures as KPIs during the TOC development phase crafted through the five values. These KPIs serve as the 'workhorse' to help guide performance. As explained in this section, the alliancing approach uses a specific leadership mechanism, the ALT, to oversee, support and guide performance. We also discussed short-term and long-term legacy performance measures that fit the projects within the program context that is critical in delivering sustainable social value infrastructure.

We then discussed a comparative global megaproject case study example from Canada to illustrate how social value delivery and visionary, transformational project outcomes emerge as best-practice project delivery performance aims. Similarities in VfM/BV perceptions were highlighted.

We emphasise that program performance management provides opportunities for within and cross-project continuous improvement (Chapters 9 and 5) enabling a strong pipeline of AWPs to reinforce project integration within a *program* mindset motivating participants to view the LXRP as a career path. We argued that the LXRP exemplifies how continuous improvement may be achieved through its innovation and improvement KRAs. Combining this with the evolution of the JCC as an effective integrative and collaborative governance mechanism demonstrates further insights into how LXRP measures and communicates its VfM/BV delivery performance across projects in the program.

Notes

1 Internal LXRP document INTLX6086 Blueprint Brochure 2021_UPDATE_V9 "Blueprint Delivering Great Change." Kevin Devlin September 2021
2 See https://bigbuild.vic.gov.au/news/level-crossing-removal-project/major-award-for-excellence-in-community-engagement
3 https://bigbuild.vic.gov.au/jobs/training-programs/grow-program
4 https://bigbuild.vic.gov.au/jobs
5 https://bigbuild.vic.gov.au/__data/assets/pdf_file/0019/644302/TFTF-Social-Procurement-in-Practice-resource-Oct-19-final.pdf

6 www.infrastructure.gc.ca/about-apropos/jccbi-pjcci-eng.html
7 Infrastructure Canada (2019), *Samuel De Champlain Bridge – Project Objectives,* accessed on 14
   April 2022 at www.infrastructure.gc.ca/nbsl-npsl/objectives-objectifs-eng.html.
8 Institute for Sustainable Infrastructure (2018a).
9 Institute for Sustainable Infrastructure (2018b).

References

Ahiaga-Dagbui, D.D. and Smith, S.D. (2014). "Rethinking construction cost overruns: Cognition, learning
   and estimation." *Journal of Financial Management of Property and Construction.* 19 (1): 38–54.

Atkinson, R. (1999). "Project management: Cost, time and quality, two best guesses and a phenomenon, its
   time to accept other success criteria." *International Journal of Project Management.* 17 (6): 337–342.

Baden-Fuller, C. and Morgan, M.S. (2010). "Business models as models." *Long Range Planning.* 43 (2–3):
   156–171.

Barrett, P.S. (2005). *Revaluing Construction A Global CIB Agenda,* Construction I.C. f. R. a. I. i. B. a.
   Rotterdam. 84pp

Bradley, G. (2010) *Benefit Realisation Management,* Aldershot, UK, Gower.

Brunet, M. (2021). "Making sense of a governance framework for megaprojects: The challenge of finding
   equilibrium." *International Journal of Project Management.* 39 (4): 406–416.

Brunet, M. and Aubry, M. (2016). "The three dimensions of a governance framework for major public
   projects." *International Journal of Project Management.* 34 (8): 1596–1607.

Department of Infrastructure and Transport (2011a). *National Alliance Contracting Guidelines Guidance
   Note No 5: Developing the Target Outturn Cost in Alliance Contracting.* Department of Infrastructure
   and Transport A.C.G. Canberra, Commonwealth of Australia. www.infrastructure.gov.au/infrastructure/
   nacg/files/NACG_GN5.pdf

Department of Infrastructure and Transport (2011b). *National Alliance Contracting Guidelines Guide to
   Alliance Contracting.* Department of Infrastructure and Transport A.C.G. Canberra, Commonwealth of
   Australia. www.infrastructure.gov.au/infrastructure/nacg/files/National_Guide_to_Alliance_Contract-
   ing04July.pdf

Drevland, F. (2019). *Optimising Construction Projects as Value Delivery Systems: Expanding the
   Theoretical Foundation,* Trondheim, Norway, Norwegian University of Science and Technology.

Drevland, F., Lohne, J. and Klakegg, O.J. (2018). "Defining an ill-defined concept: Nine tenets on the
   nature of value." *Lean Construction Journal.* 14: 31–46.

Drouin, N. and Brunet, M. (2023). "The Samuel De Champlain Bridge Corridor Project: Sustainability and
   innovation as key success factors." In *Infrastructure Development: A Critical International Perspective
   on Value in Public-Private Partnerships,* Clegg, S., Sankaran, S., Ke,Y., Mangioni, V., Devkar, G. (eds),
   Cheltenham, UK Edward Elgar Publishing.

Drouin, N. and Turner, J.R. (2022) *Advanced Introduction to Megaprojects,* Northampton, UK, Edward
   Edgar Publishing

Engwall, M. (2003). "No project is an island: Linking projects to history and context." *Research Policy.*
   32 (5): 789–808.

Flyvbjerg, B. (2014). "What you should know about megaprojects and why: An overview." *Project
   Management Journal.* 45 (2): 6–19.

Flyvbjerg, B., Holm, M.S. and Buhl, S. (2002). "Underestimating costs in public works projects: Error or
   lie?" *Journal of the American Planning Association.* 68 (3): 279.

Gil, N. (2021). "Megaprojects: A meandering journey towards a theory of purpose, value creation and
   value distribution." *Construction Management and Economics.* 1–23.

Gil, N. and Pinto, J.K. (2018). "Polycentric organizing and performance: A contingency model and
   evidence from megaproject planning in the UK." *Research Policy.* 47 (4): 717–734.

Haaskjold, H. (2021). *The Puzzle of Project Transaction Costs. Optimising Project Transaction Costs
   Through Client-Contractor Collaboration,* Trondheim, Norway, Norwegian University of Science and
   Technology.

Ika, L.A., Love, P.E.D. and Pinto, J.K. (2021). "Moving beyond the planning fallacy: the emergence of a new principle of project behavior." *IEEE Transactions on Engineering Management.* under review.

Infrastructure Australia (2021). *Infrastructure Workforce and Skills Supply. A Report from Infrastructure Australia's Market Capacity Program*, Government A.

Institute for Sustainable Infrastructure (2018a). *Envision: Sustainable Infrastructure Framework Guidance Manual V3.* Washington, DC, Institute for Sustainable Infrastructure. https://sustainableinfrastructure.org/wp-content/uploads/EnvisionV3.9.7.2018.pdf

Institute for Sustainable Infrastructure (2018b). "Samuel De Champlain Bridge Corridor: Montréal's iconic Samuel De Champlain Bridge Corridor earns Envision Platinum Award." Washington, DC, Institute for Sustainable Infrastructure. https://sustainableinfrastructure.org/project-awards/projet-de-corridor-du-nouveau-pont-champlain-new-champlain-bridge-corridor-project/

Keniger, M. and Walker, D.H.T. (2003). "Developing a quality culture: Project Alliancing versus business as usual." In *Procurement Strategies: A Relationship Based Approach*, Walker, D.H.T. and K.D. Hampson (eds), pp. 204–235. Oxford, Blackwell Publishing.

Locatelli, G., Konstantinou, E., Geraldi, J. and Sainati, T. (2022). "The dark side of projects: Dimensionality, research methods, and agenda." *Project Management Journal.* 53 (4): 367–381.

Loosemore, M., Keast, R. and Alkilani, S. (2023). "The drivers of social procurement policy adoption in the construction industry: An Australian perspective." *Building Research & Information.* 1–13.

MacDonald, C.C. (2011). Value for Money in Project Alliances. DPM Thesis, School of Property, Construction and Project Management. Melbourne, RMIT University.

Murray, P. (2004) *The saga of Sydney Opera House: The Dramatic Story of the Design and Construction of the Icon of Modern Australia*, New York, Spon Press.

Nader, M. (2020). "Accelerated bridge construction of the new Samuel De Champlain Bridge." *Journal of Bridge Engineering.* 25 (2): 05019015.

Pinto, J.K. and Slevin, D. (1987). "Critical factors in successful project implementation." *IEEE Transactions on Engineering Management.* EM-34 (1).

Porter, M.E. and Kramer, M.R. (2011). "Creating shared value." *Harvard Business Review.* 89 (1/2): 62–77.

Raiden, A., Loosemore, M., King, A. and Gorse, C. (2019). "Introduction." In *Social Value in Construction*, Raiden, A., M. Loosemore, A. King and C. Gorse (eds), pp. 3–38. Milton, CRC Press.

Ross, J. (2003). *Introduction to Project Alliancing.* Alliance Contracting Conference, Sydney, 30 April 2003, Project Control International Pty Ltd.

Shenhar, A.J., Dvir, D., Levy, O. and Maltz, A.C. (2001). "Project success: A multidimensional strategic concept." *Long Range Planning.* 34 (6): 699–725.

Shenhar, A.J., Milosevic, D., Dvir, D. and Thamhain, H. (2007) *Linking Project Manangement to Business Strategy*, Newtown Square, PA, Project Management Institute.

Troje, D. and Gluch, P. (2020). "Populating the social realm: New roles arising from social procurement." *Construction Management and Economics.* 38 (1): 55–70.

United Nations (2016). "Global Sustianability Goals." ww.un.org/sustainabledevelopment/sustainable-development-goals/

Victorian Auditor-General's Office. (2017). *Managing the Level Crossing Removal Program*, Melbourne.

Victorian Auditor-General's Office. (2020). *Follow up of Managing the Level Crossing Removal Program*, Printer V.G. Melbourne.

Victorian Government. (2021). *Victorian Government Social Procurement Framework Annual Report 2020–21*, Government V. Melbourne, Australia.

Victorian State Government. (2017). *Level Crossing Removal Project – Program Business Case*, Government V. Melbourne.

Volden, G.H. (2019). "Assessing public projects' value for money: An empirical study of the usefulness of cost–benefit analyses in decision-making." *International Journal of Project Management.* 37 (4): 549–564.

Walker, D.H.T., Mills, A. and Harley, J. (2015). "Alliance projects in Australasia: A digest of infrastructure development from 2008 to 2013 " *Construction Economics and Building.* 15 (1): 1–18.

Walker, D.H.T., Vaz Serra, P. and Love, P.E.D. (2022). "Improved reliability in planning large-scale infrastructure project delivery through Alliancing." *International Journal of Managing Projects in Business*. 15 (8): 721–741.

Winch, G. and Leiringer, R. (2016). "Owner project capabilities for infrastructure development: A review and development of the 'strong owner' concept." *International Journal of Project Management*. 34 (2): 271–281.

Zwikael, O. (2016). *Project Benefit Management*. IPMA Conference in Rekjavik, Iceland, IPMA. http://voicesfromoxford.org/video/project-benefit-management/685

Zwikael, O. (2018). "Setting effective project benefits." In *2018 IRNOP – A Skilled Hand and a Cultivated Mind*, Walker, D. (ed.). Melbourne, RMIT

Zwikael, O. and Globerson, S. (2006). "From critical success factors to critical success processes." *International Journal of Production Research*. 44 (17): 3433–3449.

Index

Printed in the United States
by Baker & Taylor Publisher Services